BIOCHEMISTRY RESEARCH TRENDS

SYNERGISTIC INTERACTION AND CELL RESPONSES TO ENVIRONMENTAL FACTORS

BIOCHEMISTRY RESEARCH TRENDS

Additional books in this series can be found on Nova's website under the Series tab.

Additional e-books in this series can be found on Nova's website under the e-book tab.

BIOCHEMISTRY RESEARCH TRENDS

SYNERGISTIC INTERACTION AND CELL RESPONSES TO ENVIRONMENTAL FACTORS

VLADISLAV G. PETIN
AND
JIN KYU KIM

nova publishers
New York

Library of Congress Cataloging-in-Publication Data

Library of Congress Control Number: 2014930678

ISBN: 978-1-63117-084-3

Published by Nova Science Publishers, Inc. † New York

CONTENTS

PREFACE

Living organisms are exposed to numerous natural and man-made agents that can interact with each other in a synergistic manner when the biological effect is greater than the sum of the effects from separate exposures to the individual agents. Because of the worldwide expansion of damaging chemical and physical agents, a large body of diseases and even aging are thought to stem from combined action of various deleterious agents. Safety standard assessment generally performed with the simplifying assumption that the agents under consideration act mainly independently on each other. For practical applications in risk analysis, deviation from simple summation of biological effect is virtually expected not very large, the synergistic interaction being expected for acute exposures. However, this problem has not been studied in sufficient detail. Therefore it is of great interest to investigate whether or not the synergistic interaction can take place at low intensities of inactivating and/or mutagenic factors. Since such experiments are very difficult to perform at the real intensities observed in the biosphere, a feasible approach to this problem is to analyze the dependence of the synergistic interaction on the intensity of the agents applied. Several hypothetical possibilities may be realized in principle. If the level of synergism is reduced progressively with a decrease in the intensity of physical factors or concentration of chemical agents, it would mean that the synergistic interaction will hardly be of any importance at low exposure intensity. This conclusion would be also true for the case when a decrease in the intensity of one agent should be accompanied by the increased intensity of another factor to provide the same level of the synergism. The only possibility important for the problem of low doses and intensities safety is the case when a decrease in the intensity of one agent must be accompanied by a decrease in the intensity of another factor used in combination to provide the same level of synergism. This problem has been investigated carefully in our work.

Original data on synergistic effects and cell responses to environmental factors are published. The data will provide some general view regarding the mechanism of synergistic interaction after combined exposures. Before presentation of these data, it is of interest to expound our probability model of cell responses to ionizing radiation that, in fact, combines hit and target model with biological stochastic and predicts the retention of some sublesions in cells surviving after irradiation. We shall use the sublesion idea in our further mathematical modeling of synergistic interaction. Since many authors attribute the mechanism of synergistic interaction with cell recovery inhibition, the relationship of some radiobiological effects with the ability of cells to recover from radiation damage is considered. There are only few papers on the RBE of densely ionizing particles for yeast cells with different capabilities of recovery. Thereby we are going to present a lot of our new data regarding this problem. In particular, we present the data indicating that the recovery processes are not impair after high-LET radiation and the RBE are mainly due to the formation of irreversible damage.

Numerous data are known for the simplest cell objects, which indicate that mechanism of chemical radioprotectors, oxygen and hypoxic cell sensitizers may involve cellular recovery processes. These findings could contribute to the analysis of participation of recovery processes in synergistic effects display. However, similar data for eukaryotic cells are almost missing. New experimental data and theoretical calculations indicating the possible interaction of ionization and excitation of molecules due to Čerenkov emission after ionizing radiation alone are also presented.

Thus, the objective of this Monograph is to develop a unifying conception and mathematical model of synergy, which would not depend on interactive agents, biological objects and tests. The model will predict a number of general rules of interaction, the condition under which the highest synergy can be achieved, its value, and the dependence of synergy on the intensity of the agents applied. Validation of the model will be verified for various physical and chemical factors, biological objects and end points. To achieve this objective, the following subjects are considered.

1. Cell responses to low- and high-LET exposure: the recent peculiarities.
2. Novel direction in study of radioprotecting and radiosensitizing action of chemical compounds.
3. Quantitative regularities of combined action of various agents.
4. Clue for understanding the mechanism involved in synergistic effects.
5. The mathematical approach for synergistic effect: prognosis and optimization.
6. Quantitative prognosis of synergistic effects and recovery for various biological objects.
7. Synergistic effects display for cultured mammalian cells, animals and humans.

We hope the Monograph will be useful for professionals in the fields of radiation biology, medicine, ecology, mathematical biosciences, synergetic, environmental protection, space flights and for students interested in the problems of synergy.

The authors express their warm-hearted gratitude to Marina Pronkevich for help in preparing the manuscript and figures.

Chapter 1

CELL RESPONSES TO LOW- AND HIGH-LET EXPOSURE: THE RECENT PECULIARITIES

ABSTRACT

For quantitative description of variety cell responses to ionizing radiation a mathematical model is suggested. The model connects basic ideas of hit and target models with biological stochastic: different irradiated cells are damaged in a random fashion in accordance with the hit-and-target principle and the number of damages defines the probability of successful division. Theoretically expected results were compared with experimental findings. A relatively unexpected role of the specific repair pathways in the relative biological effectiveness (RBE) of high-linear energy transfer (LET) is supported by extensive experimental data demonstrating the participation of recovery processes in the RBE of densely ionizing radiation. A slower recovery after high-LET radiation is demonstrated to reflect a smaller number of reversible damage produced by high-LET radiations, with the probability of recovery independent of radiation quality. It is concluded that RBE is determined not merely by the physical characteristics of ionizing radiation but also by cell ability to recover from radiation damage on biochemical stage of radiation damage display. It is supposed that such a recovery may be mediated by a recombination-like mechanism, for which the diploid state is required. It is supposed that the development of repair systems, providing more full recovery of diploid cells from genetic damages in comparison with haploid cells, may have served as a prerequisite for the transition, during evolution processes, from haplont to diplont cells. A theoretical analysis and interpretation of different dependences of the shape of dose-effect curves after low- and high-LET radiation for various eukaryotic cells is presented. It is inferred from the data presented that it may be due to the dependence of radiosensitive site location on the growth stage, at which irradiation is occurred.

INTRODUCTION

The classical hit-and-target models of cell inactivation are known to be founded upon two suppositions: various cells of the homogeneous population irradiated with the same dose are damaged in a different manner in accordance with the hit-principle; cell survival (or inactivation) is strictly determined by the number of hits: if the number of hits inside the target/s exceeds a threshold number, the cell is inactivated with unit probability, while in any other cases the cell survives [13; 43]. However, cell survival represents only one of many complex phenomena which occur in biological material irradiated. The information presently available shows that single cells among both surviving and inactivating parts of a homogeneous population respond differently to the same dose of irradiation [37; 38; 40]. Alternatively to the target-theory an idea of biological stochastic was developed [30]. The concept of biological stochastic was shown to be convenient for the description of radiobiological effects which cannot be interpreted by means of the hit-and-target model and vice versa [32; 33]. Hence, hit theory and biological stochastic are mutually complementary models. It can be anticipated that a generalized model of cell responses to irradiation will have to combine hit-and-target principles and some ideas of biological stochastic.

A lot of investigation of the RBE of ionizing particles differing in LET have been reported for a variety of biological test systems. These studies indicate that high-LET radiation produces a greater proportion of irreversible damage in comparison with sparsely ionizing radiations [16; 18; 25; 39; 41; 52; 57; 68]. The analysis of response of radioresistant (B/r) and radiosensitive (B_{s-1}) strains of *E. coli* to irradiation with beams of stripped atomic nuclei has shown that these strains are equally sensitive to irradiation with argon nuclei [26]. Analogously, the differences between radiosensitivities of 7 strains of *E. coli* became smaller with increasing LET, and at very high-LET (~210 keV/μm) disappeared [47]. The interpretation of these results was that the fraction of repairable damage decreases, whereas the fraction of irreversible defects augments, with increasing LET [13; 26].

There are only few papers on the RBE of densely ionizing particles for yeast cells with different capabilities of recovery. Schneider and Kiefer [62] have compared the radiosensitivities of wild-type, *rad*2 and *rad*9 diploid yeasts after exposure to X-rays and α-particles. The RBE of α-particles was almost identical for all these strains. It was demonstrated for yeast cells [51] that the value of the RBE correlated with the repair capacity: RBE was greater for diploid strains of wild-type and heterozygous *rad*51 mutant, which were capable of post-irradiation recovery, than for haploid strains and the homozygous *rad*51 diploid mutant, which were incapable of recovery. Many radiosensitive mutants of yeast cells incapable of recovery are now well known [9; 26]. From the surveys [6; 51; 52] of published papers it seems that uncertainty exists about the magnitude of liquid holding recovery (LHR) after exposure to low- and high-LET radiation. Therefore, it is of interest to study radiation response to low- and high-LET radiations of *S. cerevisiae* to wild-type and different *rad* mutants.

S. cerevisiae cells are diplont: they exist as diploid cells in natural conditions. There are other yeast species which exist in natural conditions as haploids and their diploid cells have been constructed in laboratory conditions. As shown in our teamworks [39; 40], the doubling of cell genome in haplont yeast does not result in notable changes in radioresistance. Unlike diplonts, natural haplonts might be defective in diploid specific recovery. It was shown [22] that this kind of recovery in haplont yeast *Pichia pinus* was absent in haploid cells and was present in a minor extent in diploid ones. The RBE of densely ionizing radiation is influenced by both the physical patterns of energy deposition and cell ability to recover from radiation damage. Similar data for haplont yeast cells are practically unknown. Therefore, it is of interest to study radiation response to low- and high-LET radiations of haplont yeast cells with a single and double set of chromosomes.

Variation of conditions promoting repair processes and cell radiosensitivity also appears to be a determinant of the effectiveness of densely ionizing radiation, the RBE value being the highest for more efficient and complete repair [7; 18]. For additional argument in favor of this point of view, it is desirable to study the dependence of RBE on the stage of growth in which yeast cells of various genotypes are subjected to ionizing radiation.

The shape of diploid yeast cell survival curves is sigmoid independently of radiation quality [17; 25; 36; 53]. It means that the RBE is independent on survival level. On the contrary, for mammalian cells exhibiting sigmoid dose-response curves at low-LET radiation many authors have shown a purely exponential response after densely ionizing radiations [15; 57; 68]. It means that the RBE of densely ionizing radiation for such the cases is strongly dependent on the survival level chosen for comparison. It is noteworthy from a fundamental and practical point of view to provide an appropriate explanation of the observed distinction in radiation response of yeast and mammalian cells to low- and high-LET radiations resulting in various RBE–dose relationships.

For cells of various origins there is apparently slower rate of cellular recovery and DNA repair of damage induced by α-particle and other high-LET radiations than that by γ-rays [11; 31]. These results indicate that either the process of recovery may be impaired or a slower rate of recovery may be a consequence of more severe and complex damage produced by high-LET radiations. The reduction of cellular recovery as expressed in both the extent and the rate of recovery after high-LET radiation might arise either from the damage of recovery mechanisms themselves or from the formation of irreversible damage which cannot be repaired at all. Both processes may take place at the same time. However, the data distinguishing these possibilities are lacking in literature.

Thus, in this Chapter we will deal with the following tasks.

1. To develop a mathematical model for cell responses to ionizing radiation that would combine the basic idea of hit and target principles with biological stochastic.
2. To compare the RBE of densely ionizing radiation for wild-type and radiosensitive mutants of haploid and diploid yeast cells.
3. To study the radiosensitivity of haplont yeast cells irradiated with sparsely and densely ionizing radiations.
4. To estimate the RBE of α-particles for yeast cells irradiated in stationary and exponential phase of growth.
5. To elucidate the reason of different recovery kinetics in yeast cells exposed to low- and high-LET radiations.
6. To determine the RBE of densely ionizing radiation for various criteria of cell death.
7. To suggest a mathematical approach for explanation of various shape of dose-effect curves for eukaryotic cells.

1.1. Probability Model for Cell Responses to Irradiation

The classical hit-and-target models of cell inactivation are known to be founded upon two suppositions: various cells of the homogeneous population irradiated with the same dose are damaged in a different manner in accordance with the hit-principle; cell survival is strictly determined by the number of hits. If the number of hits inside the target/s exceeds a threshold number, the cell is inactivated with unit probability, while in any other cases the cell survives.

However, cell survival represents only one of many complex phenomena which occur in biological material irradiated. The information presently available shows that single cells among both surviving and inactivating parts of a homogeneous population respond differently to the same dose of irradiation. Inactivated cells produce microcolonies consisting of various numbers of cells. These are so-called inactivation forms that can be seen by microscope. Surviving cells produce macrocolonies of various sizes which appear within different times after irradiation and can be seen by unaided eye. In the course of producing of such macrocolonies "lethal sectors" are often formed. This means that such colonies contain inactivated cells. The degree of manifestation of these effects depends on irradiation dose, conditions of culture and biological peculiarities. The diversity of cell responses to irradiation cannot be interpreted by multi-hit and multi-target models because they had been developed only for survival curve description.

Alternatively to the target-theory an idea of biological stochastic was developed [30]. The concept of biological stochastic is based on the two ideas: various cells of a homogeneous population irradiated with the same dose D are uniform in their damage; cell survival (or inactivation) depends on the degree of damages in accordance with the probability low. In accordance with this conception, the development of a cell clone is a sequence of successful and unsuccessful mitoses. It follows that in experiments one should observe lethal sectors, various rates of growth and different clone sizes irrespective of the shape of dose-survival curves. However, we had previously established [33] that such heterogeneous cell responses to ionizing radiation were only observed for objects with sigmoid survival curves. These results are not consistent with predictions of the concept of biological stochastic.

Comparing hit-and-target models with the concept of biological stochastic, one should pay attention to the following peculiar circumstances.

1. It follows from the classical multi-hit model that the shape of survival curves must be either sigmoid or exponential. In contrast, from the initial idea about the probability of successful division it is impossible to draw *a priori* any conclusion concerning the shape of survival curves.
2. From the concept of biological stochastic the existence of lethal sectors, the effect of delayed appearance of clones formed by irradiated cells, the variety of clone sizes, and the diversity of cell inactivation forms may be inferred. The classical models of hit-principle do not predict any of these radiobiological effects.

It is of interest to note that the concept of biological stochastic is convenient for the description of radiobiological effects which cannot be interpreted by means of the hit-and-target model and *vice versa*. In this sense hit theory and biological stochastic are mutually complementary models. It can be anticipated that a generalized model of cell responses to irradiation will have to combine hit-and-target principles and some ideas of biological stochastic.

Let different cells from the same irradiated population receive a various number of damages in accordance with the hit-and-target principle. Suppose that the probability of manifestation of the damage $\alpha < 1$. It is necessary to emphasize the principal distinction of this condition from externally similar suppositions under the analysis of cell recovery [32; 37]. The probability r of manifestation of radiation damage is believed to be less than 1 (r <1) in modeling postirradiation recovery since these damages are considered as only potential ones. In this case, from a potential lesions only $a \cdot r$ will be realized, while $a \cdot (1-r)$ lesions will be recovered. From the classical point of view, the fate of irradiated cells is resolved during the first postirradiation division, strictly determined by the number of realized damages. We believe here that even the realized damage acts upon cell viability with a probability α <1. This probability is considered to be maintained relatively unchanged through successive divisions. The conserved damages responsible for the decrease in α result in further intracellular physical and chemical changes which can give rise to the luck of the following cell division with a probability that is less than 1.

From the classical point of view, manifestation of radiation damage is determined by the cell incapacity to develop a macroscopic colony. Single cell division cannot indicate the lack of the effect as evidenced by different inactivation forms. In our case, we postulate the probability low of manifestation of radiation damage, and a single cell division must be considered as a measure of the effect. With a certain probability radiation damages can give rise to manifestation of unviable progeny in a colony.

Thus, we arrived at the following representation. Damages arising in irradiated cells reduce the probability P of successful division. If the probability of manifestation of a single damage (the probability of refusal) is α, the probability of successful division of a cell with one damage is $P_1 = 1 - \alpha$. The larger number of damages i, the smaller probability P_i of successful cell division. For example, in the simplest case of independent action of separate damages

$$P_i = (1-\alpha)^i . \tag{1.1}$$

It was demonstrated before [33] that for yeast cells Saccharomyces cerevisiae (strain Megri 139-B) α = 0.1. Hence $P_0 = 1$, $P_1 = 0.9$, $P_2 = 0.81$, $P_3 = 0.73$ *etc*. It leads to the decrease of clone-formation rate.

We assume that the reduced probability of successful division can remain unchanged through successive divisions of *cells*. Hence, the mechanism of clone formation is the so-called "birth-and-death" process. The process operates when an entity, in our case a single cell, either gives rise to progeny like itself (birth), or is removed in some way (death), and these two events occur in a random fashion. This model, in which different irradiated cells are damaged in a random fashion in accordance with the hit-and-target principle (as in classical models) and the number of damages defines the probability of successful division P_i, we shall call the probability model of cell responses to irradiation.

This model may have different versions. In particular, the probability P_i may be decreased only from the threshold number of damages (l)

$$P_i = \begin{cases} 1 & if \quad i \le l \\ (1-\alpha)^i & if \quad i \ge l \end{cases} \quad . \tag{1.2}$$

Such a version corresponds to the supposition of limited opportunity of repair cell systems [26] or to the necessity of interactions of some sublethal damages to form an absolutely lethal damage. In addition, the probability P_i may be independent of time or increased with a number of cell generations. One can choose the concrete versions of the probability model after comparison of the model predictions with experimental data. We shall now mention the common properties of all possible versions of the probability model by which they differ from multi-hit and multi-target models. The supposition of the probability of manifestation of radiation damages permits to predict the existence of lethal sectors, the effect of delayed appearance, the variety of clone sizes, and variety of inactivation forms. Therefore, the probability model may be useful in all cases when the diversity of radiobiological effects is observed. For the purpose to demonstrate the possibilities of the probability model to describe cell responses to irradiation, we shall consider the independent version of the probability model, in which separate damages (hit events) independently affect the probability of successful division (Eq. 1.1). In addition, we assume that the reduced probability of successful division can remain unchanged through successive divisions of cells. We shall also compare such a version of the model with experimental data obtained in experiments with yeast cells.

Knowing the dependence of the division probability P_i on the number of damages i, one can calculate expected survival curves and distributions of inactivation forms. Remember that most yeast strains are inactivated after at least one division. It is conceivable that either the probability of refusal means the death idydydydyn both products of a division or the first postirradiation division is always performed with unit probability. In the former case abortive clones must always contain an even number of cells. Since this conflicts with experiments, we shall adopt the latter case which means that the process of clone formation always begins from two initial cells, i.e., only after the first postirradiation division stochastic processes are involved. In such a case, if the probability of successful division is P, the probability of cell to be survival and produce a macrocolony visible by unaided eyes, can be described as

$$S(P) = \begin{cases} 1 - \left[\frac{1-P}{P}\right]^2 & if \quad P \ge 0.5 \\ 0 & if \quad P \le 0.5 \end{cases} \tag{1.3}$$

and the probability φ_n to produce microcolony consisting of n inactivated cells

$$\varphi_n = \frac{2(2n-3)!}{(n-2)!n!} P^{(n-2)} (1-P)^n, \quad n \ge 2. \tag{1.4}$$

For independent version of the model (Eq. 1.1), the cell with i damages can either with the probability S_i to produce a macrocolony (the probability of infinite reproduction) or with the probability φ_n to form a microcolony consisting of n inactivated cells, which may be written as

$$S_i = \begin{cases} 1-\left[\dfrac{1-(1-\alpha)^i}{(1-\alpha)^i}\right]^2 & if \quad (1-\alpha)^i \geq 0.5 \\ 0 & if \quad (1-\alpha)^i \leq 0.5 \end{cases} \qquad (1.5)$$

and

$$\varphi_{ni} = \frac{2(2n-3)!}{(n-2)!n!}(1-\alpha)^{i(n-2)}[1-(1-\alpha)^i]^n. \qquad (1.6)$$

If the number of damages in a cell is determined by Poisson's distribution, the surviving fraction S and the probability $\varphi_n(a)$ to form an abortive colony consisting of n inactivated cells are given by

$$S(a) = \sum_{i=0}^{k} \frac{e^{-a}a^i}{i!}\left\{1-\left[\frac{1-(1-\alpha)^i}{(1-\alpha)^i}\right]^2\right\}, \qquad (1.7)$$

$$\varphi_n(a) = \sum_{i=0}^{\infty} e^{-a}\frac{a^i}{i!}\frac{2(2n-3)!}{(n-2)!n!}P_i^{n-2}(1-P_i)^n =$$

$$= \sum_{i=0}^{\infty} e^{-a}\frac{a^i}{i!}\frac{2(2n-3)!}{(n-2)!n!}(1-\alpha)^{i(n-2)}[1-(1-\alpha)^i]^n. \qquad (1.8)$$

In the latter equations, α is the probability of "refusal" for one damage; k is the highest number of damages for which $(1-\alpha)^k \geq 0.5$ and a is the mean number of damages.

It follows from Eq. (1.8) that the relative number of microcolonies consisting of two, three and four inactivated cells can be estimated by the following equations

$$\begin{aligned} N_2 &= 1-2e^{-\alpha a} + e^{-a[1-(1-\alpha)^2]}, \\ N_3 &= 2\{e^{-\alpha a} - 3e^{-a[1-(1-\alpha)^2]} + 3e^{-a[1-(1-\alpha)^3]} - e^{-a[1-(1-\alpha)^4]}\}, \\ N_4 &= 5\{e^{-a[1-(1-\alpha)^2]} - 4e^{-a[1-(1-\alpha)^3]} + 6e^{-a[1-(1-\alpha)^4]} - 4e^{-a[1-(1-\alpha)^5]} + e^{-a[1-(1-\alpha)^6]}\}. \end{aligned} \qquad (1.9)$$

If $\alpha \geq 0.5$, the calculated survival curves are exponential. In such a case only cells without any damages are viable and macrocolonies produced by these cells do not differ from those produced by unirradiated cells. If $\alpha \leq 0.5$, the calculated survival curves are sigmoid and the smaller α, the larger shoulder of the survival curve should be observed. In this case among the viable cells there are those that have some damages. These cells are characterized by a reduced probability P of successful division and they produce clones with a lower growth rate as compared with unirradiated cells. It follows that the extent of manifestation of such non-lethal effects must increase with increasing survival curve shoulder. Just that very case was observed in experiments with haploid and diploid yeast cells [32; 33].

Theoretically expected results were compared with experimental findings as follows. Representing experimental survival curve in the dependence of dimensionless relative dose units, for example $D/D_{10\%}$, where $D_{10\%}$ is a dose corresponding to 10% survival, one can compare it with a family of theoretical curves constructed in the same dimensionless relative scale of the mean number of hits $a/a_{10\%}$. With this operation, we found [33] that the survival curve of diploid yeast strain Megry 139-B is close to the theoretical curve described by Eq. (1.7) for $\alpha = 0.1$. Since the dose corresponding to 10% survival is 620 Gy, and the average number of hits at 10% survival is 9, it is easy to see that 69 Gy correspond to one hit in a cell. Knowing the conversion factor from the number of hits to the dose and using Eq. (1.7), one can construct theoretically expected survival curve for $\alpha = 0.1$ (Figure 1.1A). It is clear that calculated survival curve (solid line) is in a good agreement with the experimental data (symbols). These and subsequent data of this section have been obtained after cell irradiation at stationary phase of growth with ^{60}Co γ-rays (15 Gy/min). The dose rate was measured with a Siemens ionization chamber.

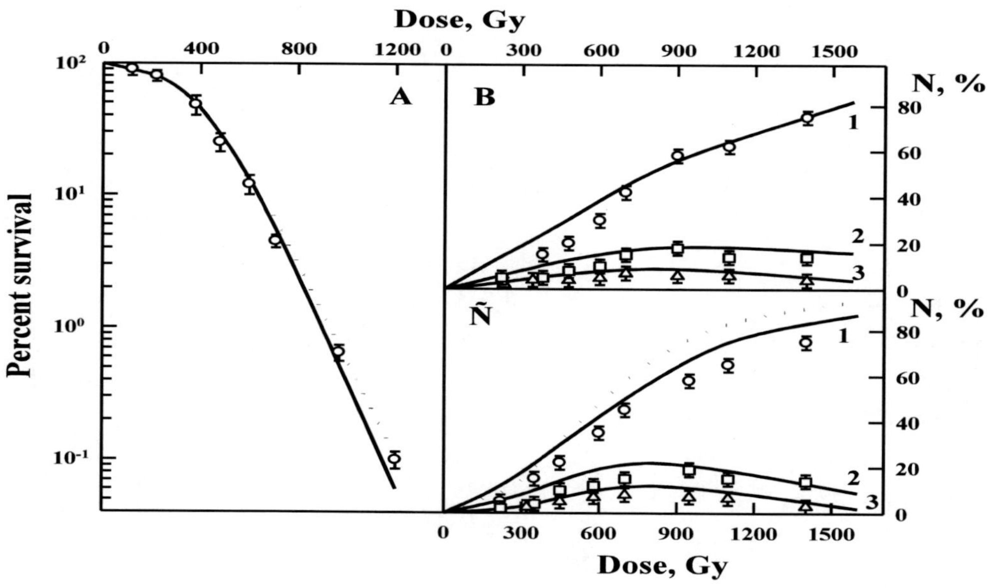

Figure 1.1. Dose dependence of cell survival (A) and the yield of microcolonies consisting of two, three and four inactivated cells (B and C, curves 1, 2 and 3, respectively) calculated according to the probability model (solid lines) and obtained in the experiments with the yeast *S. cerevisiae,* strain Megry 139-B (symbols).

If the subsequence of separate hits independently affect the probability of successful division P (Eq. 1.1), then the relative number of microcolonies consisting of two, three and four inactivated cells can be calculated by Eq. (1.9) for the same values of the basic model parameters ($\alpha = 0.1$ and one hit corresponds to 69 Gy). Figure 1.1B exhibits the results of such calculation by solid lines. Experimental findings are denoted by various symbols. One can see a good agreement between the experimental and predicted data. Similar calculations have been performed for another version of the probability model, when P_i may be decreased only after the threshold number of damages in accordance with Eq. (1.2). Figure 1.1C shows experimental data and theoretically expected curves calculated for $l = 1, \alpha = 0.15$. Dotted lines in Figure 1.1 were calculated for the case when two hits don't lead to the decrease in the probability of "refusal" ($l = 2, \alpha = 0.2$). It is evident that the survival curves are described by both versions of the model, whereas the "threshold" variant is not suitable to predict the output of various forms of inactivation. Therefore, we shall consider further only the independent version of the model, in which each hit results to the failure of cell division independently of each other.

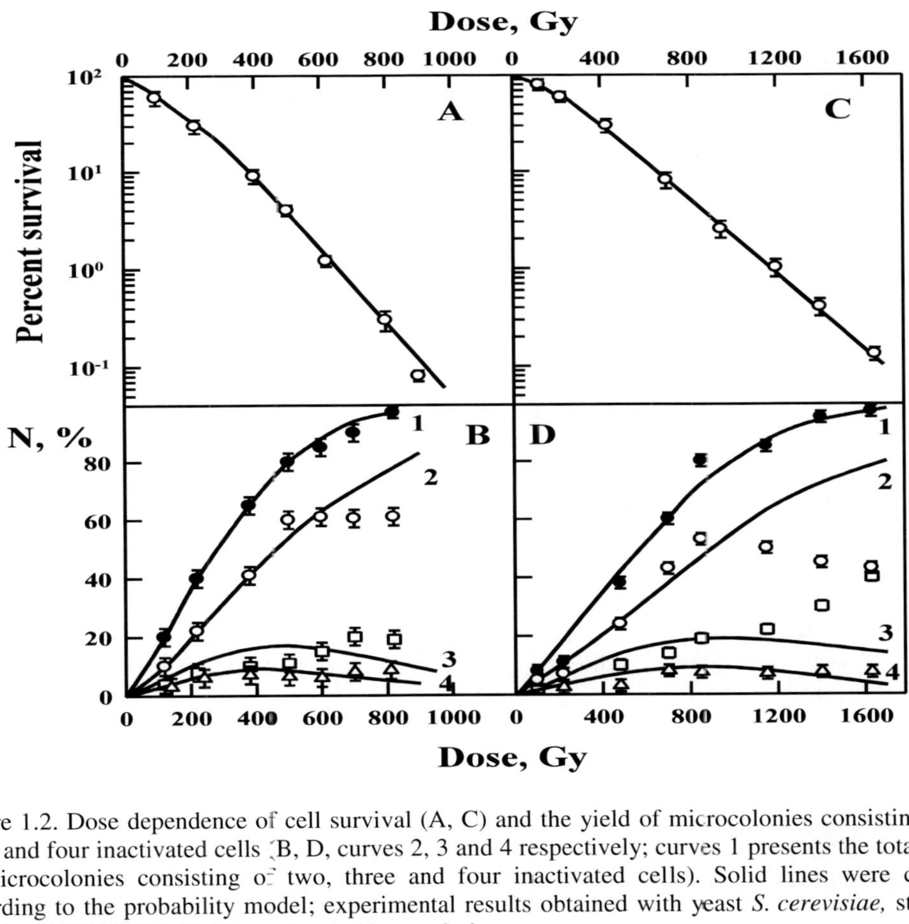

Figure 1.2. Dose dependence of cell survival (A, C) and the yield of microcolonies consisting of two, three and four inactivated cells (B, D, curves 2, 3 and 4 respectively; curves 1 presents the total number of microcolonies consisting of two, three and four inactivated cells). Solid lines were calculated according to the probability model; experimental results obtained with yeast *S. cerevisiae,* strain 39-2 (A, B) and YO2022D (C, D) are denoted by symbols.

From Figure 1.1B it follows that the relative number of microcolonies consisting of two inactivated cells is somewhat less at low doses than expected. This discrepancy between the experimental and theoretically predicted data is well reproduced for other strains. Apparently, the probability model does not take into account some factors whose influence is particularly important in small doses. One of these factors might be radiation recovery, which has almost no effect on survival, but reduces the output of this inactivation form. We tested this suggestion for two yeast strains of *S. cerevisiae* that are unable to recover from radiation damage: Megry 39-2 (Figures 1.2A and 1.2B) and YO2022D (Figures 1.2C and 1.2D).

All solid lines in these figures are the theoretically expected. For both strains $\alpha = 0.25$ and $a = D$, Gy/80 Gy for Megry 39-2 and $a = D$, Gy/160 Gy for YO2022D. In both cases there is no discrepancy between the observed and expected number of microcolonies consisting of two inactivated cells in small doses. Apparently, the observed discrepancy is due to radiation recovery of cells.

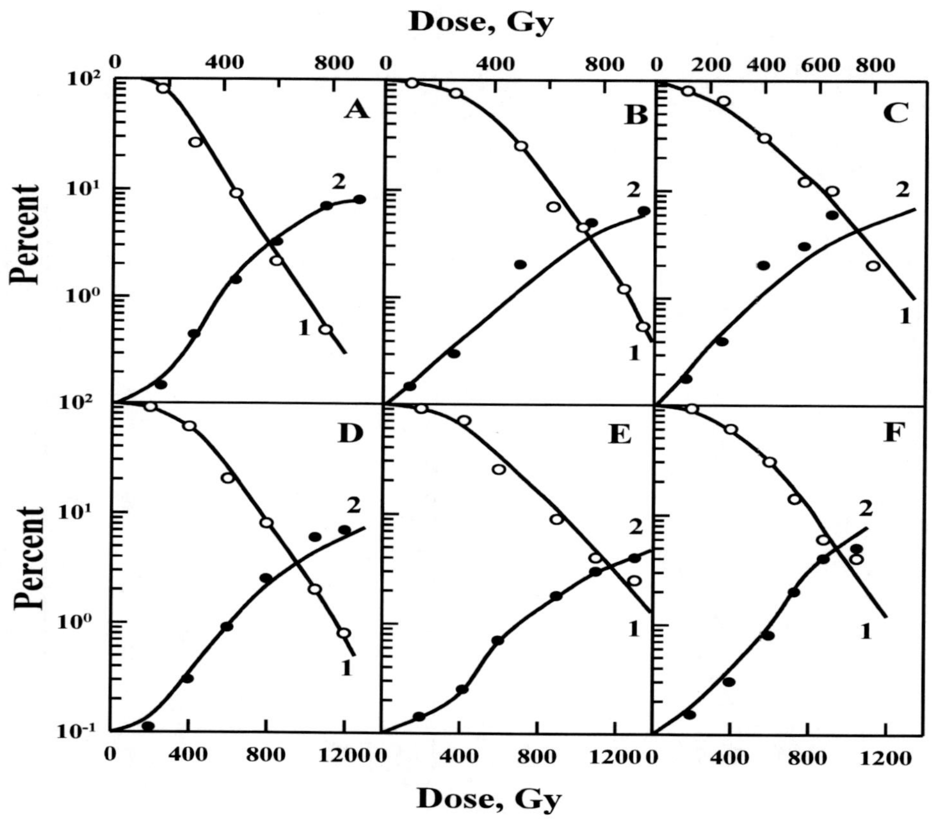

Figure 1.3. Dose dependence of cell survival (%, curves 1) and the total number of microcolonies consisting of two, three and four inactivated cells (%, curves 2) calculated by the Eqs. (7) and (9) (solid lines) and obtained in experiments (symbols) with various yeast cells: *S. cerevisiae*, strains 77A-4П3 (A), 211-1aM/2 (B), YO2587 (C) and 1306 (D); *Saccharomyces ludwgii* 3134 (E); *Saccharomyces globus* 349 (F).

Noteworthy is the decrease in the number of microcolonies consisting of two inactivated cells and corresponding increase in the number of microcolonies consisting of three inactivated cells at the region of high radiation doses (Figures 1.2B and 1.2D). It seems that at high doses some of irradiated cells that should be inactivated after the first division accomplished an additional abortive division, a peculiar radiation hormesis at high doses. It is interesting that the total number of microcolonies consisting of two, three and four inactivated cells coincides with the expected value within the entire range of doses (Figure 1.2B and 1.2D, curves 1). To be sure in these results, we present here our data obtained with six other diploid yeast strains (Figure 1.3).

In this case, we determined the basic parameters of the probability model by means of comparison of experimental and theoretical survival curves represented in dimensionless relative scale and then evaluated the yield of the total number of microcolonies consisting of two, three and four inactivated cells (curves 2) calculated by the Eq. (1.9) (solid lines) and obtained in experiments (symbols) with various yeast cells. It can be seen that the theoretical predictions are in good agreement with the experimental data.

Figure 1.4 shows the results of our experiments with haploid yeast cells of *S. cerevisiae*, strain 7D. Survival curve of this strain is exponential, so we can determine the mean dose D_0 for one hit. As concerns the magnitude of α, we can only assert that $\alpha \geq 0.5$. For this strain, the yield of the total number of microcolonies consisting from two, three and four inactivated cells, calculated by Eq. (1.9), coincides with experimental data fairly well for $\alpha = 0.55$.

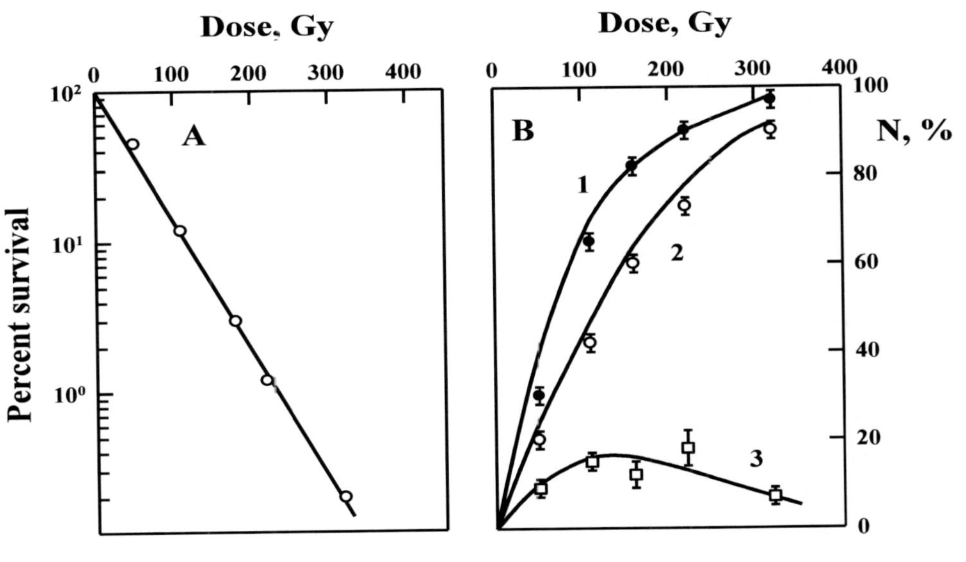

Figure 1.4. Dose dependence of cell survival (A) and the yield of microcolonies consisting of two and three inactivated cells (B, curves 2 and 3, respectively; curves 1 presents the total number of microcolonies consisting of two, three and four inactivated cells). Solid lines were calculated according to the probability model; experimental results obtained with haploid yeast cells of *S. cerevisiae,* strain 7D are denoted by symbols.

If the probability of refusal α depends on cultivation conditions, it would be possible to modify the surviving fraction as well as the shape of the survival curve by an appropriate culture condition. In particular, one may convert a sigmoid survival curve into an exponential one by a change of cultivation conditions for the worse. In terms of the model, such a treatment is equivalent to an augmentation of α. This prediction of the model was tested with diploid yeast cells growing after irradiation in a medium containing of 8% NaCl. Figure 1.5A shows the survival curves of haploid yeast cell *Z. bailii* (strain K17) containing 5% of cells with small buds and cultivated after irradiation with γ-ray of ^{60}Co at 20, 30 and 34 °C. The data indicate the relationship of temperature modification with the shape of dose-effect curve. The exponential part of this curve exhibits the radiation response of single haploid cells, for which one hit is already effective. So, the additional thermal sublesions have no influence to their radiosensitivity. The opposite situation is observed for the radioresistant tail of the survival curve, which reflects the number of cells with double number of chromosomes, but the daughter chromosomes have not yet transited to buds. This part of the population reacts similarly to diploid cells, for which several hits are needed to be inactivated. Due to this fact, the form of dose-effect curve at this dose region is sigmoid and thermal load affect the radiosensitivity.

If this reasoning is true, one may select conditions that can modify not only survival but also the shape of the dose-effect curve [10]. It is a well known fact that γ- or X-ray survival curves of haploid yeast cells taken from a growing culture consist of two components; ***a*** simple exponential portion at low doses followed by ***a*** multihit more radioresistant tail at high doses. According to the model under consideration it means that a part of cell population is defined by the enhanced probability of manifestation of radiation damages when $\alpha \geq 0.5$, while for other cells $\alpha < 0.5$. Hence, the change of cultivation conditions for the worse should cause the modification of the surviving fraction of cells only after high doses of irradiation corresponding to the radioresistant component. This prediction was confirmed in experiments with haploid yeast cells (Figure 1.5A). Figure 1.5B exhibits survival curves of diploid yeast cells of *S. ellipsoideus* (strain Megri 139-B) irradiated with γ-ray of ^{60}Co. After exposure cells were incubated on standard nutrient media at 30 and 37 °C (curves 1 and 2, respectively) and on the media additionally containing 8% NaCl also at 30 and 37 °C (curve 3, open and closed triangles, respectively). It can be seen that the salt not only reduces the survival of irradiated cells, but also transformed sigmoid shape of dose-effect curve into exponential. As expected, increased cultivation temperature result in a decline of cell survival on standard media, where dose-effect curves are sigmoid (curves 1, 2), and does not affect the cell survival when the dose-effect curve is exponential (curve 3).

Thus, the qualitative and quantitative results of radiobiological experiments with yeast cells do not contradict the probability model based on the supposition that clone formation is a random process and the probability of successful division of cells is determined by the number of damages. It is clear from the above description that the probability model is a peculiar synthesis of hit-and-target principles and biological stochastic. According to this model, cell responses to irradiation are consequences of two random factors – a random distribution of energy absorption events and biological stochastic.

It should be emphasized that for quantitative comparison of the model with experimental findings it is not necessary to know the precise dependence of the probability P on a number of damages; it is only important to know that the probability P of successful division decreases with increase in the number of damages and remains unchanged through successive divisions of cells. The preceding considerations permit to assert that the main supposition of the model under discussion (after irradiation $P<1$) seems to be consistent with the presented experimental data. Additional suppositions concerning the connection of the probability P with a number of damages and a number of cell generations are not of great importance. The specific experimental data can provide a basis for the selection of the suitable version of the probability model.

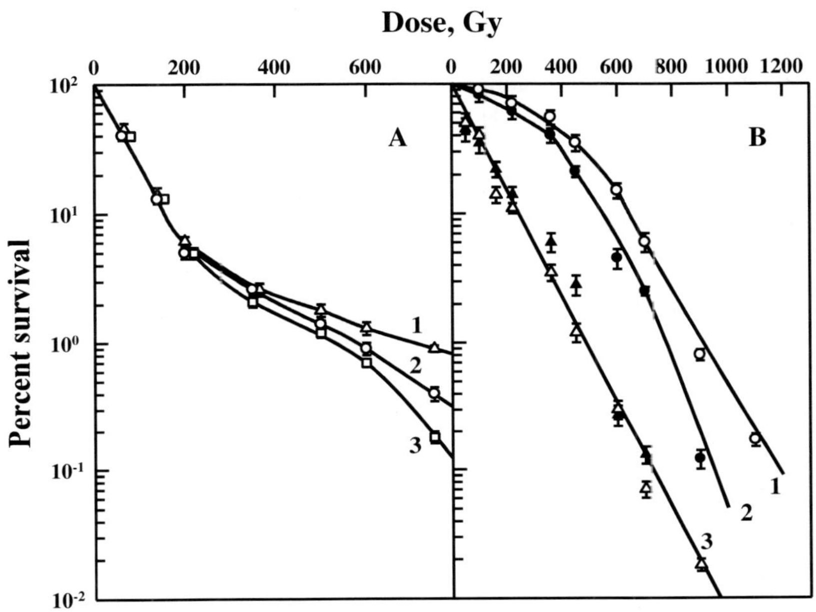

Figure 1.5. A: survival curves of haploid yeast *Z. bailii* (strain k17) obtained at various temperatures of postradiation cultivation: 20, 30 and 34 °C (curves 1, 2 and 3, respectively).
B: survival curves of diploid yeast S. ellipsoideus (vini), strain Megri 139-B. After exposure cells were incubated on standard nutrient media at 30 and 37 °C (curves 1 and 2, respectively) and on the media additionally containing 8% NaCl also at 30 and 37 °C (curve 3, open and closed triangles, respectively).

The attractiveness of the model developed consists not only in the description of the qualitative and quantitative regularities of different cells responses to radiation, but also in the fact that it can be used to predict many different effects of irradiation observed in experiments. In fact, if the behavior of irradiated cells is determined by the probability of refusal which depends on a number of damages and remains unchanged through successive divisions of cells, one can find and explain the following issues.

1. The survival curve may be either exponential or sigmoid in dependence on α, the probability of manifestation of a single damage.
2. For objects with sigmoid survival curves (diploid yeasts) one may observe lethal sectoring, the delayed appearance of clones and the diversity of macrocolony sizes after irradiation, while for objects with exponential survival curve (haploid yeasts) these non-lethal effects of radiation are negligible.
3. The inactivated cells produce microcolonies consisting of a different number of cells (different inactivation forms).

4. Among the different inactivation forms of haploid yeast cells the inactivation after a smaller number of divisions must predominate (as compared with diploid yeast at the same level of surviving fraction).
5. The viability of the cell (the reliability of the system) depends on cultivation conditions, so the probability of refusal α must also depend on cultivation conditions and consequently the degree of manifestation of different radiobiological effects must also depend on pre- and post-irradiation conditions.
6. The change of cultivation conditions for the worse has to be accompanied by the modification of the surviving fraction and the form of the survival curve for diploid and not for haploid yeast cells.
7. Cells from different clones produced *by* irradiated diploid yeast cells have a different sensitivity to cultivation conditions and irradiation.
8. *C*ells from the same small clones can produce macrocolonies of various sizes on nutrient media as well as different inactivation forms.

1.2. RBE of Densely Ionizing Radiation for Wild-Type and Radiosensitive Mutants of Yeast

A lot of investigations of RBE of ionizing particles differing in LET have been reported for a variety of biological test systems [1; 8; 35; 71]. As was also cited, studies with different biological objects indicate that high-LET radiation produces a greater proportion of irreversible damage in comparison with sparsely ionizing radiations.

There are conflicting reports about the capability of yeast cells to recover from radiation damage after low- and high-LET radiations. According to some authors, the magnitude of recovery in diploid yeast was independent of LET [46; 62[. These results contrast markedly with the aforementioned data obtained with *E. coli*. However, other investigators have shown that an irrepairable fraction of radiation damage is essentially increased for densely ionizing particles in comparison with sparsely ionizing radiation. Pohlit and Schäfer [56] have pointed out that the character of recovery is determined by a cell-energy reservoir, which may depend on pre-irradiation conditions of culture. In particular, in experiments with cells grown to the stationary state in a liquid-feeding medium, they found that the magnitude of recovery was less than that with cells grown as a dense cell layer on a nutrient agar. Thus, one may assume that the contradictions with yeast cell recovery after low- and high-LET radiation may be connected with conditions of cell cultivation. Radiosensitive mutants of yeast cells incapable of recovery are now well known. In the yeast *S. cerevisiae* there are at least 8 distinct loci that control sensitivity to ionizing radiations, while the number of independent genes capable of affecting yeast sensitivity to ultraviolet (UV) light exceeds 20 loci [9, 26]. Some of the mutant genes confer sensitivity to both UV and ionizing radiation. From the surveys of published papers it seems that uncertainty exists about the magnitude of LHR after exposure to low- and high-LET radiation. Therefore, it is of interest to study radiation response to low and high-LET radiations of haploid and diploid *S. cerevisiae* yeast cell of wild-type and different *rad* mutants.

Before irradiation strains were grown on a solid complete (wort-agar) medium for 3–5 days to stationary phase, then washed with distilled water and resuspended to make a stock solution (10^6 cells/ml). Microscopic examination of cell suspensions usually revealed less than 1–2% budding cells, no clumps and no asci. Cells from the same stock solution were irradiated with ^{60}Co γ-rays (10 Gy/min, LET ~ 0.2 keV/μm) and with ^{239}Pu α-particles (17 Gy/min, LET ~ 134 keV/μm). The γ-ray dose rate was measured with a Siemens ionization chamber. The α-particle dose rate was determined by measuring the intensity and energy of the particles with a semi-conductor silicon surface barrier detector at a distance corresponding to the cells. The mean energy of the α-particles at the surface of the cells was ~3 MeV, which corresponds to a LET of about 134 keV/μm in unit density tissue. Just at this LET value, the maximum in RBE–LET relationship was observed for most eukaryotic and some prokaryotic unicellular organisms. The dose rate was calculated according to the formula $D = 1.6 \cdot 10^{-3} \cdot LF$ (Gy/min), where L is the rate of energy loss in eV/μm and F is the flux of particles/μm^2 [4]. The γ-contamination of the α-source used was negligible, as shown by a control experiment in which the cells were shielded from the α-particles by a thin foil. The small range of α-particles necessitated the use of a monolayer of yeast cells for α-irradiation: 0.02 ml of cell suspension was placed on the surface of non-nutrient agar, and the water from this drop of suspension was evaporated. The source of α-particles was placed in a plexiglass box with fixed channels allowing the cells on the agar surface to be brought in close proximity (3–4 mm of air) of the source. After α-exposure, the cells were rinsed from the small piece of agar layer and plated on nutrient agar after appropriate dilution. The viability of yeast cells after γ-irradiation was the same for cells on the agar surface as for cells in water suspension. All irradiations were carried out in air at temperatures between 19 and 22 °C. Immediately after irradiation, samples were diluted to the appropriate cell concentration with distilled water and plated on nutrient agar plates for the assay of colony-forming ability. Colonies were scored when they reached a suitable size and the counts checked again after a further period to ensure that a final score had been reached.

Immediately after irradiation, a part of the samples was plated on nutrient agar plates for the assay of colony-forming ability. Another part of the irradiated cell suspension was placed into conditions suitable for the LHR, and in 3 days (delayed plating) their colony-forming ability was checked again. LHR was carried out in water suspension at 30 °C without constant agitation. Survival response on immediate and delayed plating was determined on the basis of the colony counts obtained at the end of 5–7 days of incubation at 30 °C. When irradiated cells were held in water at 30 °C for 3 days before plating, the survival of the wild-type diploid cells was greatly enhanced in comparison with that obtained on immediate plating. This recovery was blocked in haploid strains and some radiosensitive diploid mutants: the number of cells capable of forming visible colonies was not increased when plating was delayed for a few days.

Various parts of the experimental results related with RBE of high-LET radiation which are offered in this and subsequent sections have been published previously [51–53; 55]. Here they are published in the most complete and recalculated form.

Figure 1.6 shows the survival curves for haploid yeast of different strains irradiated with γ-rays (open circles) and α-particles (closed circles). Experimental points in this and other figures have standard errors of about 2%. All survival curves are exponential and can be described by the mean lethal dose D_0, which is the dose needed to reduce the survival S to $S \cdot e^{-1}$ or about $0.37 \cdot S$. The mean lethal doses, D_0, after γ- and α-irradiations, calculated from the data presented in Figure 1.6 are listed in Table 1.1. The RBE values of α-particles, defined as the ratio of the mean lethal doses after low- and high-LET radiations RBE = $D_0(\gamma)/D_0(\alpha)$, are also summarized for haploid strains in Table 1.1. It is evident that the RBE of α-particles was somewhat higher for wild-type than for mutant cells with the exception of *rad*54.

Figures 1.7 and 1.8 show the survival curves of wild-type diploid yeast cells and of different homozygous and heterozygous diploids mutant cells exposed to γ-rays (open circles) and α-particles (closed circles).

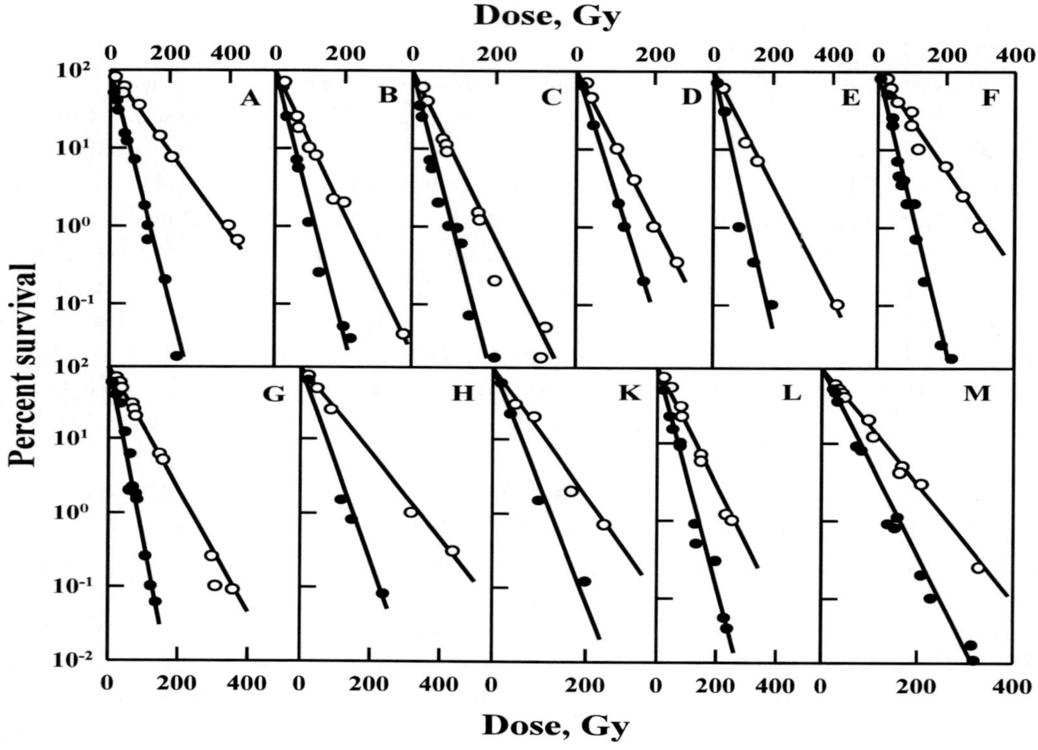

Figure 1.6. Survival curves of haploid *S. cerevisiae* yeast cells of various strains exposed to γ-rays (open circles) and α-particles (closed circles). A – S288C (RAD); B – g218/6b (a *rad*50); C – g218/4c (α *rad*50); D – g218/7a (a *rad*51); E – gl60/2d (α *rad*52); F – gl51/12c (a *rad*53); G – g218/7c (a *rad*54); H – X1687/101b (a *rad*55); K – gl51/10d (a *rad*56); L – g246/6b (a *rad*57); M – g246/27c (α *rad*57).

Table 1.2 summarizes the following radiobiological parameters of these curves: the mean lethal dose after gamma, $D_0(\gamma)$, and alpha, $D_0(\alpha)$, irradiation, the fraction of irreversible damage (K), and the RBE of α-particles. The fraction of irreversible damage was defined as the ratio of radiation doses after immediate and delayed plating to produce the same effect. This fraction was calculated at 10% survival. The survival curves for a number of homozygous *rad* mutants and their wild-type cells are collected in Figure 1.7. Radiobiological parameters of these strains are summarized in Table 1.2. After both kinds of irradiation, the wild-type cells showed sigmoid inactivation kinetics. For all mutant diploids the extrapolation number was decreased in comparison with that for wild-type cells even after low-LET radiation.

The survival curves obtained with α-particles (Figure 1.7, Table 1.2) show a smaller shoulder (extrapolation number) and a steeper exponentially decreasing part. Moreover, the wild-type cells and some of the homozygous mutant cells were capable of recovery, while others did not have this ability. Here again, the fraction of irreversible damage after LHR was greater for α-irradiation than that observed for γ-rays (Table 1.2).

As shown in Figure 1.8 and Table 1.2, the survival curves of heterozygous *rad*54 strains were somewhat lower than that of the homozygous diploid of wild-type. Figure 1.8 shows that both wild-type and heterozygous cells, in contrast with haploid and homozygous diploid mutants, were capable of recovery: their survival was greatly enhanced after delayed plating. Table 1.2 shows that the fraction of irreversible damage was greater, i.e., the magnitude of post-irradiation recovery was smaller after α-irradiation than that observed for γ-rays. Table 1.2 shows that diploid strains with the same genotype are not strongly different in their radiobiological parameters. Here again, the magnitude of post-irradiation recovery was smaller after α-irradiation than that observed for γ-rays (Table 1.2). The RBE of α-particles for heterozygous strains was a little lower than that for homozygous diploid cells of wild-type and much larger than that for the homozygous mutant strain.

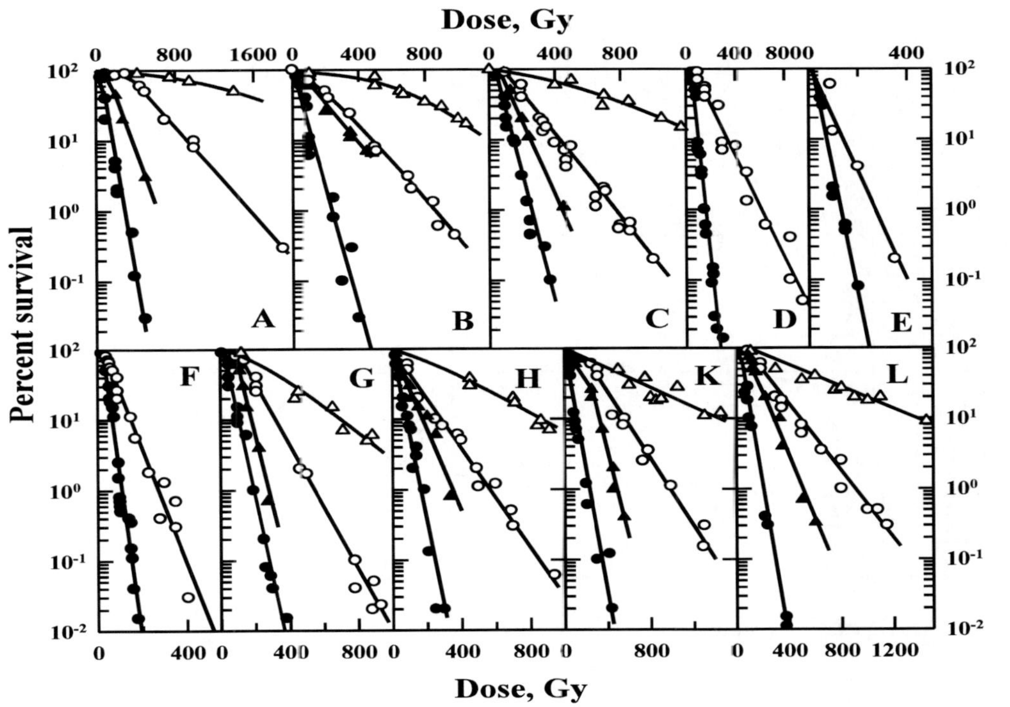

Figure 1.7. Survival curves of diploid *S. cerevisiae* yeast cells of various strains exposed to γ-rays (open circles) and α-particles (closed circles). The cells were plated immediately after irradiation (open and closed circles) and after 3 days of LHR (open and closed triangles). A – XS800 (RAD/RAD); B – XS1956 (*rad*6-1/*rad*6-1); C – XS1924 (*rad*18-2/*rad*18-2); D – XS1806 (*rad*50-1/*rad*50-1); E – XS806 (*rad*51-1/*rad*51-1); F – XS1898 (*rad*52-1/*rad*52-1); G – XS1889 (*rad*53-1/*rad*53-1); H – XS1935 (*rad*55-3/*rad*55-3); K – XS1943 (*rad*56-1/*rad*56-1); L – XS1988 (*rad*57-1/*rad*57-1).

Table 1.1. Radiobiological parameters of *S. cerevisiae* haploid yeast cells of wild-type and various radiosensitive mutants

Strain	Genotype	D_0 (γ), Gy	D_0 (α), Gy	RBE
S288C	RAD	86.0	27.0	3.2
g218/6b	a *rad*50	45.0	24.0	1.9
g218/4d	α *rad*50	35.0	18.0	1.9
g218/7a	a *rad*51	52.0	32.0	1.6
gl60/2d	α *rad*52	54.0	27.0	2.0
gl51/12c	a *rad*53	60.0	24.0	2.5
g218/7c	a *rad*54	50.0	17.0	3.1
X1687/101b	a *rad*55	60.0	27.0	2.2
gl51/10d	a *rad*56	46.0	23.0	2.0
g246/6b	a *rad*57	55.0	26.0	2.1
g246/27c	α *rad*57	55.0	32.0	1.7

Table 1.2. Radiobiological parameters of *S. cerevisiae* diploid yeast cells of wild-type and radiosensitive mutants

Strain	Genotype	γ-Rays		α-Particles		RBE
		D_0, Gy	*K*	D_0, Gy	*K*	
XS800	RAD/RAD	265	0.32	52	0.50	5.1
XS805	RAD/*rad*51-1	160	0.23	33	0.41	4.8
XS1956	*rad*6-1/*ra*d6-1	170	0.30	43	0.30	4.0
XS1924	*rad*18-2/*rad* 18-2	120	0.30	46	0.52	3.3
XS1806	*rad*50-1/*rad*50-1	120	1.00	30	1.00	3.9
XS806	*rad*51-1/*rad*51-1	60	1.00	26	1.00	2.3
XS1898	*rad*52-1/*rad*52-1	50	1.00	16	1.00	3.1
XS1889	*rad*53-1/*rad*53-1	130	0.50	30	0.68	4.3
XS1935	*rad*55-3/*rad*55-3	130	0.37	34	0.54	3.9
XS1943	*rad*56-1/*rad*56-1	200	0.43	45	0.44	4.4
XS1988	*rad*57-1/*rad*57-1	186	0.36	43	0.44	4.3
L5	RAD/RAD	220	0.44	50	0.60	4.2
L6	RAD/RAD	230	0.34	45	0.37	5.1
L1	RAD/*rad*54-1	150	0.48	33	0.44	4.5
L2	RAD/*rad*54-2	185	0.62	55	0.63	3.7
L8	*rad*54-1/*rad*54-1	32	1.00	20	1.00	1.6
L9	*rad*54-2/*rad*54-2	59	1.00	22	1.00	2.7
L10	*rad*54-3/*rad*54-3	50	1.00	26	1.00	1.9

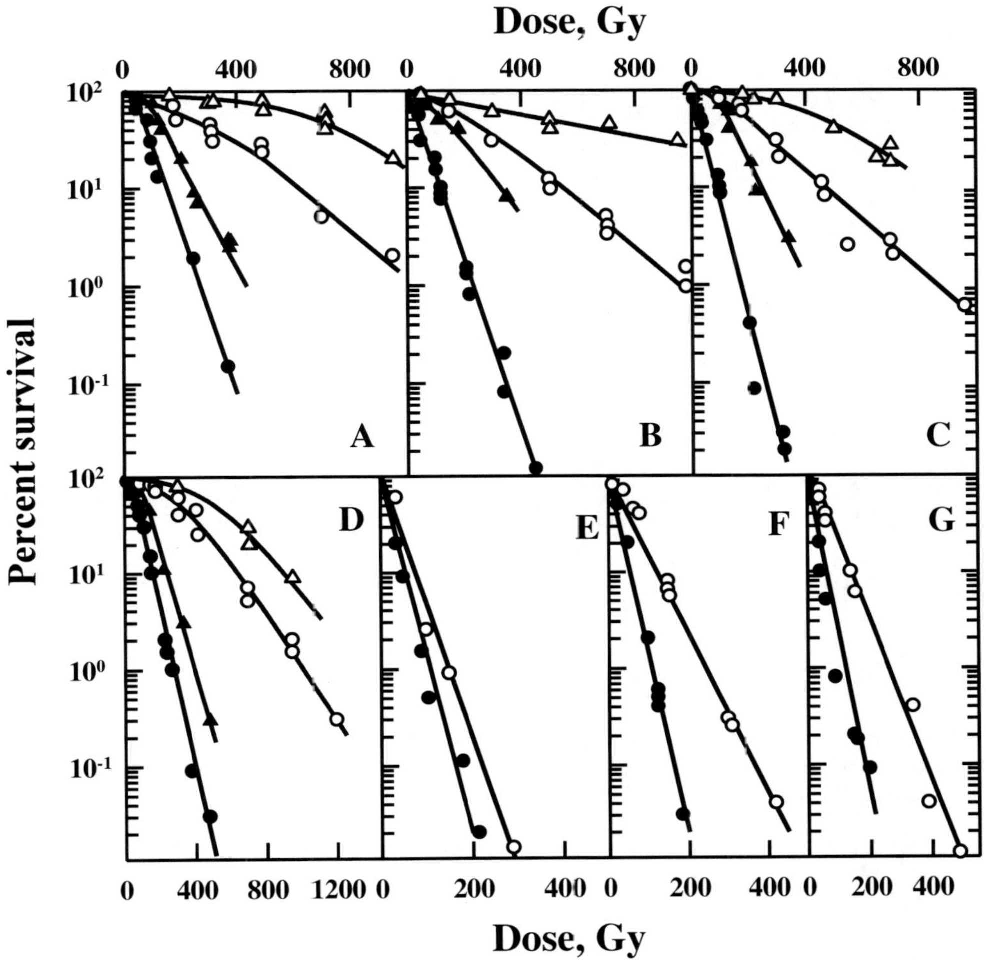

Figure 1.8. Survival curves of homozygous diploid *S. cerevisiae* yeast cells of wild-type and various yeast strains with *rad*54 mutation exposed to γ-rays (open circles) and α-particles (closed circles). The cells were plated immediately after irradiation (open and closed circles) and after 3 days of LHR (open and closed triangles). A – L5 (RAD/RAD); B – L6 (RAD/RAD); C – L1 (RAD/*rad*54-1); D – L2 (RAD/*rad*54-2); E – L8 (*rad*54-1/*rad*54-1); F – L9 (*rad*54-2/*rad*54-2); G – L10 (*rad*54-3/*rad*54-3).

To examine the dominance of *rad*51 mutation, similar experiments were performed on heterozygous diploid cells of *S. cerevisiae* (strain XS805, RAD/*rad*51). As shown in Figure 1.9, the survival curves of this strain were similar to those of the homozygous diploid of wild-type: this strain showed similar sigmoid inactivation kinetics after both kinds of radiation. Again, the magnitude of LHR was smaller after α-irradiation than that observed for γ-rays, i.e., the proportion of irreversibly damaged cells (irreversible component) was greater after exposure to densely ionizing radiation (Table 1.2).

It can be also seen that the RBE of α-particles for this strain was 4.8, a little lower than that for homozygous diploid cells of wild-type and much larger than that for homozygous mutant strain. These results show that radiobiological responses of heterozygous diploid cells are similar to those of wild-type homozygous diploid cells. This suggests that the dominance of the wild-type genes over the *rad*51 allele is almost complete.

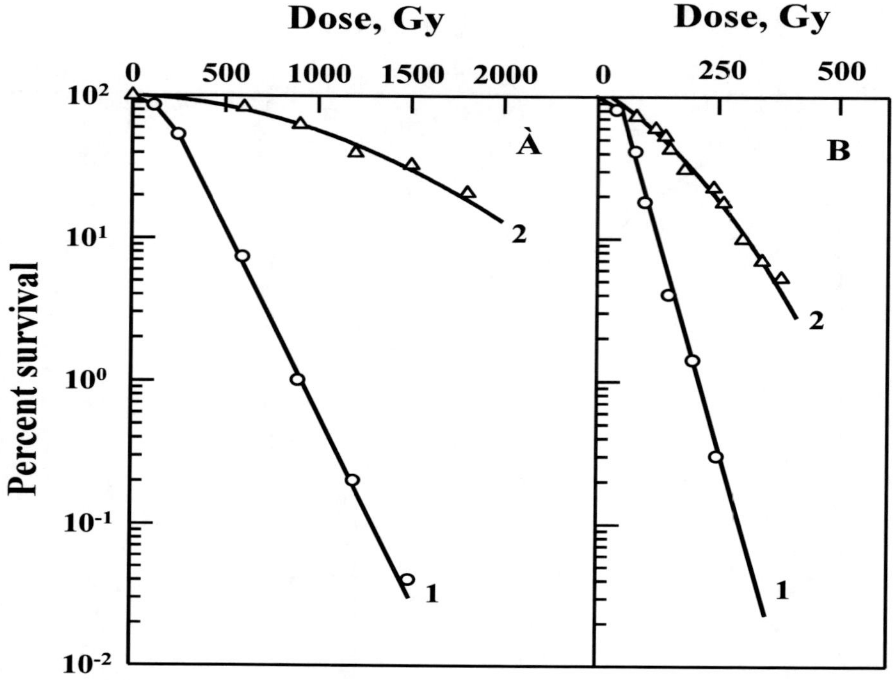

Figure 1.9. Survival-versus-dose curves of heterozygous diploid yeast cells of *S. cerevisiae*, strain XS805 (RAD/*rad*51) exposed to γ-rays (A) and α-particles (B). The cells were plated immediately after irradiation (curves 1) and after 3 days of LHR (curves 2).

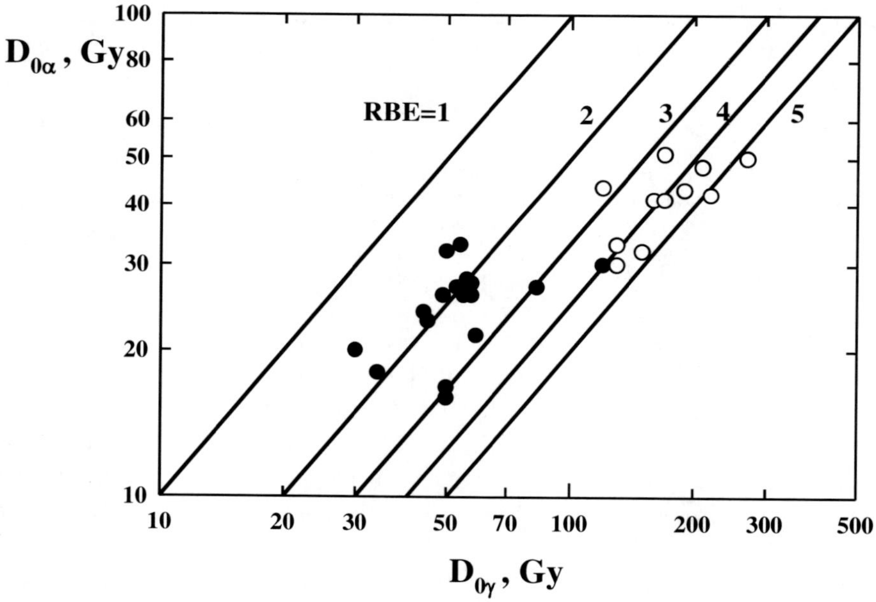

Figure 1.10. The relation of D_0 values for γ-rays and α-particles for various *S. cerevisiae* yeast strains. Open circles denote the results for cells that were able to recover radiation damage; the closed circles denote the results for cells incapable of recovery.

Figure 1.10 shows how D_0 values for γ- and α-irradiation were related for all strains studied. Open circles denote the data for cells that were able to recover radiation damage, and closed circles denote the data for cells incapable of recovery. The cells were considered to be capable of recovery if their colony-forming ability was enhanced after delayed plating. In this figure the lines of equal values of the RBE are also graphed. The RBE is seen to increase rather smoothly with the value of D_0. Figure 1.10 also shows that the variation of the mean lethal doses was more pronounced after γ- than after α-irradiation.

Thus, we compared the RBE of densely ionizing radiation for wild-type and radiosensitive mutants of yeast cells of various ploidies and genotypes. Results reported here reveal the correlation between RBE values and cell repair capacity: strains that were most resistant, i.e., were more "capable of recovery", had the highest RBE values, whereas those strains that were most sensitive, i.e., were deficient in "repair capacity", had significantly lower values of RBE.

So, the values of the RBE of α-particles were only slightly different for wild-type cells and for mutants that retained an ability to increase their survival after LHR ($K < 1$). For mutant cells incapable of recovery ($K = 1$), the RBE of densely ionizing radiation was markedly smaller. It follows that the increase in the RBE may be related to cell-repair ability. Our data showed that the RBE of α-particles was always greater for diploid than for haploid yeast cells. These data also support the discussed hypothesis that the RBE of densely ionizing radiation may be correlated with cell-repair capacity, because it is well known that the ability of diploid cells to recover radiation damage is higher than that of haploid cells. It can be concluded that the recovery, which may be responsible for a greater value of the RBE or for the decrease in the difference of radiosensitivities between wild-type and mutant cells with increasing LET, is relevant in greater extent to diploid cells.

1.3. Radiosensitivity of Haplont Yeast Cells Irradiated with Sparsely and Densely Ionizing Radiations

In the previous section, we have given our summarized data with *S. cerevisiae* yeast cells showing that the RBE of α-particles is defined by both the primary process of radiation interaction with biological objects and the process of cell recovery at the biochemical stage of radiation damage formation. *S. cerevisiae* cells are diplont: they exist as diploid cells in natural conditions. There are other yeast species which exist in natural conditions as haploids and their diploid cells have been constructed in laboratory conditions. As shown in our teamwork [40], the doubling of cell genome in haplont yeast does not result in notable changes in radioresistance. The relation between radiosensitivities of haploid and diploid strains of haplont yeasts (*Pichia pinus, Pichia guilliermondii*) was almost the same as for some radiosensitive *rad* mutants in *S. cerevisiae*, which were unable of diploid specific recovery. It was shown [22] that LHR of yeast *P. pinus* was absent (the irreversible component $K = 1$) in haploid cells and was present in a minor extent in diploid cells ($K = 0.8$). As was given above, the RBE of densely ionizing radiation was considerably higher for diploid (3.5–5.0) than for haploid (about 2.0) cells of *S. cerevisiae*. These data can be considered as an indication of the importance of postirradiation recovery in the RBE of high-

LET radiation. More cogent arguments supporting this view came from experiments with radiosensitive cells defective in recovery processes (section 1.2). Thus, the RBE of densely ionizing radiation is influenced by both the physical patterns of energy deposition and cell ability to recover from radiation damage. Similar data for haplont yeast cells are practically unknown. Therefore, it is of interest from a fundamental point of view to study radiation response to low- and high-LET radiations of haplont yeast cells with a single and double set of chromosomes.

Different haploid and diploid strains of haplont yeast cells were used in this investigation. Haploid strain of *P. pinus* (MH4) was obtained from Dr. L. Wünsche (Germany). This strain was established to be homothallic; diploid strain (MH-4D) has been constructed by autodiploidization [69]. Heterothallic haploid strains of *P. guilliermondii* of both mating types (NRRLY-2075, NRRLY-2076) were kindly supplied by Dr. C.P. Kurtzman (USA). Diploid strain (NRRLY 2075 x 2076) was constructed through mating of haploid cells [64]. *Pichia* yeast cells of all strains were incubated at 30 °C. Before irradiation, the cells were grown for 3–5 days on solid medium (yeast extract, 1%; bacto-peptone, 2%; dextrose, 2%; agar, 2%). The duration of preirradiation cultivation of various strains was determined by the time needed to attain stationary phase of growth when the number of small budding cells was less than 2%. The cells were then resuspended in sterile water and after appropriate dilution were irradiated from the same stock solution with ^{60}Co γ-rays (12 Gy/min) and with ^{239}Pu α-particles (25 Gy/min).

The typical survival curves of all strains studied exposed to both γ-rays and α-particles are presented in Figures 1.11–1.12. These curves were constructed on the basis of averaging of our previously published results [38; 40]. Experimental points in all figures have errors of 2% or less, that is approximately the size of the point. Dose-effect curves have been drawn by visually fitting the experimental points.

Table 1.3. Radiobiological parameters of yeasts of different species and ploidy

Strains (ploidy)	$D_0(\gamma)$, Gy	$D_0(\alpha)$, Gy	RBE	DNA content, 10^{-14} g/cell
Pichia pinus				
MH-4 (haploid)	51.3 ± 4.8	26.1 ± 2.9	1.96 ± 0.11	2.4 ± 0.3
MH-4D (diploid)	58.1 ± 6.1	27.7 ± 3.1	2.10 ± 0.10	4.9 ± 0.5
Pichia guilliermondii				
NRRLY-2075 (haploid)	145.0 ± 17.0	41.3 ± 4.6	3.51 ± 0.27	1.4 ± 0.2
NRRLY-2076 (haploid)	167.0 ± 15.0	60.8 ± 7.1	2.75 ± 0.25	1.4 ± 0.2
2075 X 2076 (diploid)	115.2 ± 11.0	36.5 ± 4.0	3.16 ± 0.32	2.3 ± 0.3

Each experiment was repeated 3 times as minimum. Radiobiological parameters of these curves are summarized in Table 1.3. This Table also includes the RBE values, which were calculated as the ratio of D_0 at low- and high-LET radiation: RBE = $D_0(\gamma)/D_0(\alpha)$. The errors involved in the corresponding D_0 and the RBE have been calculated from inter-experimental variations. As illustrated by the data presented, the doubling of the chromosome set for haplont yeast *P. pinus*, and *P. guilliermondii* does not result in a great increase in cell radioresistance. The γ-ray sensitivity of diploid cells of these yeast species was nearly equal or even somewhat larger to that of their haploid cells.

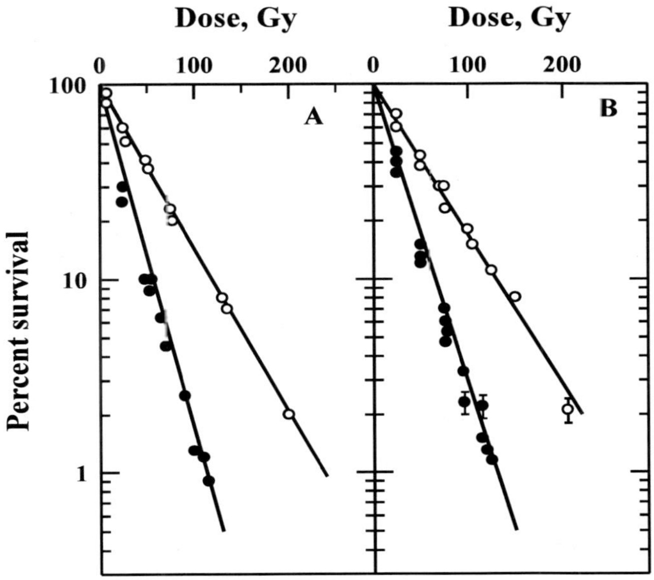

Figure 1.11. The survival curves of haploid (MH-4, A) and diploid (MH-4D, B) yeast *P. pinus* exposed to γ-rays of ^{60}Co (open circles) and α-particles of ^{239}Pu (closed circles).

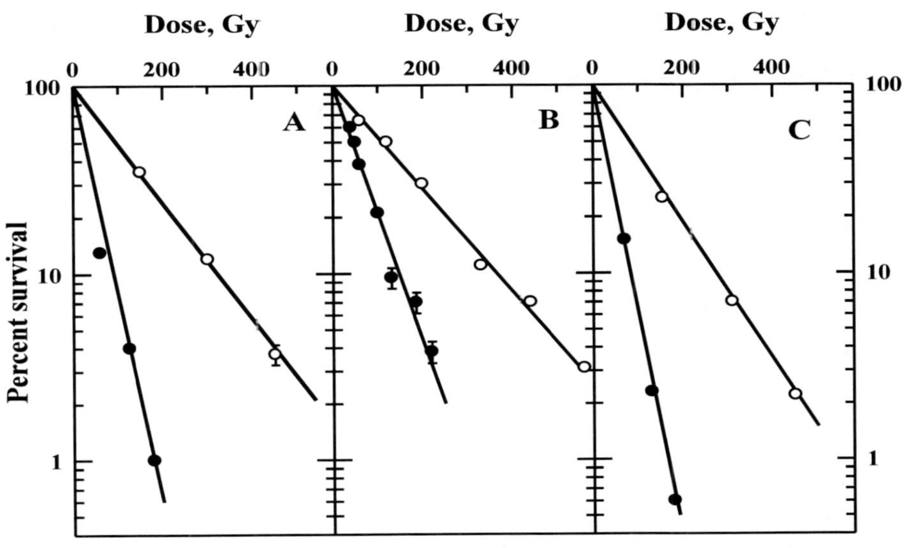

Figure 1.12. The survival curves of haploid (NRRLY-2075, A; NRRLY-2076, B) and diploid (NRRLY 2075x2076, C) yeast *P. guilliermondii* exposed to γ-rays of ^{60}Co (open circles) and α-particles of ^{239}Pu (closed circles).

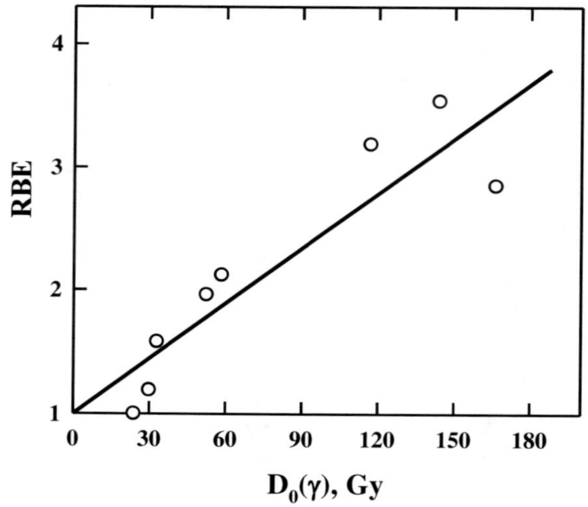

Figure 1.13. The relation of $D_0(\gamma)$ values with RBE of α-particles for various haplont yeast strains.

To distinguish the diploid state from the haploid one, we used the possibility of diploid strains to undergo meiosis and sporulation on sporulation media and produce tetrads consisting of 4 viable ascospores [38; 39]. Moreover, *P. piniis* diploid was able to sporulate even on usual complete media after a great duration of cultivation (10–14 days), the colonies being rose colored in contrast to white colonies produced by haploid cells. The diploids used in these experiments were heterozygous for mating type. It was controlled by tetrade analysis and the ability to crossing of the corresponding spores. In addition, DNA content was estimated for *Pichia* yeasts (Table 1.3) by colorimetric diphenylamine and UV-spectrophotometric methods [39]. One can believe that repair processes of haplont yeast are manifested to a small extent if they are not absent. As a consequence, the effect of ploidy on survival of these yeast species was similar to that of *S. cerevisiae* radiosensitive mutants in which defects of recombinational repair was due to *rad*50, *rad*51, *rad*52 and *rad*54 mutations responsible for the absence of the diploid specific recovery [19; 20; 61].

Figure 1.13 shows the relation between the RBE and $D_0(\gamma)$ values for haplont yeast strains studied here and presented in our previous publications [40; 55]. An almost linear relationship is found between these parameters. The regression line is RBE = 0.964 ± 0.015 D_0 and the correlation coefficient is 0.93. The data indicate that the RBE correlates well with yeast cell radioresistance (D_0) and further provide good evidence for the role of both physical (LET) and biological (radioresistance, reparation) mechanisms in the determination of RBE for high-LET radiation. The data presented in Figure 1.13 are in agreement with the conclusion that the RBE of densely ionizing radiation is greatly determined by cell radiosensitivity and cell ability to recover from damage produced by radiation.

As already mentioned, the repair capacity including diploid specific recovery is an important determinant for the RBE of high-LET ionizing radiation. From this, it can be expected that the difference in the RBE of α-particles for haploid and diploid yeast cells of the investigated species will vary to a smaller extent than for wild-type. Indeed, this was found. The doubling of cell genome was shown not to be accompanied by a large increase in the RBE of α-particles for all haplont yeast strains studied. For haploid and the corresponding diploid strains, the following RBE values were obtained: 1.96 and 2.10 for *P. pinus*; 2.75–3.51 and 3.16 for *P. guilliermondii* (Table 1.3). Comparing these data with results presented in the previous section, one can conclude that these RBE values were close to those for haploids of wild-type and all radiosensitive mutants as well as for a number of homozygous diploid mutants (*rad*18/*rad*18, *rad*50/*rad*50, *rad*51/*rad*51, *rad*52/*rad*52, and *rad*54/*rad*54) of *S. cerevisiae* exposed at the stationary phase of growth.

The following conclusions concerning the RBE for *P. guilliermondii* may be deduced. The RBE for haploid and diploid yeast cells of this species are not very different from each other, although these RBE values are somewhat higher than those for radiosensitive diploid mutants, they still did not reach the RBE values typical for wild-type diploid yeast cells.

1.4. Genetic Control of RBE of α-Particles for Yeast Cells Irradiated in Stationary and Exponential Phase of Growth

It was clear from the results presented the RBE of densely ionizing radiation is influenced by both physical characteristics of ionizing radiation and physiological cell properties controlling their repair systems. Variation of conditions promoting repair processes and cell radiosensitivity also appears to be a determinant of the effectiveness of densely ionizing radiation, the RBE value being highest for more efficient and complete repair. It is known that reparability of a cell is a function of its stage of growth at the time of irradiation. In particular, exponentially growing culture of yeast cells could not undergo the LHR. Moreover, it was demonstrated an additional cell death during their storage during postirradiation period in innutrient medium [52]. It is known that DNA helix in the stationary phase of growth is closely packed, occupying a much smaller volume compared with the exponentially growing culture, when cell chromosomes are almost homogeneously distributed within the cell nucleus. Both these factors should influence the RBE of high-LET radiations. We describe here the lethal response of some haploid and diploid yeast cells of the wild-type and radiosensitive mutants exposed to sparsely and densely ionizing radiation in stationary and exponential phase of growth.

To attain stationary phase of growth, cells were incubated before irradiation at 30 °C on a solid nutrient agar during 3–5 days for diploid and during 7–10 days for haploid cells, then washed with distilled water and resuspended to make a stock solution (10^6 cells/ml). Microscopic examination of this suspension usually revealed less than 1–2% budding cells. Exponential phase cells were obtained by inoculating a culture with $5 \cdot 10^3$ cells into a liquid nutrient medium and allowing them to grow at 30 °C overnight to a cell concentration of 10^6 cells/ml. More than 30% small-budded cells were observed in these cultures. The method of irradiation and other procedures were presented above.

Figure 1.14 shows the survival curves for diploid yeast *S. cerevisiae*, strains Megri 139-B, 211, and *Endomyces magnusii*, strain K7, exposed at stationary phase of growth to γ-rays and α-particles. All survival curves were sigmoid. There is a slight increase in the RBE in the low-dose of ionizing radiation. This means that the extrapolation number of the survival curves is somewhat lesser after exposure to densely than sparsely ionizing radiation.

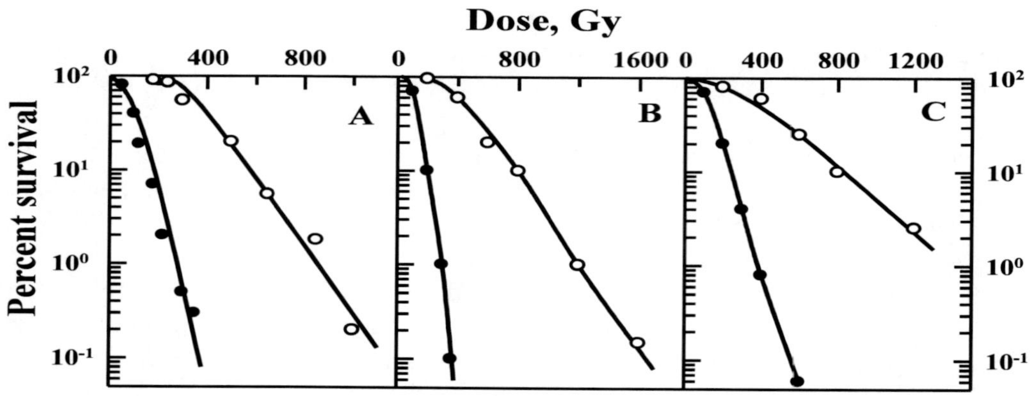

Figure 1.14. Survival curves of diploid yeast cells of *S. ellipsoideus (vini)*, strain Megri 139-B (A), *S. cerevisiae*, strain 211 (B), and *Endomyces magnusii*, strain K7 (C), exposed at stationary phase of growth to γ-rays (open circles) and α-particles (closed circles).

Similar effect was more pronounced for mammalian cells which, as a rule, are irradiated in exponential phase of growth. Therefore it would be of interest to irradiate these yeast strains in the exponential phase of growth. The results are presented in Figure 1.15. The survival of the log-phase culture of the wild-type strain exposed to graded doses of γ-rays and α-particles was typical for a population heterogeneous with respect to radiosensitivity. If the resistant tail is extrapolated to zero dose, it meets the survival axis at about the 3–20% level. This was in agreement with the percentage of cells with small buds, supporting the notion that cells with small buds are the most resistant. It can be seen that the initial sigmoid parts of the survival curves of strains Megri 139-B and 211 after gamma-irradiation (Figure 1.15A and 1.15B) transformed into exponential after exposure to alpha particles. Without taking into account the presence of resistant tails, this yeast cell response resembles mammalian cells answer to low- and high-LET radiation. Figure 1.15C exhibit survival curves for *Endomyces magnusii* cells which are multiplied not by budding like *Saccharomycetes* but by the division as mammalian cells. Like the analogous response of mammalian cells, the sigmoid survival curve of *Endomyces magnusii* after irradiation with sparsely ionizing radiation transformed into purely exponential curve after exposure to densely ionizing radiation. These results are closer to those observed for mammalian cells.

It is of interest to compare the RBE values for various haploid and diploid radiosensitive mutants irradiated at stationary and exponential phase of growth. Various fragments of this investigation have been published by us before [52, 53]. Figure 1.16 exhibits survival curves of haploid yeast cells of various strains exposed to γ-rays and α-particles in stationary and exponential phase of growth. The survival of the log-phase culture of the wild-type strain shows a resistant tail (Figure 1.16, right panel) that was connected with a number of cells with a certain size of buds.

To avoid difficulties related with radioresistant tails, the RBE for wild-type strain was estimated as the ratio $D_0(\gamma)/D_0(\alpha)$, D_0 being estimated at the initial exponential part of dose-effect curve. Table 1.4 lists the radiobiological parameters of all these haploid yeasts. The data confirm well-known results indicating that after γ-irradiation all haploid *rad* mutants exhibit slightly enhanced radiation sensitivity at stationary phase of growth compared with the wild-type strain.

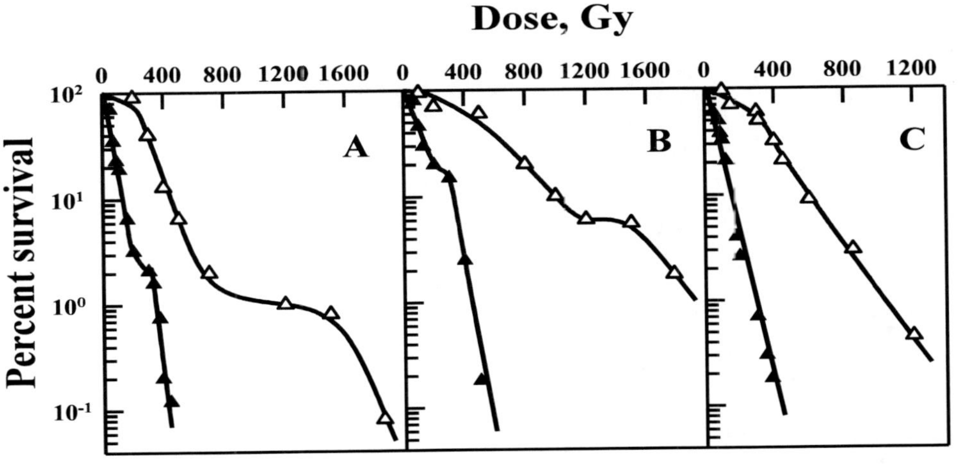

Figure 1.15. Survival curves of diploid yeast cells of *S. ellipsoideus (vini)*, strain Megri 139-B (A), *S. cerevisiae*, strain 211 (B), and *Endomyces magnusii*, strain K7 (C), exposed at exponential phase of growth to γ-rays (open circles) and α-particles (closed circles).

It is worth to note that the mutant cells used in these investigations were the most sensitive among the others. Our new results show that the mean lethal dose is influenced by the stage of growth, being decreased for all strains exposed in exponential phase of growth. From the data it is evident that the efficiency of densely ionizing particles, as might be judged by the D_0 values for α-particles, was almost the same for both the wild-type (S288C) and all mutants studied (*rad*50, *rad*51, *rad*52 and *rad*54) independently of the stage of growth. Due to these facts, the RBE of α-particles was slightly higher for wild-type than for mutant cells exposed at both phase of growth. For exponential phase of growth, the RBE of α-irradiation was found to be greatly reduced for all haploid strains. As can be seen, this is mainly related to the increase of cell radiosensitivity after γ-irradiation in exponential phase of growth.

Figure 1.17 shows the survival-versus-dose curves of wild-type diploid yeast cells and various homozygous radiosensitive mutants exposed to γ-rays and α-particles in stationary and exponential phase of growth. The cells were plated immediately after irradiation. All mutant cells showed exponential survival curves. Figure 1.17A shows quite different shapes of dose-effect curves for wild-type cells exposed to low- and high-LET radiation in stationary and exponential phase of growth. For stationary culture, the shape of the survival curve was sigmoid independent of radiation quality, and their exponential parts can be extrapolated to the same extrapolation number. This means that the RBE of α-particles is constant for any survival level and equal to the ratio of slopes of the survival curves, i.e., the ratio $D_0(\gamma)/D_0(\alpha)$. Here again, for exponential culture, the shapes of the survival curves exhibit radioresistant tails.

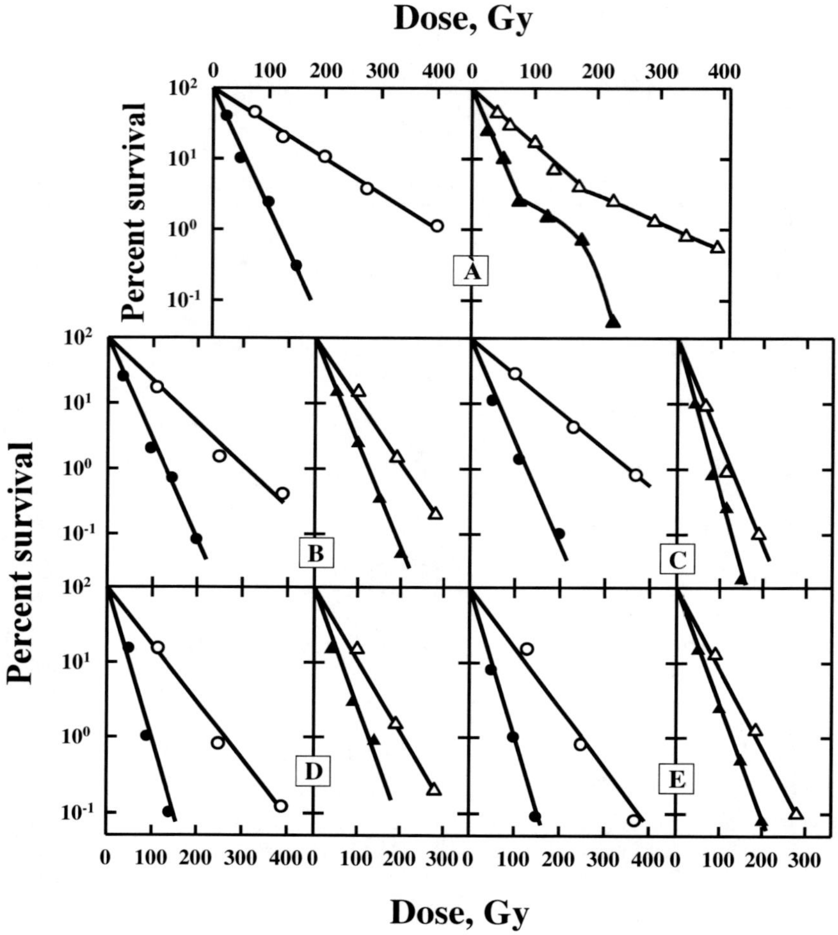

Figure 1.16. Survival curves of haploid *S. cerevisiae* yeast cells of various strains exposed to γ-rays (open symbols) and α-particles (closed symbols) in stationary (left panels, circles) and exponential (right panels, triangles) phase of growth: A, S288C (RAD); B, g218-6b (*rad*50); C, XS774-4d (*rad*51); D, g160-2d (*rad*52); E, g218-7c (*rad*54).

Table 1.4. Radiobiological parameters of *S. cerevisiae* haploid yeast cells of various genotypes exposed to γ-rays and α-particles at stationary and exponential phase of growth

Strain	Genotype	Stationary phase			Exponential phase		
		$D_0(\gamma)$, Gy	$D_0(\alpha)$, Gy	RBE	$D_0(\gamma)$, Gy	$D_0(\alpha)$, Gy	RBE
S288C	RAD	80.1	25.6	3.1	44.8	23.7	1.9
g218-6b	*rad*50	63.5	28.2	2.2	44.0	26.5	1.7
XS774-4d	*rad*51	75.0	27.2	2.7	27.4	18.0	1.5
g160-2d	*rad*52	57.4	23.6	2.4	44.4	27.0	1.6
g218-7c	*rad*54	52.8	21.5	2.4	40.4	26.9	1.5

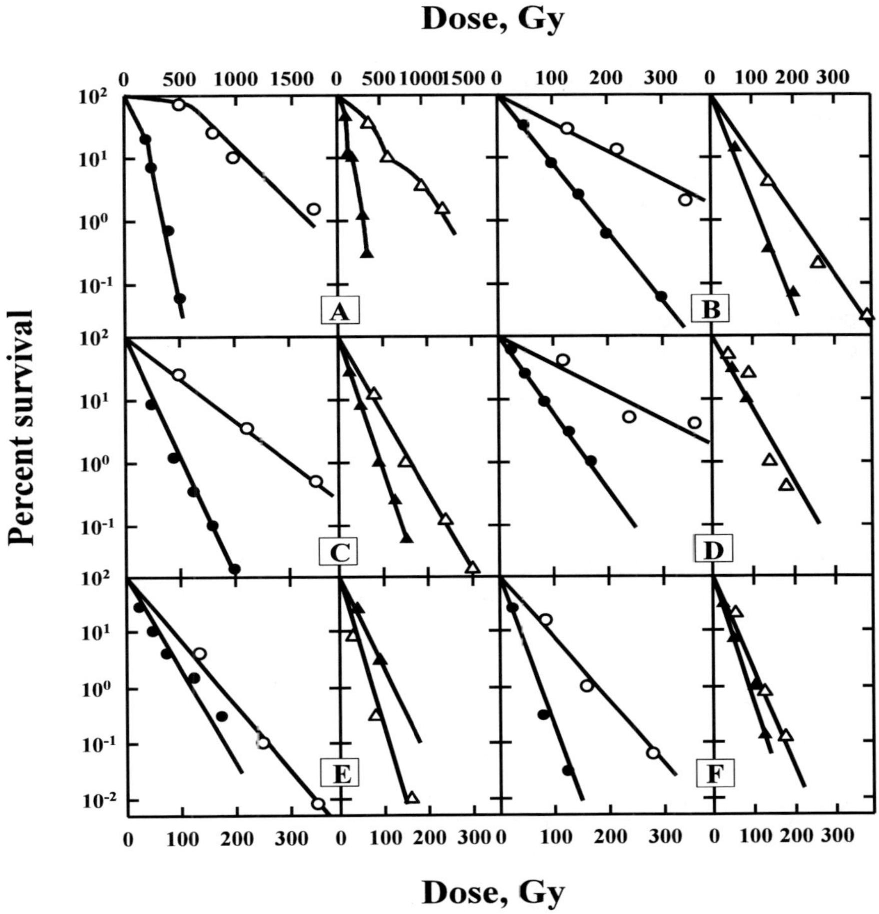

Figure 1.17. Survival curves of diploid yeast cells of various strains exposed to γ-rays (open symbols) and α-particles (closed symbols) in stationary (left panels, circles) and exponential (right panels, triangles) phase of growth. The cells were plated immediately after irradiation; A, XS800 (RAD/RAD; B, XS1806 (*rad*50/*rad*50); C, XS806 (*rad*51/*rad*51); D, XS1893 (*rad*52/*rad*52); E, T3 (*rad*54/*rad*54); F, T4 (*rad*2*rad*54/*rad*2*rad*54).

Not taking into consideration the radioresistant tails, it is evident that, for exponential culture, the sigmoid shape of the dose-effect curve after low-LET radiation transforms into an exponential curve after high LET radiation. Similar response has just been presented above for diploid yeast cells of *S. ellipsoideus* (*vini*), strain Megri 139-B (A), and *S. cerevisiae*, strain 211 and *Endomyces magnusii*, strain K7 (Figures 1.14 and 1.15). It is of interest that very similar data have been published for another wild-type diploid yeast cell by Murthy et al. [48]. For such a case, RBE values are very large at low doses and approach the ratio of slopes as the level of the effect observed increases. Therefore, here we calculated the RBE as the ratio $D_0(\gamma)/D_0(\alpha)$, the mean lethal dose being estimated at exponential parts of the survival curve (before the radioresistant tails are expressed).

Table 1.5 summarizes the radiobiological parameters of homozygous diploid yeast *S. cerevisiae* of the same genotypes as in Table 1.4. Additionally, we used the double mutant (strain T4, *rad*2*rad*54/*rad*2*rad*54). Both strain T3 and T4 were obtained from Dr. I.A.

Zakharov (Russia). It is clear that in stationary phase of growth all the *rad* mutants exhibit greatly enhanced sensitivity to low-LET radiation compared with the wild-type strain. This fact has been well known for a long time. As we and others authors have shown previously, the relative sensitivity varies more widely in diploids than in haploids.

Our new results show that $D_0(\gamma)$ values are dependent on stage of growth, being strongly decreased for all strains irradiated in exponential phase of growth. Here again, the relative sensitivity of various diploid mutants varies more widely than in haploids. Also it can be seen from the data presented that $D_0(\alpha)$ values vary not so widely as $D_0(\gamma)$, being slightly dependent on stage of growth only for two strains (RAD/RAD and *rad*50/*rad*50). Due to these facts, RBE values were the highest for wild-type diploid yeast cells and greatly decreased for all mutant cells irradiated in stationary phase of growth. It suggests a correlation between the RBE values and cell radiosensitivity: the strain that was more resistant had the greater RBE value, whereas those strains that were more sensitive had significantly lower values of RBE.

Table 1.5. Radiobiological parameters of diploid yeast *S. cerevisiae* of various genotypes exposed to γ-rays and α-particles at stationary and exponential phase of growth

Strain	Genotype	Stationary phase			Exponential phase		
		$D_0(\gamma)$, Gy	$D_0(\alpha)$, Gy	RBE	$D_0(\gamma)$, Gy	$D_0(\alpha)$, Gy	RBE
XS800	RAD/RAD	310.3	54.1	5.7	172.0	40.1	4.3
XS1806	*rad*50/*rad*50	95.5	37.2	2.6	43.1	25.5	1.7
XS806	*rad*51/*rad*51	62.3	22.9	2.7	32.7	20.9	1.6
XS1893	*rad*52/*rad*52	88.0	32.5	2.7	36.8	36.8	1.0
T3	*rad*54/*rad*54	36.1	23.9	1.5	15.8	24.0	0.7
T4	*rad*2*rad*54/*rad*2*rad*54	37.0	17.3	2.1	22.9	18.9	1.2

The most interesting new information following from these data pertains to the dependence of RBE values on stage of growth. It can be seen that for all mutant diploid cells, exposed in exponential phase of growth, the RBE values were close to unity (Table 1.5), i.e., cell radiosensitivity was almost identical for both sparsely and densely ionizing radiation. These data are an additional argument supporting the idea that the RBE of densely ionizing radiation is determined by both physical characteristics of ionizing radiation and physiological cell properties controlling their repair systems. Moreover, these data indicate that the role of repair in the determination of RBE may prevail over LET radiation for the most radiosensitive yeast cells.

Other X-ray-sensitive haploid and diploid yeast cells were not used in these experiments because of the following reasons. X-ray-sensitive diploid mutants (*rad*53/*rad*53, *rad*55/*rad*55; *rad*56/*rad*56, *rad*57/*rad*57) were comparatively more resistant, exhibited sigmoid survival curves and even retained the ability for LHR during the postirradiation period [53]. This means that, in principle, the radiobiological responses should be similar to those obtained for wild-type cells. The haploid strains (*rad*53, *rad*55, *rad*56, *rad*57) irradiated in exponential phase of growth showed survival curves with radioresistant tails corresponding to a number of budding cells with a certain size of bud [61]. Raju et al. [58] have been investigated in detail the radiation response of budding cells to low- and high-LET radiation. They showed that the relative resistance of budding cells decreases rapidly with increasing LET. Therefore there was no need to repeat their experiments with other strains. Our results with haploid and diploid wild-type cells exposed in exponential phase of growth show a similar radiation response: radioresistant tails were more expressed after γ- than after α-irradiation.

DNA double-strand breaks seem to be the critical lesions for cell killing by ionizing radiation. It has been shown for yeast cells that one double-strand break per cell corresponds to a lethal event [17]. The influence of double-strand breaks on the RBE of high-LET radiation was also described [7; 18; 25]. In particular, it was shown that the RBE value is low and independent of dose when no repair of double-strand breaks is involved, whereas it becomes gradually larger with a gradual increase in the level of double-strand break repair [18]. The repair of DNA double-strand breaks requires the availability of two homologous DNA duplexes [45]. It means that the postulated diploid specific recovery has a certain real genetic and molecular basis.

It should be pointed out that, unlike bacterial strains [1; 73], the RBE of high-LET radiation in the *rad* yeast strains irradiated in stationary phase of growth was not reduced to unity, indicating that other repair processes or factors other than repair are also responsible for the determination of RBE [25]. The most important new results obtained in this study pertain to the RBE of α-particles for wild-type and mutant cells irradiated in exponential phase of growth. It was shown that, during exponential phase of growth, the RBE of α-particles for all strains studied was decreased in comparison with that for the stationary phase of growth. For haploid cells, the RBE varied from 2.4 to 3.1 for stationary phase cultures and from 1.5 to 1.9 for exponential phase cultures. As can be seen from the results obtained, compared with haploids, the RBE varies more widely in diploids. For diploid wild-type cells, the RBE value was found to be the highest (5.7) for stationary phase culture, being slightly decreased for exponential phase culture (4.3). For diploid mutant cells, the corresponding RBE values varied within either 1.5–2.7 for stationary or 0.7–1.7 for exponential cultures. The last RBE values were close to unity, i.e., cell radiosensitivity was almost the same for sparsely and densely ionizing radiations.

The major point to be inferred from the data presented is that the reduction of the RBE of α-particles in exponential phase of growth was mainly due to the increase of cell radiosensitivity after γ-irradiation. Alpha-particle sensitivity was shown to be not greatly dependent on phase of growth. Thus, in addition to variation due to genetic defects, the physiological state of the cells can influence their sensitivity and RBE of densely ionizing radiation. This may be due to the expression of different repair processes, thereby altering their radiosensitivity. Studies with various biological objects indicate that high-LET radiation produces a greater proportion of irreversible damage in comparison with sparsely ionizing radiation. It follows that recovery processes play a less important role after high-LET radiation. It may perhaps be the reason why the sensitivity to α-particles was largely independent of stage of growth.

1.5. Liquid Holding Recovery Kinetics in Wild-type and Radiosensitive Mutant Yeast Cells Exposed to Low- and High-LET Radiations

A relatively unexpected role of the specific repair pathways in the RBE of high-LET radiation was demonstrated for bacterial [1; 26; 41], yeast [18; 25; 53] and mammalian cells [11; 67; 72]. It seems exist a common agreement that high-LET radiations produce more portion of damage that are considered to be irreversible [27; 65] compared with low-LET radiation such as photons.

Cellular repair of radiation-induced DNA double-strand breaks (DSB) could be also dependent upon radiation quality. Studies concerning the rate of the recovery from radiation damage produced with low- and high-LET radiations in cells of various origins on the survival and macromolecular level have also revealed that in generalб at a high ionization density, these processes may be reduced or even absent [6; 18; 23; 36]. It was shown that in diploid cells of the yeast *S. cerevisiae* the cellular recovery and the rejoining of DSB slow down with increasing LET [18]. In yeast cells more than 100 loci influence cell sensitivity to ionizing radiation [20]. Since the cellular recovery is a result of multistep enzymatic repair process realized on macromolecular level, it is not excluded that the degree of lesion of the recovery process itself may be increased with increasing LET. For cultured mammalian cells there was also apparently a slower rate of cellular recovery and DNA repair of damage induced by α-particle and other high-LET radiations than that by γ-rays. These results indicate that either the process of recovery may be impaired or a slower rate of recovery may be a consequence of more severe and complex damage produced by high-LET radiations. In other words, the reduction of cellular recovery as expressed in both the extent and the rate of recovery after high-LET irradiation might arise either from damage of the recovery mechanism itself or from the formation of irreversible damage which cannot be repaired at all. Both processes may take place at the same time. However, data distinguishing these possibilities are lacking in literature.

The ability of yeast cells to undergo LHR requires many DNA-repair and checkpoint genes [5; 20]. In the yeast *S. cerevisiae* there are at least eight well established distinct loci that control its sensitivity to ionizing radiation, while the number of independent genes capable of affecting the yeast's sensitivity to ultraviolet radiation exceeds 20 loci; some of the mutant genes confer sensitivity for both UV and ionizing radiation. It is thought that the mechanism of their enhanced radiosensitivity is related to the reduced capacity of cellular recovery. However, it was not quantitatively elucidated to what extent the reduced recovery would be due to damage in the recovery mechanism itself and/or to the formation of irreversible unrepairable damage. This question remains.

Here we tried to answer the question whether or not high-LET radiation affects the recovery process itself or if it only produces a higher level of severe irreversible damage that cannot be repaired at all as well as to elucidate the role of irreversible damage and the probability of recovery in some *rad* mutants in yeast *S. cerevisiae*. Getting a better insight into the relationship between radiation quality and cell recovery from potentially lethal radiation damage is of interest both from a fundamental standpoint and with respect to the practical applications of densely ionizing radiation in tumor therapy.

The diploid yeast cells of *S. ellipsoideus* (*vini*), strain Megri 139-B (RAD/RAD), the diploid yeast cells of *S. cerevisiae*: 211 (RAD/RAD), XS800 (RAD/RAD), XS1924 (*rad*18/*rad*18), XS1889 (*rad*53/*rad*53), XS1935 (*rad*55/*rad*55), XS1943 (*rad*56/*rad*56), XS1988 (*rad*57/*rad*57), were used for these experiments. Strain 211 was initially obtained from Dr. W. Laskowski (Germany), a parent strain of wild-type (XS800) and its radiosensitive mutants have been obtained from Dr. S. Nakai (Japan). All radiosensitive mutants used here were heterozygous in mating type, other genetic markers have also been published [61]. It was demonstrated earlier that these mutant strains exhibited a higher radiosensitivity as compared with their parent strain, sigmoid survival curves and preserve their capability for LHR to a certain extent, whereas radiosensitive strains with other mutant loci *rad*50, *rad*51, *rad*52, *rad*54 lost this capability. The cells were grown before irradiation up to the stationary phase of growth on a solid growth medium.

Aliquots from the same cell suspension were irradiated with ^{60}Co γ-rays (10^6 cells/ml, 10 Gy/min, LET ~ 0.2 keV/μm) and with ^{239}Pu α-particles (17 Gy/min). The LET of the particles reaching a cell monolayer was estimated to be 134 keV/μm. Exactly at about this LET value the maximum RBE–LET relationship was observed for most eukaryotic and some prokaryotic unicellular organisms.

Immediately after irradiation, a part of each sample was plated onto nutrient agar plates (immediate plating) for the assay of the colony-forming ability. To observe the recovery kinetics, another part of the irradiated or controlled cell suspension was placed in suitable conditions for the LHR and their colony-forming ability was determined as a function of the storage time in the recovery conditions (delayed plating). LHR was carried out in a water suspension at 30 °C without constant agitation (the cell concentration was 10^6 cells/ml). The dose-response curves and recovery kinetics were independent of whether or not the cell suspension was prepared with a 0.07 M phosphate buffer or with distilled water. Survival response on immediate and delayed plating was determined on the basis of the colony counts obtained at the end of 5–7 days of incubation at 30 °C and the counts checked again after a further period to ensure that the final score had been reached.

During the LHR process a number of the primary radiation damage is eliminated, resulting in an increased cell survival. This can be considered as a reduction of the initial dose D_1 to a certain effective dose $D_{eff}(t)$ which is proportional to the mean number of the residual damages, both reparable and irreversible, after a recovery for t hours. It has been demonstrated for yeast cells [37; 38] that the decrease in the effective dose $D_{eff}(t)$ with the recovery time t could be fitted by an equation of the form

$$D_{eff}(t) = D_1\left[K + (1-K)\cdot e^{-\beta\cdot t}\right], \quad (1.10)$$

where K is the irreversible component of radiation damage, i.e., the fraction of radiation damage resistant to recovery, and β is the recovery constant that characterizes the probability of the recovery per unit time. In other words, the recovery constant is approximately equal to the fraction of the reversible radiation damage recovering per unit time. This equation was used previously to fit the recovery kinetics of various biological organisms after low-LET ionizing radiation applied alone [38] or combined with high temperatures [54] but it was not widely tested for high-LET radiation. The ratio

$$K(t) = D_{eff}(t) / D_1 \tag{1.11}$$

reflects the relative part of the primary radiation damage that has not been repaired during t hours of recovery. If t is sufficiently large (for yeast cells it is about 2–3 days), the recovery curves reach a plateau when the ability of the cells to recover is saturated or exhausted. For this case, we can write

$$K = K(plateau) = \frac{D_{eff}(plateau)}{D_1}. \tag{1.12}$$

In this expression, $D_{eff}(plateau)$ is the effective dose corresponding to the plateau of the recovery curve, which is proportional to the mean number of the irreversible damage. It can be easily shown that

$$e^{-\beta \cdot t} = \frac{D_{eff}(t) - D_{eff}(plateau)}{D_1 - D_{eff}(plateau)}. \tag{1.13}$$

By inserting

$$A(t) = \frac{D_{eff}(t) - D_{eff}(plateau)}{D_1 - D_{eff}(plateau)}, \tag{1.14}$$

we have

$$\beta = -\frac{\ln A(t)}{t}. \tag{1.15}$$

In biological terms, $A(t)$ reflects the relative part of the reparable damage that has not been repaired after t hours of recovery. Thus, knowing the survival and recovery curves after cell exposure to low- and high-LET radiation, one can calculate the corresponding values of $D_{eff}(t)$, $D_{eff}(plateau)$, $K(t)$, K, and β.

Figure 1.18 exhibits the survival curves (A) and LHR recovery kinetics (B) of the diploid yeast cells *S. ellipsoideus* (*vini*), strain Megry 139-B. The cells were exposed to graded doses of γ-rays (curve 1) or α-particles (curve 2). As can be seen, the survival curve obtained after α-irradiation shows a smaller shoulder (extrapolation number) and a steeper exponentially decreasing part. The RBE of the α-particles, defined as the ratio of the mean lethal doses after low- and high-LET radiations RBE = $D_0(\gamma)/D_0(\alpha)$, was found to be of 3.7. The recovery kinetics after both kinds of radiation were compared at approximately equal levels of survival (i.e., at equal amounts of lethal lesions). When irradiated cells were held before plating in conditions favoring the LHR, their survival was greatly enhanced, the number of viable cells being increased as a function of time, reaching a plateau after about 2–3 days. Differences on repair kinetics are also evident, with slower recovery for α-particle irradiation. Thus, the data presented suggest that recovery decreases with increasing LET of the radiation. The decrease becomes apparent in both the rate and the volume of recovery.

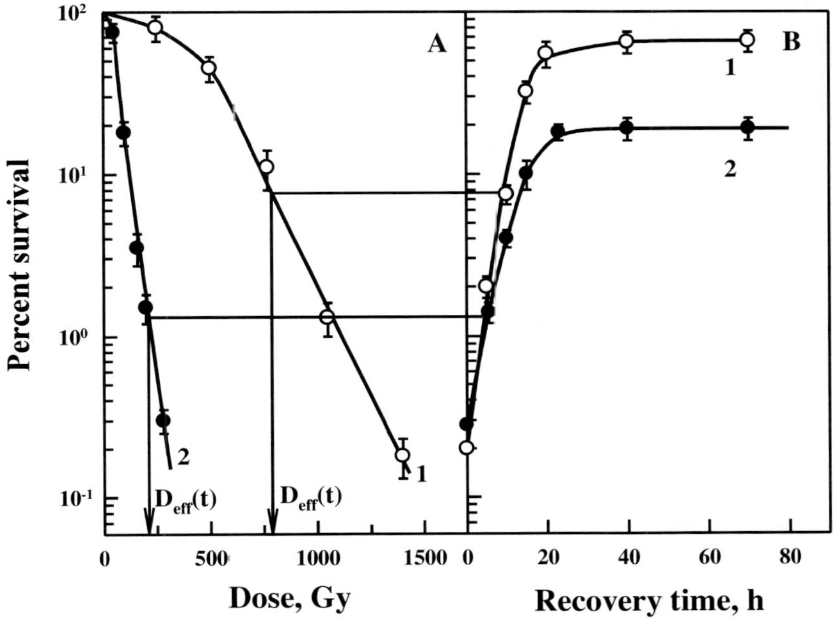

Figure 1.18. Survival curves (A) and LHR recovery kinetics (B) of the diploid yeast cells *S. ellipsoideus* (vini), strain Megry 139-B. Cells were exposed to γ-rays (curves 1) or α-particles (curves 2) and plated on a nutrient agar immediately after irradiation (A) or after various recovery times (B). The arrows indicate the examples of the effective dose, $D_{eff}(t)$, determination following 5 (α-irradiation) and 9 (γ-irradiation) hours of recovery. Here and in the following figures data are the mean values calculated from at least three independent experiments. Error bars indicate the standard error of the mean. Curves were fitted to the data points by eye.

Having used the results presented in Figure 1.18 and Eq. (1.11), we estimated the decrement of the relative fraction of radiation damage resistant to recovery, $K(t)$, as a function of the recovery time for a low- and high-LET radiation (Figure 1.19). These data demonstrate that the values of $K(t)$ decreased steadily, reaching a plateau after 2–3 days of recovery for both sparsely and densely ionizing radiation. It also appears from these data that the least quantities of $K(t)$, i.e., the values of the irreversible component of radiation damage, $K(plateau)$, increased with the ionizing density. The experimental data presented make it possible to calculate the relative part of the reparable damage $A(t)$, defined by Eq. (1.14).

The outcome is shown in Figure 1.20 for both low- and high-LET radiation. It is evident that this function decreases exponentially with the recovery time and does not depend on the quality of the radiation. Using Eq. (1.15) and the results shown in Figure 1.20, we obtained that the recovery constant $\beta = 0.1\ h^{-1}$ for both the low- and high-LET radiation. It means that about 10% of the residual reparable damage is recovered every hour independent of the radiation quality.

Similar experimental data have been carried out for all the other yeast strains tested. For all of them, it was shown that both the extent and rate of recovery decreased with increase in LET of radiation. The final results are presented in Table 1.6. It shows the mean lethal doses D_0 after γ- and α-irradiation, the RBE values, and both parameters of postirradiation recovery: the irreversible component K and the probability of the recovery per unit time β after γ- and α-irradiation. As it was obtained in the previous section, *rad* mutant cells have lower RBE values in comparison with that for wild-type parent strain (XS800).

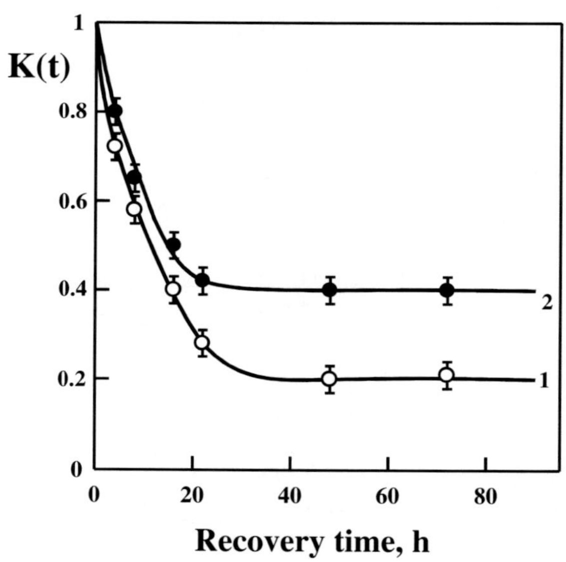

Figure 1.19. The decrement of the relative part of the unrecovered radiation damage, $K(t)$, as a function of recovery time of the diploid yeast cells *S. ellipsoideus* (*vini*) strain Megry 139-B after treatment with γ-rays (curve 1) and α-particles (curve 2).

Thus, it was shown that the irreversible component was enhanced for the densely ionizing radiation in comparison to the low-LET radiation while the probability of the recovery was identical for both the low- and high-LET radiations for all the strains investigated. It means that the recovery process itself is not damaged after densely ionizing radiation and the enhanced RBE of the high-LET radiation may be caused by the enhanced probability to produce primary radiation damage and the increased yield of the irreversible damage. A parent diploid strain (XS800) and all its radiosensitive mutants (XS1924, XS1889, XS1935, XS1943, XS1988) showed the same probability for recovery from radiation damage.

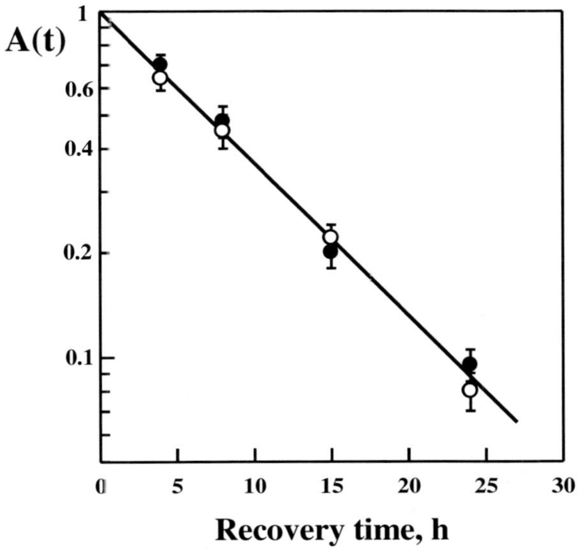

Figure 1.20. The decrement of the relative part of the recovered radiation damage, $A(t)$, as a function of recovery time of the diploid yeast cells *S. ellipsoideus* (*vini*) strain Megry 139-B after treatment with γ-rays (open circles) and α-particles (closed circles).

Table 1.6. Radiobiological parameters of the diploid yeast cells of different species and strains exposed to γ-rays and α-particles

Strain Genotype	$D_0(\gamma)$, Gy	$D_0(\alpha)$, Gy	RBE	$K(\gamma)$	$K(\alpha)$	$\beta(\gamma)$, h^{-1}	$\beta(\alpha)$, h^{-1}
Megry 139-B RAD/RAD	150 ± 12	41 ± 5	3.7 ± 0.3	0.22 ± 0.02	0.40 ± 0.03	0.1 ± 0.02	0.1 ± 0.02
211 RAD/RAD	270 ±25	60 ± 6	4.5 ± 0.5	0.45 ± 0.05	0.54 ± 0.03	0.06 ± 0.009	0.06 ± 0.009
XS800 RAD/RAD	265 ±28	52 ± 7	5.1 ± 0.6	0.32 ± 0.03	0.50 ± 0.04	0.05 ± 0.007	0.05 ± 0.007
XS1924 *rad*18/*rad*18	120 ±14	46 ± 6	3.3 ± 0.2	0.30 ± 0.02	0.52 ± 0.05	0.05 ± 0.007	0.05 ± 0.007
XS1889 *rad*53/*rad*53	130 ±15	30 ± 4	4.3 ± 0.3	0.50 ± 0.04	0.68 ± 0.06	0.05 ± 0.007	0.05 ± 0.007
XS1935 *rad*55/*rad*55	130 ±16	34 ± 3	3.9 ± 0.3	0.37 ± 0.04	0.54 ± 0.04	0.05 ± 0.007	0.05 ± 0.007
XS1943 *rad*56/*rad*56	200 ±21	45 ± 5	4.4 ± 0.3	0.43 ± 0.04	0.45 ± 0.05	0.05 ± 0.007	0.05 ± 0.007
XS1988 *rad*57/*rad*57	186 ±15	43 ± 5	4.3 ± 0.4	0.36 ± 0.04	0.44 ± 0.04	0.05 ± 0.007	0.05 ± 0.007

It follows that the mechanism of the enhanced radiosensitivity of the mutant cells might not be related to the damage of the repair systems themselves but with the production of some kind of radiation damage which cells are incapable of recovery. The data presented here indicate that the irreversible component was enhanced for the densely ionizing radiation in comparison to the low-LET radiation while the probability of the recovery was identical for both kinds of radiation for all the strains investigated. The latter fact means that the slowing down of yeast cells to undergo LHR after high-LET radiation is associated with a greater number of irreversible damage rather than with the impairment of recovery process itself.

1.6. RBE of α-Particles and Postradiation Recovery in a Temperature Conditional Diploid Yeast Mutant for Various Criteria of Cell Death

It is known that dying cells can be inactivated after ionizing radiation exposure either without any division (interphase killing) or after one or several reproduction cycle (reproductive killing). During the LHR process a number of the primary radiation damage is eliminated, resulting in an increased cell survival. It is well-known fact that the interphase cell death occurs at much higher doses than reproductive one. For reproductive criterion of cell death, irradiated cells can recover from radiation damage by LHR. Conversely, interphase cells do not possess such ability. Since the RBE of densely ionizing radiation correlated with the ability of cells to recover from the damage induced by ionizing radiation, it can be expected that for interphase killing the RBE will be considerably less than that for reproductive killing. To test this hypothesis, we obtained dose-effect curves for a number of diploid yeast cells irradiated with gamma-rays ^{60}Co (dose rate 40 Gy/min) and alpha-particles (dose rate 25 Gy/min) at the stationary phase of growth. The following yeast cells were used in this study: S288C (RAD), g160/2d (*rad*52), XS800 (RAD/RAD), XS1898 (*rad*52-1/*rad*52-1), g580L (*rad*54-3/*rad*54-3). Last strain was obtained from Dr. D. Frankenberg (Germany). It is known as the diploid temperature conditional radiosensitive mutant of *S. cerevisiae* cells that is capable of repairing of radiation-induced DNA double strand breaks (DSB) if they are cultured at the permissive temperature of 23 °C during postirradiation period [10]. The RBE of α-particle was higher for this postirradiation temperature while it was smaller at the restrictive temperature of 36 °C due to a considerable depression of DNA double strand breaks repair [18].

Survival of reproductive dying cells was assessed by counting macrocolonies visible by the unaided eye after 4 days of cultivation on the solid nutrient medium. For interphase death, irradiated cells were considered as killed if they could not perform even one division after 36 hours of incubation on a nutrient medium. For this criterion, cells were considered as survival if they were able to perform at least one division and their survival was estimated mictroscopically. Cells that were capable of producing a microcolony consisting of more than 40 cells were considered to be viable (reproductive death). It is worth noting that the cell survivals estimated with macrocolony and microcolony counting coincided well.

As an example, Figure 1.21 shows the survival curves for reproductive cell death of wild-type diploid yeast *S. cevevisiae* (strain XS800), irradiated with ^{60}Co γ-rays (40 Gy/min) and α- particles (25 Gy/min) at a stationary phase of growth. Cells were plated on a nutrient medium immediately after the irradiation and after 48-hour LHR at 30 °C.

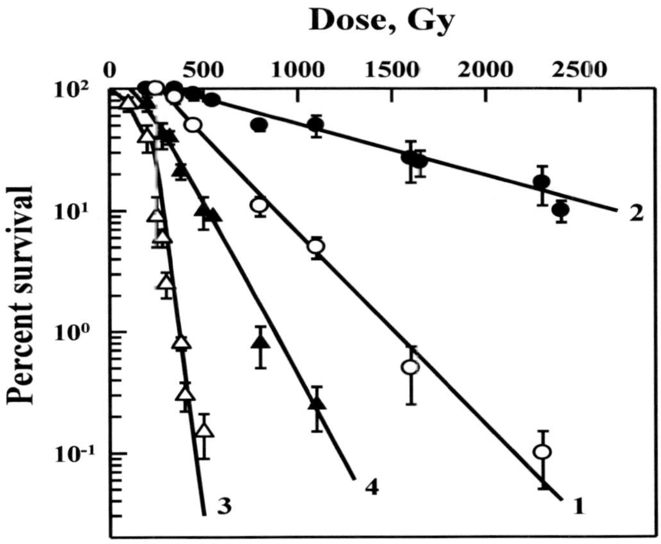

Figure 1.21. The dependence of cell survival for reproductive cell death on radiation dose for diploid yeast cell (*S. cerevisiae*, strain XS800), irradiated with ^{60}Co γ-rays (1, 2) and α-particles (3, 4) at a stationary phase of growth and plated on a nutrient medium immediately after the irradiation (1, 3) and after 48-hour LHR at 30 °C (2, 4). Data are the mean values calculated from at least three independent experiments. Error bars indicate the standard error of the mean. Curves were fitted to the data points by eye.

Figure 1.22 shows survival curves of the same strain and conditions of irradiation and recovery, but for interphase death. Similar curves were obtained for all the strains of yeast cells used, but their dose-effect curves were obtained only for immediate cell plating. Radiobiological parameters of all the strains studied are summarized in Table 1.7. This table also includes the RBE values estimated as the ratio of the corresponding mean lethal dose RBE = $D_0(\gamma)/D_0(\alpha)$.

Analysis of all these data leads to the following conclusions regarding the radiation resistance, recovery and RBE of α-particles for various forms of yeast cell death. differing in ploidy and genotype, controlling their ability to recover from radiation damage. As it can be expected, the interphase death occurred after much higher doses than reproductive killing after both kinds of radiation for all strains studied. This effect was more pronounced for the radiosensitive mutants. Actually, cell resistance increased in about 5–20 and 90–210 times for wild-type cells and radiosensitive mutants, respectively. Due to this fact, radiosensitive mutants became even more resistant to the damaging effects of ionizing radiation than their parent strains of wild-type. These data show that the mechanism of interphase death is not related with cell ability to recover from radiation damage. An additional argument in favor of this view is the luck of the LHR for interphase death: after liquid holding of irradiated cells even further death took place for both kinds of radiation (Figure 1.22). It is evident from Table 1.7 that the RBE of densely ionizing radiation was significantly higher for reproductive cell killing of diploid yeast cells of wild-type (4.6) than for interphase one (1.1). For reproductive cell death the RBE of alpha particles was smaller for haploid strain of wild-type and the radiosensitive *rad*52 and *rad*52/*rad*52 mutants, varying in the range of 2.3–3.1 (Table 1.7). These data are consistent with results presented above. For interphase killing the RBE of α-particles for these strains varied in a narrow range of 2.0–2.2. This means that RBE of densely ionizing particles for cells incapable of LHR depends on the form of cell killing in a lesser degree.

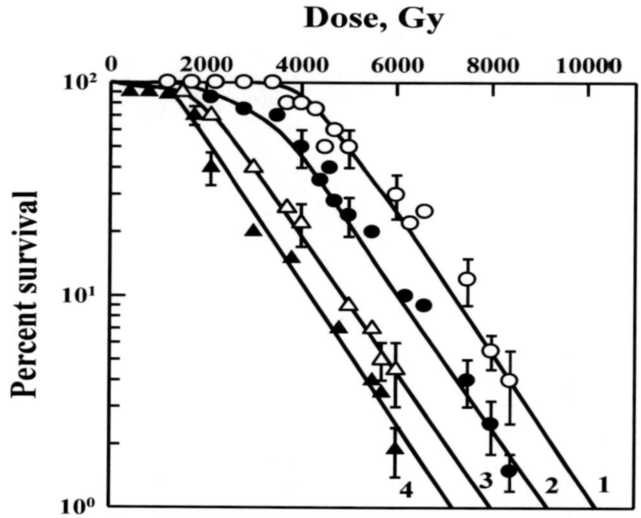

Figure 1.22. The same as in Figure 1.21, but for interphase cell death.

Consider in more detail the response of g580L strain incubated after irradiation at 23 and 36 °C. Cells incubated at the temperature permissive for DSB repair (23 ºC) yielded shoulder survival curves with a small extrapolation number. The RBE of α-particles was found to be of 4.7 (Table 1.7). This value is close to that observed for wild-type yeast cells. It means that this mutant cells being incubated at 23 ºC shows similar radiobiological responses as wild-type cells capable of recovery. The survival of irradiated cells held before plating in conditions favoring LHR was greatly enhanced in comparison with that obtained after immediate plating for both the low- and high-LET radiations. The number of viable cells increased as a function of time, reaching a plateau after about 2–3 days. The fraction of irreversible damage K was calculated as the ratio of the mean lethal doses after immediate and delayed plating. It was shown that the fraction of irreversible damage was markedly greater, i.e., the magnitude of postirrradiation recovery was smaller, after α-irradiation than that observed for γ-rays. The probability of recovery per unit time β was estimated using the survival and recovery curves after cell exposure to low- and high-LET radiation and Eq. (1.15). The probability of recovery was shown to be identical for both kinds of radiations. Moreover, the probability of recovery of these mutant cells was identical with that presented above for the wild-type strain. Thus, the data presented suggest that recovery decreases with increasing LET of radiation. The decrease becomes apparent in both the rate and the volume of recovery. The independence of the probability of recovery on radiation quality means that the process of recovery itself stayed unchanged and a slower recovery may reflect a smaller yield of reversible damages produced by high-LET radiation. It can be concluded on this basis that the LHR process itself is not damaged after densely ionizing radiation and the enhanced RBE of the high-LET radiation may be related both with the enhanced probability to produce primary radiation damage and the increased yield of the irreversible damage. Strong evidence for a causal relationship between rejoining of DSB, LHR and RBE of α-particles has been reported [18; 45]. We can suppose on this basis that the identical probability of the LHR observed in this section may signify the identical probability of DSB rejoining after low- and high-LET radiations while slowing down and smaller volume of DSB rejoining may reflect the reduction in a number of DSB which are capable to rejoining.

Table 1.7. Radiobiological parameters of the survival curves of different strains of yeast cells *S. cerevisiae*

Strain Genotype	Reproductive cell killing			Interphase cell killing		
	$D_0(\gamma)$, Gy	$D_0(\alpha)$, Gy	RBE	$D_0(\gamma)$, Gy	$D_0(\alpha)$, Gy	RBE
XS800 (RAD/RAD)	274	60	4.6	1465	1304	1.1
XS1898 (*rad*52/*rad*52)	45	15	3.0	3900	1950	2.0
S288C (RAD)	102	33	3.1	1500	670	2.2
g160/2d (*rad*52)	25	11	2.3	4800	2300	2.1
g580L at 23 °C (*rad*54-3/*rad*54-3)	57	12	4.7	4350	1575	2.8
g580L at 36°C (*rad*54-3/*rad*54-3)	22	8	2.8	3100	1250	2.5

The data described here prompted the similar types of experiments using restrictive (36 °C) temperature for postirradiation cultivation. Table 1.7 exhibits the results of these experiments. It is evident that at the restrictive temperature of 36 °C the *rad*54-3/*rad*54-3 radiobiological response is quite different from that observed at the temperature (23 °C) permissive for DSB rejoining. The mutant cells became more sensitive to low-LET radiation and their ability to undergo the LHR was also decreased. The decrease was apparent in both the rate and the volume of recovery. It was shown that the probability of recovery was identical for low- and high-LET radiation at both the permissive (23 °C) and restrictive (36 °C) for DSB rejoining temperatures. It means that an identical fraction of the reversible radiation damage is recovered per unit time independently of radiation quality at both temperatures investigated. The irreversible component was enhanced for restrictive temperature even for low-LET radiation and was similar to that observed for high-LET radiation. Therefore, we can conclude again that the recovery process itself is not damaged after densely ionizing radiation and the enhanced RBE value of the high-LET radiation as well as the high radiosensitivity of mutant cells at the restrictive temperature may be due to the increased yield of the irreversible radiation damage.

The LHR process was shown to fail for interphase cell killing when they are inactivated immediately after irradiation without any postradiation division (Figure 1.22). It can be seen that unusually tremendous radiation doses were used in these experiments after which a degradation of main cell compounds was observed [37]. Due to this fact no any recovery was recorded. Moreover, in all cases some decrease in cell survival was observed after LHR procedure (Figure 1.22). This fact can be considered as an indication to the failure of the connection between interphase cell killing and cell ability to recover radiation damage and perhaps may be related with unrepairable radiation damages of vitally important cell structure. The RBE of α-particles was smaller in comparison with that for the reproductive cell killing. This prediction is justified by the data presented in Table 1.7. It is evident that the RBE of α-particles for cell killing without division was smaller than that for reproductive cell killing, lowering from 4.7 to 2.8 for 23 °C and from 2.8 to 2.5 for 36 °C. These reduced RBE values are closed to those observed for radiosensitive yeast mutants incapable of recovery.

Thus, the new experimental data obtained in this work can be served as an additional argument in favor of the view that the RBE of densely radiation is largely determined by the ability of cells to recovery from radiation damage. In other words, the RBE of densely ionizing radiation is controlled by a certain biochemical properties of cells, but not only by the physical processes of energy absorption at microgeometrical level. The data of this section can be considered as an indication to the fact that the observed RBE value for unrecoverable condition (RBE = 2.5–3.0) may be caused by the physical property of high-LET radiation, while further increase of this value (up to 4.7) may be due to the cell ability to recover from radiation damage inflicted by low-LET radiation.

1.7. Radiation Quality and the Shape of Dose-effect Curves for Eukaryotic Cells: Theoretical Consideration, Recent Data and Interpretation

Cell inactivation is the radiation response which is of interest for radiation therapy and sterilization. The use of high-LET radiation for these purposes is based on the higher values of the RBE and a lesser cell ability to sublethal damage recovery after fractionated irrfadiation. The problem of the RBE and low dose effect is now very important not only for radiation therapy and sterilization but also for aerospace flights and ecology problems newly arising after the Chernobyl and Fukushima reactor accidents. Variation of the RBE as a function of dose is also a subject of considerable current interest because of the radiation safety and the value of radiation quality factor for natural radon daughter alpha-particles.

Radiobiological responses of simple eukaryote such as yeast cells are qualitatively identical to those of mammalian cells. This is one of the reasons why yeast cells are considered suitable model system for wide radiobiological investigations. Yeast cells as well as mammalian cells reveal an increase of the RBE with LET until it reaches a maximum around 120–130 keV/μm from where it declines again. To explain a rising RBE with LET, numerous models have been invoked in which multiple ionizations are deposited in the sensitive target(s). The decreasing of the RBE with increasing LET implies that a particle passing through a critical structure will dissipate more energy in this structure than it is required for damage. However, from a survey of published data it seems there appear to be inconsistencies in the findings with regard of radiation quality on the shape of survival curves of yeast and mammalian cells. For mammalian cells exhibiting sigmoid dose-response curves at low LET radiation many authors have shown a purely exponential response after densely ionizing radiations [4; 15; 57; 68]. The change from shouldered to exponential type of survival curves has been also seen when chlorella cells were irradiated with high-LET radiation [70]. It means that the RBE of densely ionizing radiation for such the cases is strongly dependent on the survival level chosen for comparison, RBE being very large (10–15) for low doses and approaches ratio of slopes of exponential parts of survival curves (3–5) as level of the survival fraction decreases. Conversely, for diploid yeast cells which have a double set of chromosomes like mammalian cells the shape of survival curves is sigmoid independently of radiation quality [17; 25; 36; 42; 46; 51; 53; 56; 62]. It means that for this case the RBE is independent on survival level. We have quoted here a lot of works performed with yeast and mammalian cells irradiated with low- and high-LET radiation to emphasize that the observed differences in the responses of these cells are not random and systematically obtained by different researchers. Thus, there appear to be a considerable uncertainty in the findings with regard to the LET dependence on survival curve shape for mammalian and yeast cells at the low dose of ionizing radiation. It must be concluded, therefore, that theoretical model for RBE effects based on the changes in the shapes of survival curves, when radiation quality changes, are not sufficient general to be applicable to various biological objects. When we take up this problem we are immediately confronted with some awkward questions. Why the RBE of densely ionizing particles is increasing with dose decreasing for cultured mammalian cells and is independent on dose for yeast cells? In other words, why radiation response to low dose of high-LET radiation is qualitatively different for yeast and mammalian cells?

The purpose of this section is to provide an appropriate explanation of the observed distinction in radiation response of yeast and mammalian cells to low- and high-LET radiations resulting in various RBE–dose relationships and test it by means of mathematical calculations. To answer the foregoing questions, we shall consider a possible dependence of internal structure of complex target consisting of genes and chromosomes on the stage of cell growth. Target theory was the first theoretical model that explained an intriguing paradox of ionizing radiation action – a negligible total energy absorbed resulted in a significant damage effect [1; 43]. And although after the target theory other mathematical models were developed, which took into account many of the newly discovered cell responses to radiation, they, in fact, took into consideration the main idea of the target theory – the existence of unique radiosensitive cell micro volumes.

Interpretation of mammalian cell data. As far back as 1958, Howard-Flanders [29] had shown in his "track segment theory" that for objects with exponential survival curves the increase of the RBE with LET increasing may be explained by a simple one-hit model if hit-event is several "ionizations" produced simultaneously by a single particle within the sensitive target. It is reasonable to suggest such a kind of hit-event for both diploid yeast and mammalian cells to explain the similar RBE-LET dependence. The change of the survival curve shape from sigmoid into exponential with increasing LET may be explained by the following suggestions. Cell sensitive volume may be presented as a certain region filled by sensitive sites. The number of sites is very large and every high-LET particle hitting this region can cross many such sites. The site may be damaged only by several ionizations produced by the same particle (dimensions of the site are very small and the probability that two independent particles hit the site simultaneously is negligible). The cell is inactivated when the number of damaged sites reaches a critical value. According to current knowledge, one would expect this region to be identical with vulnerable sections of chromosomes which are crossed by ionizing particles. It is obviously that every chromosome consists of a large number of sections; therefore cell nucleus can contain a lot of sensitive sites even if the number of chromosomes is limited.

This idea may be proved with simple calculations. A complete mathematical description of physical stochastic must yield the probability of incidence of each pattern of ionizations, which depends on radiation quality. For the simplest quantitative analysis of this problem we do not need consider all details of interaction of ionizing particles with a matter, in particular, we may do not take into account the distribution of energy (or a number of ion pairs) in the primary events of interaction (primary ionizations). Most dose-effect curves in radiobiology are consistent with the idea that biological effect results from essentially random transfer of energy from the ionizing particle to critical target structures. Without going into details, we use a number of primary ionizations as a measure of site damage and take into consideration only the facts that a number of primary ionizations per part of particle track and a number of particles per unit of area subjected to random statistical fluctuation and can be described by Poisson distribution.

If L is the overall number of primary ionizations per unit length of track and t is the thickness of a sensitive site, then the mean number of primary ionizations produced by a particle inside the sensitive site is $a = Lt$. If the site needs μ primary ionizations to be damaged, then the probability to damage the site is given by

$$\varphi = \sum_{n=\mu}^{\infty} e^{-a} a^n / n! = 1 - \sum_{n=0}^{\mu-1} e^{-a} a^n / n!, \tag{1.16}$$

where n is the Poisson variate. If an ionizing particle crosses l such sites, the probability to damage i sites by a single particle traversing is given as follows

$$P_{1,i} = \frac{l!}{i!(1-i)!} \varphi^i (1-\varphi)^{l-1} = \binom{l}{i} \varphi^i (1-\varphi)^{l-i}. \tag{1.17}$$

Assuming that two ionizing particles transit the sensitive volume, it would be expected that the probability to find a cell with i damages is

$$P_{2,i} = \sum_{n=0}^{i} P_{1,n} \cdot P_{1,i-n} = \sum_{n=0}^{i} \binom{l}{n} \cdot \varphi^n (1-\varphi)^{l-n} \cdot \binom{l}{i-n} \varphi^{i-n} (1-\varphi)^{l-i+n} =$$

$$= \varphi^i (1-\varphi)^{2l-i} \cdot \sum_{n=0}^{i} \binom{l}{n} \cdot \binom{l}{i-n} = \binom{2l}{i} \cdot \varphi^i (1-\varphi)^{2l-i}. \tag{1.18}$$

Correspondingly, if j particles cross the sensitive volume, the following relation must be valid

$$P_{j,i} = \frac{(jl)!}{i!(jl-i)!} \cdot \varphi^i (1-\varphi)^{jl-i}. \tag{1.19}$$

Let the mean number of particles which hit the sensitive volume after irradiation with dose D is m, i.e., $D \sim am$. Then the probability to find a cell with i damaged sites is given by

$$P_i = \sum_{j=0}^{\infty} \frac{e^{-m} \cdot m^j}{j!} \cdot \frac{(jl)!}{i!(jl-i)!} \cdot \varphi^i (1-\varphi)^{jl-i}. \tag{1.20}$$

Let cell inactivation needs at least k damaged sites. It leads to a survival relationship

$$S = \sum_{i=0}^{k-1} P_i = \sum_{i=0}^{k-1} \sum_{j=0}^{\infty} \frac{e^{-m} m^j}{j!} \cdot \frac{(jl)!}{i!(jl-i)!} \cdot \varphi^i (1-\varphi)^{jl-i}, \tag{1.21}$$

where S is the proportion of the cell population surviving a dose D. For two-hit process Eq. (6) may be written as

$$S = \exp[-m \cdot (1-(1-\varphi)^l)] \cdot [1 + ml\varphi(1-\varphi)^{l-1}]. \tag{1.22}$$

The calculated two-hit curves for S in dependence of am, which is proportional to irradiation dose are presented in Figures 1.23–1.25 for different values of m, which is proportional to LET. Values of the critical number of primary ionizations μ needed to cause site damage were chosen to be 1, 2, 3, 4, 5 and 6, l = 10. If hit-event is one primary ionization (Figure 1.23), the calculated survival curves change their shapes with LET from sigmoid to exponential and simultaneously the effectiveness of radiation decrease. Such a behavior can be easily explained. Indeed, for low-LET radiation the probability of the site to be damaged (Eq. 1.16) is low, correspondingly the probability to damage several sites by a single particle traversing is also low and several particles are needed to damage several sites, therefore the survival curve is sigmoid. With LET increasing these probabilities increase too and due to this fact the shape of the survival curve trends to exponential. But in this case, when even one primary ionization is effective to damage a site, the probability of site damage or by other words the efficiency of particle to produce damage $\varphi = 1 - e^{-a}$ increases with a (i.e., with LET growing) not faster than a dose delivered by a single particle which is proportional to a. That is why the effectiveness of irradiation per unit dose decreases. Thus the considered model with hit-event assuming to be one primary ionization is inconsistent with experimental data. Injury must, therefore, be the result of more than one ionization.

Earlier Günther and Schulz [24] validated the same results by means of other calculations and we presented survival curves here (Figure 1.23) only for comparison with other versions of the model.

On the contrary, if the critical number of primary ionizations equals 2 or 4, then the transformation of sigmoid survival curves into exponential takes place with a increasing and now it is accompanied by a rising of radiation efficiency (Figures 1.24 and 1.25) and only after very high a the RBE of radiation decreases. In these cases the requirement of several primary ionizations productions by the same particle results in the decreasing of RBE with a (i.e., with LET) initially faster than a dose delivered by a single particle which increases with a linearly so the efficiency of radiation per unit dose (RBE) raises. Then the particle efficiency approaches saturation but a dose per particle continues to increase, therefore the efficiency of irradiation per dose unit decreases. This pattern looks like results of radiobiological experiments with mammalian cells. We may conclude that the radiobiological response of mammalian cell is in qualitatively accordance with the observed mathematical model if (i) hit-event is a production of several ionizations by the same single particle in a sensitive site, and (ii) the total number of sensitive sites is large enough and they are distributed relatively homogeneously throughout the volume of nucleus so that every ionizing particle hitting the sensitive volume crosses several sensitive sites. Similar conclusions have been made by Kellerer and Rossi [34] on the basis of much more complex microdosimetry considerations. They suggested the existence of a gross sensitive volume which contains the sensitive loci homogeneously dispersed, and at least two inactivated loci are necessary for the effect.

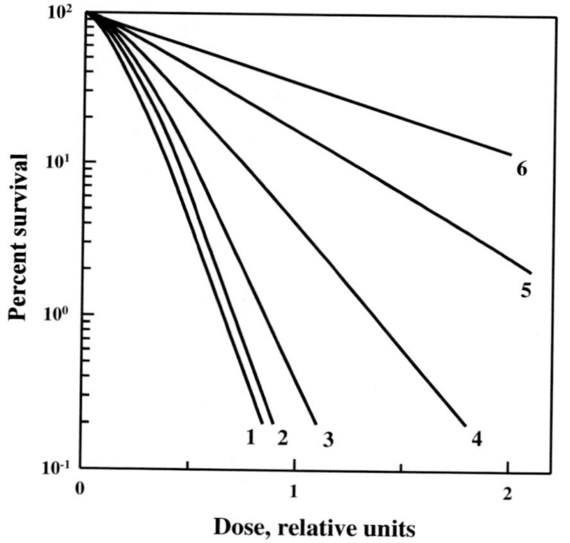

Figure 1.23. Two-hit dose-survival curves calculated for different meanings of a (LET). Target structure is a thick layer; hit-event is one primary ionization. Curves 1, 2, 3, 4, 5, and 6 were calculated for the following values of a : 0.001, 0.01, 0.05, 0.2, 0.5, and 1.0, respectively.

Interpretation of yeast cell data. Independence of survival curve shape on radiation quality suggests that no observable biological effect is produced until a certain amount of biologically effective energy transfer (hit-event) has accumulated both after low- and high-

LET radiations The persistence of the shoulder at high-LET radiation would indicate that a single densely ionizing particle is unable to inactivate a number of targets or induce a number of units of sublethal damages.

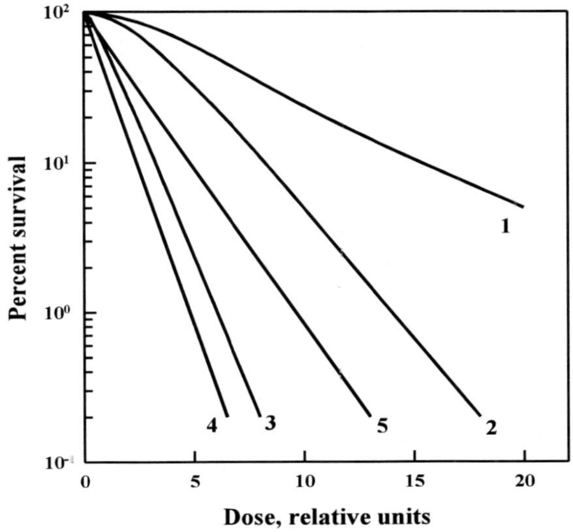

Figure 1.24. Two-hit dose-survival curves calculated for different meanings of a (LET). Target structure is a thick layer; hit-event is two primary ionizations. Curves 1, 2, 3, 4 and 5 were calculated for the following values of a : 0.05, 0.1, 0.3, 0.8, and 2.0, respectively.

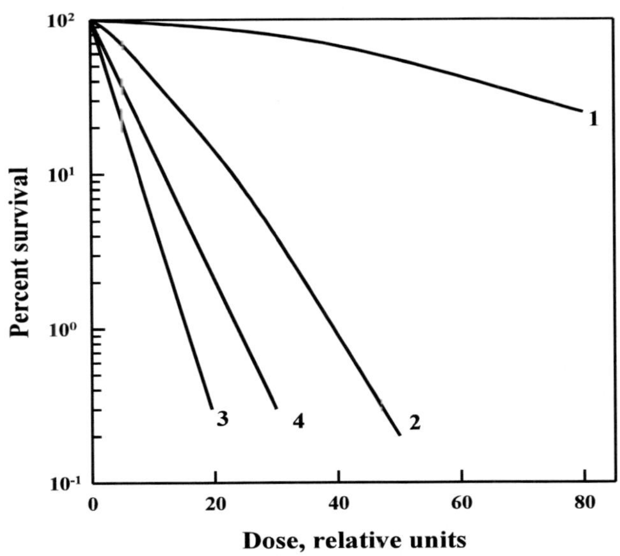

Figure 1.25. Two-hit dose-survival curves calculated for different meanings of a (LET). Target structure is a thick layer; hit-event is four primary ionizations. Curves 1, 2, 3 and 4 were calculated for the following values of a : 0.5, 1.0, 2.0, and 5.0, respectively.

Hence, the resulting shape of the survival curve can be used to deduce information concerning the structure of radiosensitive sites in the cell. For yeast cells it is also possible to describe the radiation response by the suggestion that hit-event should be some group of ionizations produced by the same ionizing particle. But in this case some peculiar composition of the target has to provide the damage of a limiting number of sensitive sites, for example one, after traversing of ionizing particle even with high-LET. Then the probability of sensitive site to be damaged increases with LET, while the mean number of sites damaged by the same particle is independent of LET. In such a case several ionizing particles have to traverse the sensitive volume to inactivate the cell even after exposure with high-LET radiation and correspondingly the sigmoid shape of survival curve would independent of LET. These considerations suggest that the target of diploid yeast cell may be imagined as a very thin layer, for example a shell-like region.

It is of interest that track-segment theory [29] shows that experiments with haploid yeast cells (in contrast to diploids, these cells after any LET radiations exhibit exponential survival curves, which may be analyzed with this theory) are agreed with the supposition in accord with the observed effect is resulted from 8–10 ion pairs being formed by a single ionizing particle within a track segment of 40–70 angstroms and the effective cross section of the target is about 10^{-8} cm^2, i.e., approximately the cross section of cell nucleus. Hence, by analogy with haploid yeast cell one may believe that the target of diploid yeast cell is a very thin layer or spherical shell with diameter to be equal to that of cell nucleus.

Our next task is to calculate an expected survival curves for such a specific fine target structure for several of a (LET). To have a possibility to compare results with those for mammalian cells, we consider the target is a thin layer with a thickness which is equal the thickness of a sensitive site of mammalian cells and with cross section being a l-fold greater size than for the previous case. It means that the total volume of sensitive sites, the mean number of primary ionizations a and the probability of a site to be damaged φ (Eq. 1.16) are the same in both cases, but the mean number of particles which hit the sensitive volume after irradiation of cells with dose D now is lm. If cell inactivation needs at least k damaged sites, then the probability of survival is given by

$$S = \sum_{i=0}^{k=1} \exp(-ml\varphi) \cdot \frac{(ml\varphi)^i}{i!}. \tag{1.23}$$

For two-hit process Eq. (1.23) may be written as follows

$$S = \exp(-ml\varphi) \cdot (1 + ml\varphi). \tag{1.24}$$

Calculated in accordance with Eq. (1.24) two-hit curves for S in dependence of am which is proportional to irradiation dose D are presented in Figure 1.26 for different values of a which is proportional to LET. The meaning of μ (the critical number of primary ionizations needed for a site damage) was chosen to be 2 (solid lines). As can be seen, the dependence of radiation efficiency on a is similar to that of the preceding case for mammalian cells when more than one primary ionizations were needed to damage a site (Figures 1.24 and 1.25) but now the sigmoid shape of survival curves is independent on a.

In order to facilitate a comparison of both sets of calculations, the corresponding survival curves for the preceding target geometry are presented in Figure 1.26 by the dotted lines. One can see that if a is small enough the survival curves for both target geometry are identical. It is a natural consequence of our initial suggestion that the total volumes of sensitive sites are equal in both cases. Indeed, if a is small, the sensitive volume rather than target geometry is essential. It is of interest that with a growing thin-layer-target version reveals both more cell sensitivity and higher RBE achieved at a higher a (LET). It would thus appear that in the case when one particle can cross many sensitive sites the decrease of the RBE with a growing is related not only with saturation of the particle efficiency φ but also with the saturation of the probability to damage the necessary number of sites by one particle. If a is too high, a single particle produces not only excessive ionizations to damage a site but also an excessive number of damaged sites to kill a cell.

The mathematical description presented may be called phenomenological in that it seeks to account for one class of observed radiobiological phenomena, but it is not based on the molecular mechanisms of cell death which are beyond the main scope of this work. Results of our calculations reveal quantitative agreement with experimental data if we assume that hit-event for both mammalian and yeast cells is a group of ionizations produced by the same ionizing particle but a fine structure of a sensitive target is quite different for these objects. For mammalian cells the target is expected to be the cell nucleus packed up by sensitive sites which are assumed to be vulnerable sections of chromosomes. For yeast cells, the target is a thin layer which is expected to be nuclear membrane.

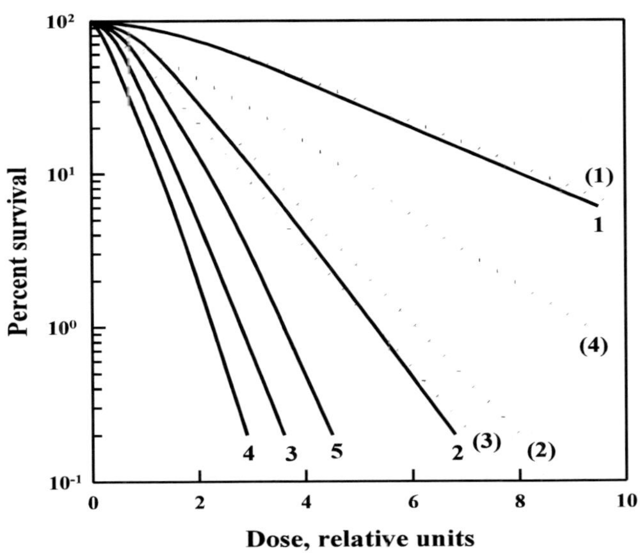

Figure 1.26. Two-hit dose-survival curves calculated for different meanings of a (or LET). Target structure is a thin (solid lines) or thick (dotted lines) layer. Hit event is two primary ionizations. Curves 1 (1), 2 (2), 3 (3), 4 (4), and 5 were calculated for the following values of a : 0.1, 0.3, 0.8, 2.0, and 5.0, respectively, for thin (solid lines) and thick (dotted lines, indexes in brackets) target structure.

However, the last conclusion seems to be wrong. Frankenberg-Scwager et al. [18] have shown a good correlation of cell killing, repair of DNA double-strand breaks and RBE of α-particles for diploid yeast cells: the increased effectiveness of densely ionizing radiations was reflected in decreased double-strand breaks induction. It means that sensitive structures in yeast have to be vulnerable sections of chromosome too. Obviously, the difference in the supposed target structures would not related with various nature of sensitive sites but to some other reasons. In particular, it would be noted that in radiobiological experiments mammalian cells, as a rule, are used at exponential phase of culture growth, while yeast cells – in stationary phase, when most of cells are in interphase state. Electron micrographs of interphase cell indicate [12] that heterochromatin contains a well-defined structural unit which can form ordered regions, namely layers in contact with nucleus envelope. Autography, light microscopy investigations and other data have shown the distribution of DNA in the nucleus was characteristically peripheral as cell moved through the interphase [2; 3; 14]. It appears reasonable to assume that even if the target is a thin spherical shell the sensitive sites may be recognized as chromosome sections. Hence, one can believe that in stationary phase of culture growth (yeast cells) chromosome are arranged at a periphery of a cell nucleus as a thin layer so that several particle traversals through the cell nucleus are needed for the production of several chromosome damages even at high-LET radiation. At exponential phase of culture growth (mammalian cells), different cells are in different phase of cell cycle and in most cells chromosomes are distributed throughout the volume of the nucleus so that every particle can cross several sections of chromosome and damage several sites for one traversing.

If the different shape of survival curves for mammalian and yeast cells irradiated with high-LET radiation is related with the dependence of the target structure on phase of culture growth, it would be expected that (i) the shape of the survival curve for mammalian cells in stationary phase of growth has to be independent on LET, and (ii) the shape of the survival curve for exponentially growing yeast cells should be changed from sigmoid after low-LET radiation into exponential one after high-LET radiation.

Unfortunately, the survival curves of mammalian cells irradiated in plateau phase of growth are exponential even after low-LET radiation. This is connected with the lowered viability (the plating efficiency) of mammalian cell growth to stationary phase. In some papers the independence of sigmoid survival curves of LET for mammalian cells was noted, however, we were in difficulties under determination whether or not this effect was connected with the phase of growth.

We have shown above that when *S. cerevisiae* diploid yeast strains were irradiated with sparsely and densely ionizing radiations in exponential phase of growth the initial sigmoid part of survival curves after low-LET radiation was transformed into exponential one after high-LET radiation. In addition, it would be expected from our calculation that the RBE of densely ionizing radiations has to be higher for stationary than for exponentially growing cells that was observed in our experiments described above.

As a final comment, the work of this section has been done deliberately in the hope that it may turn attention to the areas where further work would help to clarify our understanding of different cell responses to low- and high-LET radiation.

CONCLUSION

A mathematical model for cell responses to ionizing radiation was developed that would combine the basic idea of hit and target principles with biological stochastic. The model is based upon the two suppositions: different irradiated cells are damaged in a random fashion in accordance with the hit-and-target principle (as in classical models) and the number of damages defines the probability of successful division. Theoretically expected results were compared with experimental findings. Radiobiological experiments with yeast cells do not contradict the probability model based on the supposition that clone formation is a random process and the probability of successful division of cells is determined by the number of damages. This model is a peculiar synthesis of hit-and-target principles and biological stochastic. According to this model, cell responses to irradiation are consequences of two random factors – a random distribution of energy absorption events and biological stochastic.

A multiversity of cell responses to irradiation predicted by the probability model seems to be typical for bacterial cells, water-plants, some protozoa as well as for yeast and mammalian cells. In particular, different inactivation forms, lethal sectoring and heterogeneous in size of colonies formed by irradiated cells were observed for mammalian cells [68]. These facts indicate that the probability model may be useful for interpretation of radiation experiments with mammalian cells. Unfortunately, the distribution of mammalian cells on inactivation forms has not been yet studied in detail. With this in mind, we cannot choose the specific version of the model to interpret mammalian cell responses to irradiation quantitatively. Nevertheless, one may assert that the information presently available must be considered as not contradicting the probability model in its common presentation.

The probability model is based on hit-principle and operates with the discrete primary damages. Since in our experiments we used sparsely ionizing radiation for testing of the model, we believed that primary radiation damages in cells are determined by Poisson's distribution. The pattern of this distribution is not of great importance for the probability model and one can always take into account the radiation quality with help of microdosimetry ideas. If DNA lesions account for the radiation effect under consideration, one may use calculations made by Günther and Schulz [24]. The survival curve of diploid yeast cells irradiated with α-particles has almost the same shape as after γ-irradiation. It follows from the probability model point of view that the yield of inactivation forms and the effect of delayed appearance of clone must be identical after these kinds of irradiation. In fact, the yield of inactivation forms of diploid yeast cells depicted in the dependence of cell survival was the same after both γ- and α-irradiation [51]. Hence, the probability model can be suitable for the description of experiments with different LET.

It is often assumed that the effectiveness in reducing the colony-forming ability by high-LET radiation may be due to the quality of damage inflicted. Beginning with Lea [43] for many years it was widely accepted that an increase in effectiveness of ionizing radiation with increase in LET was due to merely an increase in the probability of the production of primary radiation damage, i.e., interaction between energy-loss events at the physical level. Neary [50] was the first to suggest, and give some evidence in flavor of, the view that, when radiation is more effective at higher LET, it may be connected with biochemical interaction between sites of damage. First information on this problem has been obtained for bacterial cells [1; 26; 47]. Some years later, we published our first data supporting the idea related with cell recovery role in the RBE of densely radiation for yeast cells, the simplest model of eukaryotic cells [52]. Somewhat later, similar data have been published for cultured mammalian cells [11]. In this chapter, we have given our summarized data with yeast cells showing that the RBE of α-particles is defined by both the primary process of radiation interaction with biological objects and the process of cell recovery at the biochemical stage of the radiation damage.

Double-stranded breaks are probably important lesions induced by ionizing radiation because, in a variety of organisms, single-strand breaks are repaired efficiently. Yeast cells are capable of repairing of double-strand breaks [16; 18]. The repair of DNA double-strand breaks requires the availability of 2 homologous DNA duplexes [45]. Resnick [60] has proposed a model for the repair of double-strand breaks involving recombination processes. It is interesting that mutant diploid strain rad51 is deficient in intragenic and intergenic recombination processes [49], and mutant diploid strain rad52 is deficient in the ability of rejoining radiation-induced double-strand breaks [60]. As was pointed out by Rao et al. [59], absence of the LHR in rad51 and *rad*52 strains may also suggest the possible role of recombination repair in this recovery. On the other hand, the absence of the LHR in rad50 and *rad*54 strains suggests possible defects in the recombination-repair pathway. These data throw some new light on the conflicting results in regard with the capability of yeast cells to recover from radiation damage after low- and high-LET radiations, which support the view that the damage caused by high-LET particles is also less repairable in yeast cells.

Participation of recovery systems in the RBE display was also confirmed by the new data demonstrating the dependence of RBE on the stage of growth, on the criteria of cell death, and on of yeast cell species. In particular, a common feature of the haplont yeasts (*P. pinus* and *P. guilliermondi*) was that, in contrast to diplont cells of *S. cerevisiae,* they have a haploid stage as a normal vegetative state in natural conditions. The experimental data presented here may be explained by the key role of repair for radioresistance of haplont yeast cells of different ploidy. Apparently, haplont yeast cells are deficient in a repair system, the effect of which in diplont yeast cells are related to the presence of the double genome allowing recombination processes to occur. If radiation injuries in haploid and diploid cells involve damage leading to chromosome aberrations it can be expected that such lesions may arise at a higher frequency in diploid than in haploid cells because of differences in target volume (number of chromosomes, DNA content, *etc.*). In such a case, diploid cells have to be more radiosensitive for cells incapable of recovery in comparison with haploid cells. If dark recovery is not completely inhibited, then an intermediate situation may apply and haploid and diploid cells may show very similar radiosensitivity. Then the RBE values are expected to be nearly identical for haploid and diploid yeast cells as was obtained here. It is known that there was no remarkable difference in sensitivity to both ionizing radiation and UV-light between haploid and diploid cells of the fission yeast *Schizosaccharomyces pombe* [63] and the yeast-like eukaryote *Ustilago maydis* [44]. Shahin et al. [63] have shown that haploid and diploid wild-type strains of S. pombe exhibited no liquid-holding effect after both γ- and UV-irradiation. It is not excluded that all haplont yeast cells were defective in diploid-specific recovery including DNA double-strand break repair which was found to be responsible for the LHR in *S. cerevisiae*. If the cell radioresistance to ionizing radiation may, to some extent, serve as a measure of cell sensitivity to other environmental factor inflicted genetic damage, one may suppose that only the yeasts capable of diploid cell recovery will have a selective advantage. On the contrary, for yeasts defective in repair systems, haploid cells may possess a selective advantage. It is not excluded that owing to this fact, the vegetative phase in natural conditions is presented by diploid cells in *S. cerevisiae* and by haploid cells in *P. pinus* and *P. guilliermondii*. It follows that the development of repair systems, providing more full recovery of diploid cells from genetic damages in comparison with haploid cells, may have served as a prerequisite for the transition, during evolution processes, from haplont to diplont cells.

Here, differences in recovery kinetics were demonstrated for all strains studied, with slower recovering for increasing LET. Several possible explanations for the slower recovery observed when the LET is increased have been widely discussed. (1) The primary radiation damage are more severe and complex and therefore requires a long time to be repaired [23]. This is an interesting hypothesis, likely to be true. (2) The number of small and intermediate-size DNA fragments is larger, and a longer time is needed for repair and recovery [28]. (3) Impairment of recovery process at high-LET could also be considered although it was not really proven. (4) It is more likely that the repair systems involved are overwhelmed by the presence of more complex DNA damage (in the case of high-LET radiation) as already suggested [18]. These authors showed that in diploid cells of the yeast *S. cerevisiae* the rejoining of double-strand breaks (DSB) slows down with increasing LET, DSB induced by electrons were rejoined more rapidly than DSB induced by α-particles. The authors proposed to interpret their data as being due to accumulating DNA damage which impairs the operation of enzymes involved in the process of DSB rejoining. (5) A slower recovery may reflect a smaller number of reversible damage produced by high-LET radiations, with the probability of recovery independent of radiation quality.

It was demonstrated in this Chapter that the probability of the recovery was identical for both a low- and high-LET radiation for all the strains investigated. It means that an identical fraction of the reversible radiation damage is recovered per unit time for both the sparsely and densely ionizing radiations. The molecular mechanisms involved in LHR are not clear; however, some aspects could be mentioned. It is known that several DNA repair pathways are involved in the recovery from radiation damage. Among these, in stationary phase diploid yeast cells, homologous recombination is the pathway that is most frequently employed. Apparently, in yeast cells, a recombinational type of repair involving (or leading to) gene conversion can be favored by LHR [18]. This type of repair is obviously absent in mutants defective in homologous recombination [61]. In addition, base excision repair, mismatch repair and even nucleotide excision repair pathways contribute [5; 20; 36]. The *rad*18/*rad*18 is known to be defective in the repair of radiation-induced single-strand breaks [21] and the *rad*53/*rad*53 is responsible for the genome integrity of checkpoint kinase [66]. It may be concluded on this basis that neither all these pathways nor repair enzymes are damaged more seriously after a densely ionizing radiation. Moreover, a slower recovery after high-LET radiation does not reflect the deceleration of recovery processes itself and is due to a smaller number of repairable damage inflicted by high-LET radiation. In other words, the slowing down of recovery kinetics and therefore the enhanced RBE of a high-LET radiation may be exclusively caused by the increased yield of irreversible damage that cells are incapable of recovery. Thus, the role of ionization density in the repair system itself may be excluded. It may be suggested therefore that the impact of a high ionization density is connected with a change in the quality of the lesions rather than with the quantity or the properties of the repair systems themselves.

In this Chapter the recovery kinetics was estimated for wild-type diploid yeast cells and different mutants: *rad*18/*rad*18 is deficient in postreplication repair but competent in homologous recombination; *rad*53/*rad*53 is affected in an important signal induction pathway (kinase); *rad*55/*rad*55, *rad*56/*rad*56, *rad*57/*rad*57 are partially defective in homologous recombination repair [5; 20; 66]. The probability of the recovery was shown to be identical (β = 0.05 h^{-1}) for the parent strain (XS800) and all *rad* mutants investigated. And here again, the probability of the recovery in the parent and *rad* mutant cells does not depend on the ionizing radiation quality. It means that about 5% of the residual reparable damage is recovered every hour for both wild-type and radiosensitive mutants independently of the radiation quality. It corresponds to the data that in some mutants the genetically defined defect can at least partially be compensated by other repair protein (channeling of repair). The lower RBE values for mutant cells may be due to the slight deficiency of the mutants to cope already with γ-ray induced damage. The fact that the radiosensitive mutants tested showed the same probability of recovery during LHR as the parent strain may indicate that the deficiency of the mutants does not affect the repair system facilitating the repair of radiation-induced lesion by LHR.

Some experimental results for mammalian cells obtained in molecular level are in agreement with this conclusion: the kinetics of those DSBs which were actually repaired did not show any significant differences for three cell lines studied [66]. It was suggested by this author that the higher number of non-reparable breaks in *xrs* cells may not be attributable to a deficient enzymatic repair system. These data indicate that the mechanism of the enhanced radiosensitivity of the radiosensitive yeast and mammalian cells may be related to the production of more severe damage from which cells are incapable to recover. It is likely that the inability of other *rad* mutants (*rad*50, *rad*51, *rad*52, *rad*54) to recover from radiation damage has to be considered as an indication for the absence of repair and corresponds to the incapacity to repair certain types of radiation lesions.

The interesting point can be inferred from the data presented: the action of sparsely and densely ionizing radiations on both yeast and mammalian cells can be interpreted with help of multi-hit model with hit-event to be a local production of several ions by the same single particle in a sensitive site which is presumably a section of chromosome. In the dependence of cell cycle phase, sensitive sites can be located either in periphery of a cell nucleus forming a thin spherical layer or occupy inner region of the nucleus, both cases being not contradict to cytological observations [14]. Published autoradiographic studies of Chinese hamster cell have suggested a peripheral distribution of cellular chromatin within the nucleus of the cell during interphase [14]. Such rearrangement of the target leads to the alteration of the dependence of both survival curve shape and RBE value on radiation quality.

It is of interest to note that Zermeno and Cole [74] in their studying of radiosensitive structure with low-voltage electrons of different penetration capacity also concluded that the major radiosensitive region was a thin peripheral shell close to nuclear membrane in the interpose cells and a considerably thicker shell in metaphase cells. In addition, the metaphase cells exhibited a four-fold sensitivity.

Now it is almost generally assumed that a shape of survival curve is an expression of cell capacity to recover. If a survival curve is exponential, then a cell is not capable to repair radiation damage at all, but the sigmoid shape of a survival curve means that a cell is capable to repair a certain number of radiation damages. According to the repair idea, the sigmoid shape of survival curve results from failure of a repair mechanism with increasing dose. Not discussing whether this viewpoint is universal or not we note only that our interpretation does not contradict it. Indeed, the multi-hit model in itself contains a recovery idea in a concealed form. The fact that multi-hit model requires energy to be accumulated in successive events leads to obvious possibility of recovery processes. Then one may believe that many damage in a cell should be inactivated since it is not understandable why cells in stationary phase of culture growth repair both types of damages induced by low and high-LET radiations but cells in exponential phase of growth repair only damages induced by low-LET radiations and are not capable to repair damages induced by high-LET radiations. But taking into account the target geometry it becomes clear: when cells are in exponential phase of growth every densely ionizing particle produces too many damages so the cell is unable to repair all of them.

Briefly, the above mentioned new results may be summarized as follows.

1. A mathematical model for cell responses to ionizing radiation has been developed. The model combines the basic idea of hit and target principles with biological stochastic. The attractiveness of the model developed consists in the fact that it can be used to predict diversity effects of irradiation observed in experiments.
2. The RBE of densely ionizing radiation is determined both by the physical characteristics of radiation and by cell biochemical property. In particular the cell recovery ability is partly responsible for RBE values. It is concluded that such a recovery is mediated by a recombination-like mechanism, for which the diploid state is required.
3. The reduction of the RBE of α-particles in exponential phase of growth was mainly due to the increase of cell radiosensitivity after γ-irradiation. Alpha-particle sensitivity was shown to be not greatly dependent on phase of growth. Thus, the physiological state of cells can influence their sensitivity and the RBE of densely ionizing radiation.
4. The irreversible component of radiation damage was enhanced for the densely ionizing radiation in comparison to the low-LET radiation while the probability of the recovery per unit time was identical for both kinds of radiation. This means that the slowing down of yeast cells recovery after high-LET radiation is associated with a greater number of irreversible damage rather than with the impairment of recovery process itself.
5. On basis of the dependence of the of RBE on ploidy of diplont and haplont yeast cell it is affirmed that the development of repair systems in diplont cells, providing more full recovery of diploid cells from ionizing radiation damages in comparison with haploid cells, may have served as a prerequisite for the transition, during evolution processes, from haplont to diplont eukaryotic cells.
6. A theoretical analysis of the dependence of the shape of dose-effect curves for eukaryotic cells is presented. It is inferred that the independence of sigmoid survival curve on radiation quality for diploid yeast cells and its transformation from sigmoid into exponential one for mammalian cells with LET radiation increase may be due to the dependence of radiosensitive site location on the growth stage, at which irradiation is occurred.

REFERENCES

[1] Alper, T. *Cellular Radiobiology*. Cambridge University Press, 1979.

[2] Ballarini, F; Biaggi, M; Ottolenghi, A. Nuclear architecture and radiation induced chromosome aberrations: models and simulation. *Radiat. Protect. Dosimetry*, 2001; 99(1–4), 175-182.

[3] Ballarini, F; Friedland, W; Jacob, P. Role of DNA organization and environmental scavenging capacity in the evolution of radiobiological damage models and simulation. *Radiotherapy and Oncology*, 2004; 73(2), 170-172.

[4] Barendsen, GW; Beusker, TLJ. Effects of different ionizing radiations on human cells in tissue culture. 1. Irradiation techniques and dosimetry. *Radiat. Res.*, 1960; 13, 832-840.

[5] Bennett, CB; Lewis, LK; Karthikeyan, G; Lobachev, KS; Jin, YH; Sterling, JF; Snipe, JR; Resnick, MA. Genes required for ionizing radiation resistance in yeast. *Nature (Genetics)*, 2001; 29, 426-434.

[6] Bertsche, U; Iliakis, G. Radiation quality and repair of potentially lethal damage. In: Booz J., Ebert H.G., editors. *Radiation Protection*. Brussels: Commission of the European Communities, 1983; 383-395.

[7] Blöcher, D. DNA double-strand break repair determines the RBE of α-particles. *Int. J. Radiat. Biol.*, 1988; 54, 761-771.

[8] Brahme, A. Accurate description of the cell survival and biological effect at low and high doses and LET's. *J. Radiat. Res.*, 2011; 52(4), 389-407.

[9] Brendel, M; Haynes, RH. Interaction among genes controlling sensitivity to radiation and alkylation in yeast. *Mol. Gen. Genet.*, 1973; 125, 197-216.

[10] Budd, M; Mortimer, RK. Repair of double-strand breaks in a temperature conditional radiation-sensitive mutant of *Saccharomyces cerevisiae*. *Mutat. Res.*, 1982; 103, 19-24.

[11] Coquerelle, TM; Weibezahn, KF; Lucke-Huhle, C. Rejoining of double strand breaks in normal human and ataxia- telangiectasia fibroblasts after exposure to ^{60}Co gamma-rays, ^{241}Am alpha-particle or bleomycin. *Int. J. Radiat. Biol.*, 1987; 51, 209-218.

[12] Davies, HG. Electron-microscope observations on the organizations of heterochromatin in certain cells. *J. Cell Sci.*, 1968; 3, 129-133.

[13] Dertinger, H; Jung, H. *Molekulare Strahlenbiologie*. Berlin, Heidelberg, New York: Springer-Verlag, 1969.

[14] Dewey, WC; Noel, JS; Dettor, CM. Changes in radiosensitivity and dispersion of chromatin during the cell cycle of synchronous Chinese hamster cells. *Radiat. Res.*, 1972; 52, 373-394.

[15] Fox, JC; McNally, NJ. Cell survival and DNA double-strand break repair following X–ray or neutron irradiation of V79 cells. *Int. J. Radiat. Biol.*, 1988; 54, 1021-1030.

[16] Frankenberg, D. Repair of DNA double-strand breaks and its effect on RBE. *Adv. Space Res.*, 1994; 14, 235-248.

[17] Frankenberg, D; Frankenberg-Schwager, M; Blöcher, D; Harbich, R. Evidence for DNA double-strand breaks as the critical lesions in yeast cells irradiated with sparsely or densely ionizing radiation under oxic or anoxic conditions. *Radiat Res.*, 1981; 88(3), 524-32.

[18] Frankenberg-Schwager, M; Frankenberg, D; Harbich, R. Repair of DNA double-strand breaks as a determinant of RBE of alpha-particles. *Br. J. Cancer*, 1984; 49, Suppl. VI, 169-173.

[19] Friedberg, EC; Siede, W; Cooper, AC. Cellular responses to DNA damage in yeast. In: Broach JR, Pringle JR, Jones EW editors. *The Molecular and Cellular Biology of the yeast Saccharomyces. Vol. I, Genome Dynamics, Protein Synthesis and Energetics.* Cold Spring Harbor, NY: Cold Spring Harbor Laboratory Press, 1991; 147-192.

[20] Game, JC. The Saccharomyces repair genes at the end of the century. *Mutat. Res.*, 2000; 451, 277-293.

[21] Geigl, EM; Eckardt-Schupp, F. The repair of double-strand breaks and S1 nuclease-sensitive sites can be monitored chromosome-specifically in *Saccharomyces cerevisiae* using pulse-field gel electrophoresis. *Mol. Microbiol.*, 1991; 5(7), 1615-1620.

[22] Glasunov, AV; Lobachevsky, PN. Postradiation recovery of *Pichia pinus* yeast cells. *Radiobiologiya*, 1983; 23, 409-411.

[23] Goodhead, DT. Initial events in the cellular effects of ionizing radiations: clustered damage in DNA. *Int. J. Radiat. Biol.*, 1994; 65, 7-17.

[24] Günther, K; Schulz, W. *Biophysical Theory of Radiation Action. A. Treatise on Relative Biological Effectiveness*. Berlin: Akademie-Verlag, 1983.

[25] Hannan, MA; Paul, M; Phillips, RL. Fast neutron r.b.e. for lethahty and genotoxicity in a wild-type and repair-deficient strain of yeast. *Int. J. Radiat. Biol.*, 1986; 50, 811-824.

[26] Haynes, RH. The interpretation of microbial inactivation and recovery phenomena. *Radiat. Res.,* 1966; Suppl. 6, 1-29.

[27] Heilmann, J; Rink, H; Taucher-Scholz, G; Kraft, G. DNA strand break induction and rejoining and cellular recovery in mammalian cells after heavy-ion irradiation. *Radiat. Res.*, 1993; 135, 46-55.

[28] Höglund, E; Stenerlöw, B. Induction and rejoining of DNA double-strand breaks in normal human skin fibroblasts after exposure to radiation of different linear energy transfer: Possible roles of track structure and chromatin organization. *Radiat. Res.*, 2001; 155, 818-825.

[29] Howard-Flanders, P. Physical and chemical mechanisms in the injury of cells by ionizing radiations. *Adv. Biol. Med. Phys.*, 1958; 6, 553-603.

[30] Hug, O; Kellerer, AM. *Stochastik der Strahlenwirkung*. Berlin-Heidelberg-New York: Springer-Verlag, 1966.

[31] Jenner, TJ; deLara, CM; O'Neil, P; Stevens, DL. Induction and rejoining of DNA double-strand breaks in V79-4 mammalian cells following γ- and α-irradiation. *Int. J. Radiat. Biol.*, 1993; 64, 265-273.

[32] Kapultcevich, YuG. *Quantitative Regularities of Cell Radiation Injury*. Moscow: Atomizdat, 1978.

[33] Kapultcevich, YuG; Petin, VG. The analysis of survival curves and inactivation forms of irradiated yeast cells with the help of the probable model. *Radiobiology,* 1972; 12, 864-872.

[34] Kellerer, AM; Rossi, HH. RBE and the primary mechanism of radiation action. *Radiat. Res.*, 1971; 47, 15-34.

[35] Keszenman, DJ; Sutherland, BM. Yields of clustered DNA damage induced by charged-particle radiations of similar kinetic energy per nucleon: LET dependence in different DNA microenvironments. *Radiat. Res.*, 2010; 174(2), 238-250.

[36] Kiefer, J; Egenolf, R; Ikpeme, S. Heavy ion-induced DNA double-strand breaks in yeast. *Radiat. Res.*, 2002; 157, 141-148.

[37] Korogodin, VI. *The problems of Postirradiation Recovery*. Moscow: Atomizdat, 1966.

[38] Korogodin, VI. The study of post-irradiation recovery of yeast: the premolecular period. *Mutat. Res.*, 1993; 289, 17-26.

[39] Korogodin, VI; Kapultcevich, YuG; Petin, VG; Blisnik, KM. Radiosensitivity of haplont yeast cells irradiated with sparsely and densely ionizing radiations. *Mutat. Res.*, 1996; 357, 67-74.

[40] Korogodin, VI; Bliznik, KM; Kapultcevich, YuG; Korogodina, VL; Korolev, VG; Mezhevaya, EV; Petin, VG; Simonyan, NV; Tolstorukov, II; Tsyb, TS. Cascade mutagenesis: regularities and mechanisms. In: Korogodina VL, Cigna AA, Durante M editors. *Modern Problems of Genetics, Radiobiology, Radioecology and Evolution*. Dubna: JINR, 2007; 419-447.

[41] Krasavin, EA. *Problem of RBE and Repair of DNA*. Moscow: Energoatomizdat, 1989.

[42] Laskowski, W. Inaktivierungsversuche mit homozygoten Hefestammen verschidenen Ploidiegrades. 1. Aufban homozygoter Stamme und Dosiseffektkurven fur ionisierende Strahlen, UV and organische Peroxyde. *Z. Naturforschung*, 1960; 15b, 495-506.

[43] Lea, DE. *Action of Radiation on Living Cells*. Cambridge University Press, 1946.

[44] Lemontt, JF. Induced mutagenesis in *Ustilago maydis*. I. Isolation and characterization of a radiation-reversible allele of the structural gene for nitrate reductase. *Mol. Gen. Genet.*, 1976; 145, 125-132.

[45] Luchnik, AN; Glaser, VM; Shestakov, SV. Repair of DNA double-strand breaks requires two homologous DNA duplexes. *Mol. Biol. Rep.*, 1977; 3, 437-442.

[46] Lyman, JT; Haynes, RH. Recovery of yeast after exposure to densely ionizing radiation. *Radiat. Res.*, 1967; Suppl. 7, 222-230.

[47] Munson, RJ; Neary, GJ; Bridges, BA; Preston, RJ. The sensitivity of *Escherichia coli* to ionizing particles of different LETs. *Int. J. Radiat. Biol.*, 1967; 13, 205-224.

[48] Murthy, MS; Reddy, NM; Rao, BS; Subrahmanyam, P; Madhvanath, U. Dependence of radiation response of diploid yeast upon growth phase and LET. *Int. J. Radiat. Biol.*, 1976; 29, 85-91.

[49] Nakai, S; Matsumoto, S. Two types of radiation-sensitive mutants in yeast. *Mutat. Res.*, 1967; 4, 129-136.

[50] Neary, GJ. Cell killing, repair, and radiation quality. *Stud. Biophys.*, 1969; 18, 11-25.

[51] Petin, VG. *Genetic Control of Cell Radiosensitivity Modification*. Moscow: Energoatomizdat, 1987.

[52] Petin, VG; Kabakova, NM. RBE of densely ionizing radiation for wild-type and radiosensitive mutants of yeast. *Mutat. Res.*, 1981; 82, 285-294.

[53] Petin, VG; Kabakova, NM; Wrigglesworth, VI et al. RBE of α-particles and postradiation recovery in a temperature conditional diploid yeast mutant for various criteria of cell death. *Physics of Elementary Particles and Atomic Nuclei, Letters,* **3** №4 (133) (2006), 241-245.

[54] Petin, VG; Kim, JK. Survival and recovery of yeast cells after combined treatments with ionizing radiation and heat. *Radiat. Res.*, 2004; 161, 56-63.

[55] Petin, VG; Komarov, VP. *Quantitative Description of Radiosensitivity Modification*. Moscow: Energoatomizdat, 1989.

[56] Pohlit, WE; Schafer, M. Recovery and repair in yeast cells after irradiation with densely ionizing particles. In: *Biological Effects of Neutron Irradiation*. Vienna: IAEA, 1974; 177–184.

[57] Raju, MR. *Heavy Particle Radiotherapy*. New-York: Academic Press, 1980.

[58] Raju, MR; Gnanapurani, M; Stackler, B; Madhvananth, U; Howard, J; Lyman, JT; Manney, TR; Tobias, CA. Influence of linear energy transfer on the radioresistance of budding haploid yeast cells. *Radiat. Res*., 1972; 51, 310-317.

[59] Rao, BS; Reddy, NM; Madhvanath, U. Gamma radiation response and recovery studies in radiation sensitive mutants of diploid yeast. *Int. J. Radiat. Biol*., 1980; 37, 701-705.

[60] Resnick, MA. The repair of double-strand breaks in DNA: a model involving recombination. *J. Theor. Biol*., 1976; 59, 97-106.

[61] Saeki, T; Machida, I; Nakai, S. Genetic control of diploid recovery after gamma-irradiation in the yeast *Saccharomyces cerevisiae*. *Mutat. Res*., 1980; 73, 251-265.

[62] Schneider, E; Kiefer, J. Delayed plating recovery in diploid yeast of different sensitivities after X-ray and alpha-particle exposure. *Int. J. Radiat. Biol.,* 1976; 29, 77-84.

[63] Shahin, MM; Centner, NE; Nasim, A. The effect of liquid holding in *Schizosaccharomyces pombe* strains after gamma and ultraviolet irradiation. *Radiat. Res*., 1973; 53, 216-225.

[64] Sibirny, AA; Shavlovsky, GM; Kshanovskaya, BV; Naumov, GI. Hybridization and meiotic segregation in the paraffin-utilizing yeast *Pichia guilliermondii* Wickerham. *Genetica*, 1977; 13, 314-321.

[65] Stenerlöw, B; Höglund, E; Carlsson, J; Blomquist, E. Rejoining of DNA fragments produced by radiations of different linear energy transfer. *Int. J. Radiat. Biol*., 2000; 76, 549-557.

[66] Symington, LS. Role of RAD52 epistasis group genes in homologous recombination and double-strand break repair. *Microbiol. Mol. Biol. Rev*., 2002; 66(4), 630-670.

[67] Tobias, CA; Blakely, EA; Chang, EP; Lommel, L; Roots, R. Response of sensitive human ataxia and resistant T-1 cell lines to accelerated heavy ions. *Br. J. Cancer*, 1984; 49, Suppl. VI, 175-185.

[68] Todd, P. Heavy-ion irradiation of cultured human cells. *Radiat. Res*., 1967; Suppl. 7, 196-207.

[69] Tolstorukov, II; Dutova, TA; Benevolensky, SV; Soom, YaO. Hybridization and genetic analysis of methanol-utilizing yeast *Pichia pinus*. *Genetica*, 1977; 13, 322-329.

[70] Vekshina, LK; Sakovich, IS; Shevchenko, VA. The effect of densely ionizing radiation on chlorella. *Radiobiologiya*, 1974; 14, 386-389.

[71] Wang, JF; Lu, D; Wu, X et al. Inactive and mutagenic effects induced by carbon beams of different LET values in a red yeast strain. *Nuclear Instruments & Methods in Physics Research*, 2010; 268(17-18), 2719-2723.

[72] Weyrather, WK; Ritte, S; Scholz, M; Kraft, G. RBE for carbon track-segment irradiation in cell lines of differing repair capacity. *Int. J. Radiat. Biol*., 1999; 75, 1357-1364.

[73] Yatagai, F; Matsuyama, A. LET-dependent radiosensitivity of Escherichia coli K-12 rec and uvr mutants. *Radiat. Res*., 1977; 71, 259-263.

[74] Zermeno, A; Cole, A. Radiosensitive structure of metaphase and interphase hamster cells as studied by low-voltage electron beam irradiation. *Radiat. Res.*, 1969; 39, 669-684.

Chapter 2

NOVEL DIRECTION IN STUDY OF RADIOPROTECTING AND RADIOSENSITIZING ACTION OF CHEMICAL COMPOUNDS

ABSTRACT

The correlation between the radioprotecting action of mexamine, cystamine, cysteine and cysteamine and cell repair capacity was demonstrated for diploid yeasts; such a correlation was not expressed for wild-type and mutant haploid yeast cells. Radiosensitization of hypoxic yeast cells by oxygen, metronidazole and misonidazole was compared in wild-type and various X-ray radiosensitive mutants. The oxygen enhancement ratio (OER) and enhancement ratio (ER) of hypoxic sensitizers depended markedly on cell genotypes. Metronidazole was found to enhance radiosensitivity of hypoxic cells only of those strains that showed an OER more than 2.2. A correlation between oxygen and misonidazole sentsitization was observed. The data confirm that misonidazole mimics the sensitizing effect of oxygen. Yeast strains which were deficient in ability to repair radiation damage had lower OER and ER values than their more radioresistant parents. Some possible explanations of these effects are discussed. Modification of cell radiosensitivity was expressed mathematically in terms of OER for different yeast cell strains. It was shown that the parameters used permit to describe the dependence of the efficiency of radioprotectors and radiosensitizers upon yeast cell genotype. It was concluded that the radioprotective action of chemical compounds and radiosensitizing effect of oxygen and hypoxic sensitizers may involve cellular recovery processes, which may be mediated by a recombination-like mechanism, for which the diploid state is required.

INTRODUCTION

Modification of cell radiosensitivity by chemical agents influencing both protecting and sensitizing action upon the effects produced by ionizing radiation stays an actual task. Many hypotheses have been suggested for mechanisms of chemical protection against ionizing radiation [6; 16; 36]. The radioprotective effects of chemical compounds are explicable by their hydroxyl radical scavenging action and competition with oxygen. Protection against the direct action of radiation has also been proposed [3]. A new hypothesis involving specific repair mechanisms has been widely discussed for prokaryotic cell systems. The effect of radioprotection of indolylalkylamines and aminothiols was shown to be almost absent if bacterial cells *Escherichia coli* have genetic defects in their repair systems [7; 35]. On this basis, it was postulated that the repair mechanism of radioprotection is the main mechanism in bacterial cells [7]. The role of repair in the decreased radiosensitivity of *E. coli* because of metabolic radioprotective effect of different chemical compounds was also analyzed by Hernadi and co-workers [15]. Maisin has studied the length of the mitotic cycle of the small intestinal mucosa cells and has shown that the first complete cell cycle lasts about 17 h after treatment by mixtures of radioprotectors compared to 11 h in untreated mice [21]. There is a good reason to believe that this delay allows more time for repair of radiation-induced damage. On the basis of the radioprotective effect of some chemical compounds on the yield of chromosome aberrations in seedlings *Vicia faba* the conclusion was drawn [4] that protection ratio is caused by an increase in the repair volume and not by a decrease of the initial lesion number. It is of interest to study the radioprotective action of a well-known chemical compounds on *S. cerevisiae* yeast cells of wild-type and various radiation sensitive mutants.

The increase in the radiosensitivity of biological objects exposed to ionizing radiation in oxic than in hypoxic or anoxic conditions is well known fact. This phenomenon is known in radiobiology, inherent to most biological objects and calls the oxygen effect. A possible participation of recovery processes in the mechanism of the oxygen effect inspires a renewed interest to this problem. Bacterial and mammalian cells have been extensively studied concerning the oxygen effect. Comparative studies of the survival characteristics of different mutants of *E. coli* under oxic and anoxic conditions showed that the oxygen enhancement ratio (OER) is negatively correlated with cell sensitivity [1; 17]. Wild-type forms of *Chlamydomonas reinhard* [10] and the slime mould *Dictyostelium discoideum* [11] have values of OER of about 2.2 and 1.8 respectively whereas these values for radiosensitive mutants were 1.0 or less. These data may indicate that oxygen may involve cellular recovery processes. Similar data for eukaryotic cells are not abundant.

Several chemicals have been found to be hypoxic cell sensitizers. Amongst these, the trichomonacide, metronidazole (Flagyl) has been investigated intensively in many cellular systems and animal tumors to establish its radiosensitizing efficiency, its mechanisms of action at molecular and cellular levels and its usefulness in clinical radiotherapy. The radiosensitizing effect of metronidazole with respect to survival was observed only in wild-type diploid yeast cells [19]. No work was reported on the effect of this compound or other radiosensitizers on yeast strains possessing different ability to recover radiation damage. It is known that there is significant radiosensitization by nitroimidazole, another typical electron affinic radiosensitizer, in *E. coli* wild-type cells. But in mutant cells, which are genetically unable to repair radiation damage in DNA, the radiosensitization was negligible [14; 32]. Similar results were found for membrane-binding drugs [37] in *E. coli B/r* and a radiosensitive mutant *E. coli* B_{s-1}.

The radiosensitizing action of chemical compounds on hypoxic cells has been studied mainly with wild-type cells of various origins. Only some experiments were performed with cells characterized by different ability to recover from radiation-induced damage. A comparative study of the effects of some radiosensitizers upon the survival of bacterial cells of wild-type and radiosensitive mutants resulted in an empirical relationship

$$ER = \alpha \cdot OER \qquad (2.1)$$

relating the enhancement ratio (ER) produced by sensitizer with oxygen effect of a bacterium [14; 32]. The highest value of α for bacterial cells and another typical electron-affinic radiosensitizer *p*-nitroacetophenone was 0.75; the sensitizer showed an ER approaching the full OER. Hence, it was considered to be worth-while to examine the effect of a more effective sensitizer (misonidazole) on wild-type yeast cells and radiosensitive mutants. It is hoped that studies with misonidazole in yeast cells with different abilities to repair radiation damage may offer additional insight into its mode of action. The compound is known to be an effective radiosensitizer of hypoxic cells systems both *in vitro* and *in vivo*. Yeasts are also a convenient model to test the relation between metronidazole and misonidazole activity and cell ability to recover from potentially lethal damage. It would be of interest to express mathematically the participation of recovery processes in cell radiosensitivity modification by radioprotectors, oxygen and hypoxic radiosesitizers.

Thus, in this Chapter we will deal with the following tasks.

1. To investigate radioprotecting action of chemical compounds on γ-irradiated yeast cells of various genotypes in *Saccharomyces* yeast cells.
2. To determine oxygen effect in DNA-proficient and -deficient yeast cells.
3. To estimate radiosensitization of hypoxic cells by chemical compounds in wild-type and radiosensitive mutants of yeast.
4. To study radiosensitivity modification of γ-irradiated haplont yeast cells.
5. To analyze participation of recovery systems in cell radiosensitivity modification.

2.1. Radioprotecting Action of Chemical Compounds on γ-irradiated Yeast Cells of Various Genotypes in *Saccharomyces* Yeast Cells

A number of hypothesis have been put forward to explain the mechanism underlying chemical protection and sensitization of biological objects against ionizing radiation. It was shown that sensibilization effect of oxygen [1; 2] and protective action of some compounds [7; 35] on *E. coli* depends on the state of the DNA repair systems. It was shown that mexamine (5-metoxitryptamine) protected mutants of *E. coli* with defects in polymerase I but did not protected mutants deficient in ligase or *rec* repair enzymes and caffeine in low doses eliminated protecting action of mexamine. These facts show that the protecting action of mexamine is entirely dependent on repair enzymatic processes. The radioprotecting effect of mexamine is practically zero to mutants with deficient ligase or the *rec* ABC repair enzymes. On the other hand, the *pol* A mutant behaves in this respect like the wild-type. The intact *rec* and *lig* genes products are absolutely required for the expression of the protecting action from γ-irradiation. If a highly dilute solution of caffeine, a well known inhibitor of repair processes, is introduced into the *E. coli* culture (this amount of caffeine is indifferent for inirradiated cells) it eliminates the protecting action of mexamine completely. It was shown that damage in ligase-dependent, *rec*-dependent or *pol*-dependent repair system drastically reduces or even completely prevents the protective action. In experiments *in vitro* it has been shown that cysteamine is an active inhibitor of isolated and purified endonuclease II. It is also known that endonuclease II is an enzyme involved in repair of radiation damage. It is possible to believe that endonuclease activity may be inactivated by cysteamine. Cell protection through such kind of mechanism may take place only when the repair systems are operating normally. Defects in reparation systems should result in the reducing of protective efficiency.

Papers indicating a possible connection between radioprotective efficiency and recovery processes in eukaryotic cells are not abundant. It would be of interest to compare radioprotecting action of chemical compounds on yeast cells with regard to their various abilities to recover from radiation damage. The results will be presented below. It is worth to note that some of these results have already been published concerning the radioprotecting action of chemical compounds [26; 27] on yeast cells of various genotype. Here, we present more complete and averaged results. The main goal was to obtain survival curves, with and without radioprotectors, of some haploid and diploid strains of *S. cerevisiae* known to have different capacities for repair.

Before irradiation cells were incubated for 4 days at 30 °C on a solid complete medium to stationary phase. We have used cysteamine (Sigma), mexamine, cystamine and cysteine (Reanal), as radioprotective compounds. The radioprotectors were added to the cell suspension (10^6 cells/ml) 30 min before irradiation. The optimal concentrations for the protecting action were about 10^{-2} M for cysteamine, cystamine, mexamine and $3 \cdot 10^{-2}$ M for cysteine. These concentrations were not toxic for any strains investigated. Irradiation of cells was performed by means of a ^{60}Co γ-ray source at a dose rate of 10 Gy/min at room temperature. Immediately after irradiation, the samples were suitably diluted and a part of cell suspension was plated on nutrient agar for assay of colony forming ability; another aliquot was placed in conditions promoting LHR (30 °C) and in 3 days (delayed plating) their colony forming ability was checked again. First, the radioprotective effect of mexamine, cystamine and cysteine on survival of haploid and diploid yeast *S. cerevisiae* of wild-type and a radiosensitive mutant (*rad51*) was investigated. As an example, Figure 2.1 shows survival curves of haploid and diploid yeast *S. cerevisiae* of wild-type and radiosensitive mutants exposed to ionizing radiation with and without 0.03 M cysteine. Experimental points in this and other figures have errors of 5 % or less. To estimate quantitatively the efficiency of radioprotection, we used the dose-modifying factor (DMF), which was calculated as the ratio of the mean lethal doses of the survival curves (D_0) obtained with and without radioprotectors. The DMF by mexamine, cystamine and cysteine of various strains of *S. cerevisiae* are listed in Table 2.1. It also includes the results obtained but not presented here with mexamine, cystamine and various concentration of cysteine.

It is apparent that the efficiency of radioprotection is dependent on cell ploidy and genotype, the magnitude of radioprotection being greater for wild-type yeast cells capable of recovery. These data can be interpreted in terms of the possibility that the mechanism of radioprotective action may involve the capacity of cells to recover from radiation damage.

In the yeast *S. cerevisiae* there are at least eight distinct loci which control sensitivity to ionizing radiations. Therefore, it is of interest to study the radioprotective action of the most effective compound (cysteamine) that we have among others on wild-type *S. cerevisiae* and various radiation sensitive mutants. The LHR served as an indicator of repair activity.

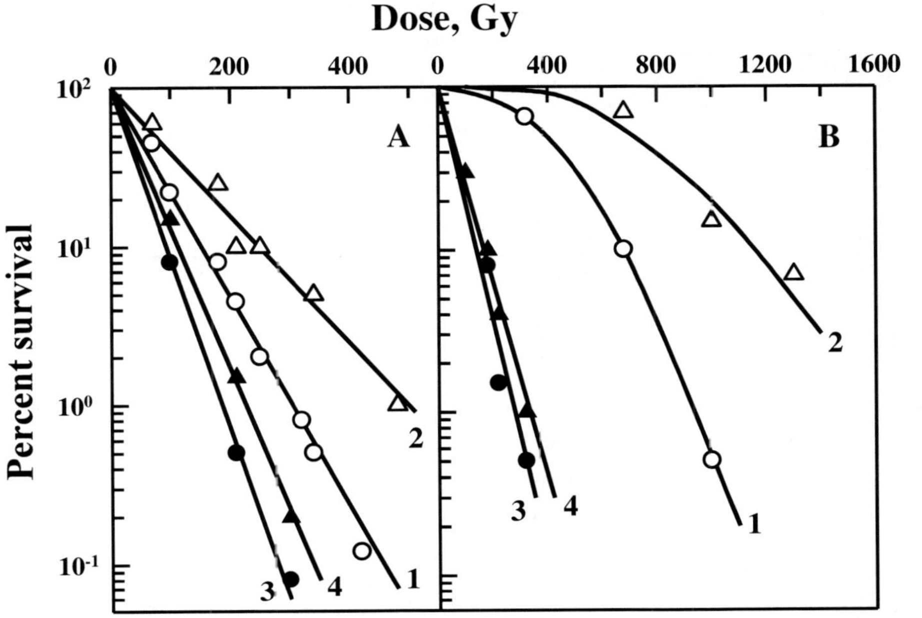

Figure 2.1. Survival-versus-dose curves of haploid yeast cells *S. cerevisiae* (A) of wild-type (strain S288C, RAD; curves 1, 2) and radiosensitive mutant (strain XS774-4d, *rad*51; curves 3, 4) and diploid cells (B) of wild-type (strain XS800, RAD/RAD; curves 1, 2) and homozygous radiosensitive mutant (strain XS806, *rad*51/*rad*51; curves 3, 4): curves 1 and 3, without cysteine; curves 2 and 4, with 0.03 M cysteine.

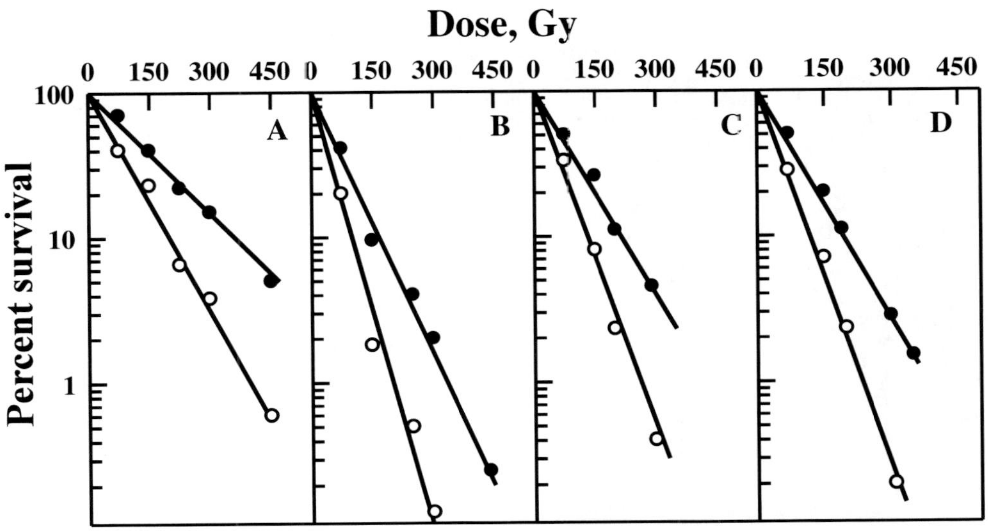

Figure 2.2. Survival curves of *S. cerevisiae* haploid yeast cells of various genotypes: A, S288C (RAD); B, g218/4d (*rad*50); C, g218/7a (*rad*51); D, gl60/2b (*rad*52); cells were irradiated with ionizing radiation without (open circles) and with (closed circles) 0.01 M cysteamine.

Table 2.1. The effectiveness of radioprotection (DMF) by mexamine, cystamine and cysteine of various strains of *S cerevisiae*

Compound concentration, M	Strain, genotype			
	S288C RAD	XS774-4d *rad*51	XS800 RAD/RAD	XS806 *rad*51/*rad*51
Mexamine 0.01 M	1.20 ± 0.02	0.96 ± 0.05	1.23 ± 0.04	1.00 ± 0.01
Cystamine 0.01 M	1.34 ± 0.03	1.00 ± 0.02	1.37 ± 0.03	1.03 ± 0.04
Cysteine: 0.01 M	1.53 ± 0.12	1.11 ± 0.10	1.48 ± 0.01	1.03 ± 0.04
0.03 M	2.16 ± 0.17	1.71 ± 0.32	2.02 ± 0.11	1.35 ± 0.09
0.05 M	2.41 ± 0.13	1.42 ± 0.07	1.98 ± 0.13	1.28 ± 0.05

Gamma-radiation survival response curves of several sensitive haploid mutants and the parent wild-type strain irradiated without and with 0.01 M cysteamine are presented in Figure 2.2. Similar curves for other haploid mutants have been also obtained, they are not presented here but they were used for DMF estimation. The mean lethal doses, calculated from the data presented in Figure 2.2, as well as the DMF values, are listed for haploid strains in Table 2.2.

It is clear that all haploid *rad* mutants exhibit enhanced radiation sensitivity compared to the wild-type strain. Among the *rad* loci investigated, *rad*50 exhibits maximum sensitivity in the haploid state. From the data it is evident that the efficiency of cysteamine protection was almost the same for wild-type and some mutant (*rad*50, *rad*53, *rad*55, *rad*57) cells, while for others it was slightly lower compared to that of wild-type cells. Survival curves for diploid yeast cells of wild-type and the most sensitive diploid mutants in homozygous state of mutant repair loci irradiated without and with 0.01 M cysteamine are presented in Figure 2.3. It shows the survival curves of wild-type yeast cells and different mutants in diploid state.

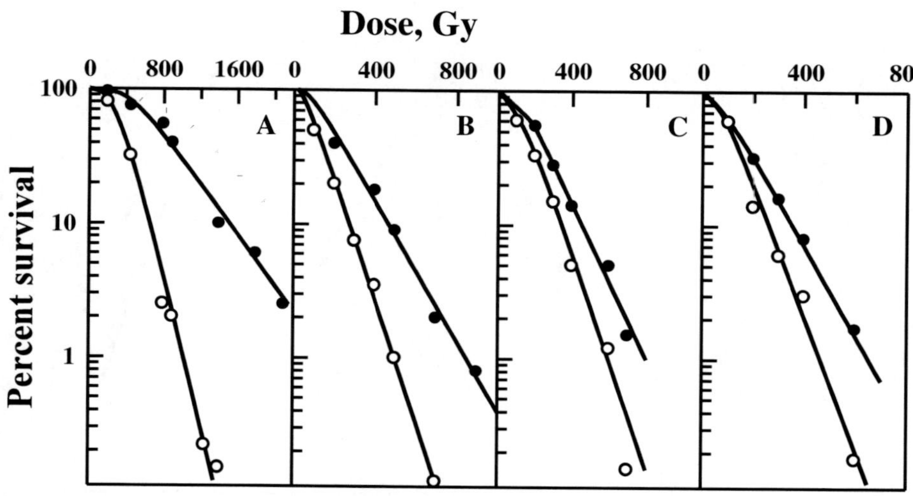

Figure 2.3. Survival curves of diploid yeast cells *S. cerevisiae* of various genotypes. A, XS800 (RAD/RAD); B, XS1806 (*rad*50/*rad*50); C, XS806 (*rad*51/*rad*51); D, XS1898 (*rad*52/*rad*52); cells were irradiated with ionizing radiation without (open circles) and with (closed circles) 0.01 M cysteamine. The cells were plated immediately after irradiation (continuous lines) and after 3 days of LHR at 30 °C (dotted line).

Table 2.2. The efficiency of radioprotecting action of cysteamine (0.01 M) against γ-irradiation of haploid cells of *S. cerevisiae* of various genotypes

Strain	Mutant repair loci	D_0, Gy without cysteamine	D_0, Gy with 0.01 M Cysteamine	DMF
S288C	RAD	84	157	1.87 ± 0.12
g2I8/4d	*rad*50	40	72	1.80 ± 0.15
g218/7a	*rad*51	63	99	1.57 ± 0.12
g160/2b	*rad*52	60	95	1.58 ± 0.17
g151/12c	*rad*53	79	137	1.73 ± 0.12
g218/3b	*rad*54	72	110	1.53 ± 0.08
X3465/3d	*rad*55	70	127	1.85 ± 0.10
30-57-rs31	*rad*56	65	100	1.54 ± 0.13
g246/6b	*rad*57	68	120	1.76 ± 0.11

It is evident that only wild-type cells were able to LHR (Figure 2.3A). Similar curves for somewhat resistant diploid mutants have been also obtained, they are not presented here but they were used for DMF estimation. The mean lethal doses calculated for these survival curves and the DMF values are listed in Table 2.3. One can see that the efficiency of cysteamine protection against γ-irradiation is correlated with cell radiosensitivity controlled by cell repair ability: it was larger for wild-type and relatively resistant mutant cells capable of recovery (irreversible component $K < 1$) and was markedly reduced for homozygous mutant cells incapable of recovery ($K = 1$).

These data indicate an important role of recovery processes in radioprotecting action of cysteamine on γ-irradiated yeast cells of various genotypes. The correlation between the effectiveness of radioprotective action and cell radiosensitivity (Figure 2.4) indicates that more radioresistant cells, which are presumably capable of recovery in a great extent, show the greater DMF values. This means the recovery processes take part in occurrence of radioprotective action. One of the well-studied processes of recovery is liquid holding recovery (LHR). This process is known to occur under nongrowth (liquid or agar-holding) conditions. One may suppose that such a recovery may take place also under growth conditions during the time until the first postradiation division or budding has been completed and the realization of potentially lethal damage has occurred. If the protectors could result in an increase in the delay of the first postradiation budding, the cells might be repaired for a longer time. To estimate a possible role of such a recovery in radioprotective action, we have compared the delay of postradiation budding of wild-type strains and radiosensitive mutants irradiated with and without protectors [27]. It was shown that cysteine slowed slightly the budding of all non-irradiated cells, did not change the delay of the first postradiation budding of all strains, shortened the delay of the second postradiation budding for haploid strains by 1.5 h and for diploid wild-type cells by 3.0 h, and does not affect the postradiation budding of the radiosensitive diploid mutant. Since the radioprotector influenced only the second postradiation budding and cannot change the duration of the first postradiation budding, one can believe that recovery processes identical to the LHR do not take part in radioprotecting action.

Table 2.3. The efficiency of radioprotective action of cysteamine (0.01 M) against γ-irradiation of diploid yeast cells of *S. cerevisiae* of various genotypes

Strain	Mutant repair loci	D_0, Gy without cysteamine	K	D_0, Gy with 0.01 M cysteamine	DMF
XS800	RAD/RAD	175	0.31	498	2.84 ± 0.21
L5	RAD/RAD	174	0.42	410	2.36 ± 0.21
XS1806	*rad*50/*rad*50	90	1.00	180	2.00 ± 0.11
XS806	*rad*51/*rad*51	36	1.00	50	1.39 ± 0.08
XS1898	*rad*52/*rad*52	32	1.00	47	1.46 ± 0.10
XS1889	*rad*53/*rad*53	70	0.33	135	1.93 ± 0.15
L9	*rad*54/*rad*54	50	1.00	65	1.30 ± 0.08
XS1935	*rad*55/*rad*55	80	0.28	170	2.12 ± 0.15
XS1943	*rad*56/*rad*56	140	0.36	310	2.21 ± 0.16
XS1988	*rad*57/*rad*57	137	0.32	271	1.98 ± 0.12
XS805	*RAD*/*rad*51	165	0.28	310	1.88 ± 0.10
L2	*RAD*/*rad*54	160	0.52	375	2.34 ± 0.17

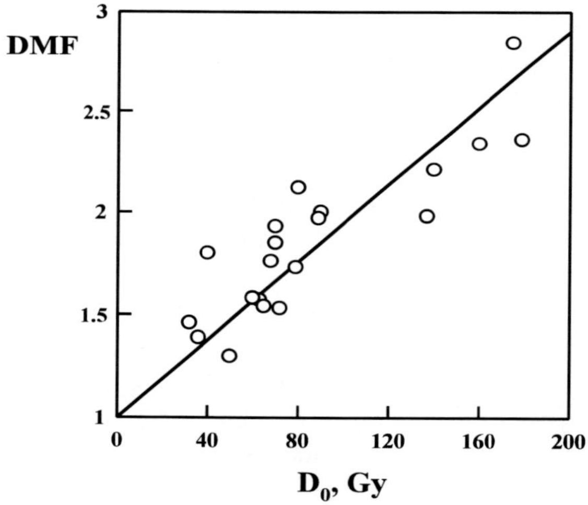

Figure 2.4. The correlation between the efficiency of radioprotective action of cysteamine (DMF) and yeast cell radiosensitivity (D_0).

For many years it was widely accepted that the mechanism of radioprotective actions of chemical compounds against ionizing radiation was connected with scavenging of radicals, competition with oxygen or, in some cases, the reducing of the direct action of radiation. A protective action of chemical compounds against γ-irradiation has been shown to depend strictly on the competence of the repair systems of *E. coli* cells [7; 15]. The data presented in this section show very similar results for various strains of yeast cells, the simplest model for eukaryotic cells.

2.2. Oxygen Effect in DNA-Proficient and -Deficient Yeast Cells

Some experimental results published by other authors [1; 10; 11; 17] indicate that oxygen may involve cellular recovery processes. Similar data for eukaryotic cells are not abundant. Only a few papers concerning the oxygen effect are known for yeast cells. It was shown that the sensitization by oxygen involves repair systems in *Saccharomyces* [5]. These authors have demonstrated a very small sensitizing effect of oxygen for the haploid strain rs_I supposedly as a consequence of the absence of a repair system in this strain. Kiefer [18] has shown an OER = 2 for a wild-type diploid yeast strain and an OER = 1.75 for a strain carrying a mutation in the *rad*9 gene. McKee and Lawrence [22] have demonstrated no consistent differences between OER values for wild-type cells and different groups of yeast mutants. These authors have no evidence for the existence of a reduced oxygen effect in strains lacking two types of repair capacity of the kind observed in *E. coli*. They also suggested that the lack of an oxygen effect for the rs_I strain might be connected with the physiological effects of extremely high anoxia. There is obviously no agreement on the correlation of the oxygen effect and repair capacity in prokaryotic and eukaryotic cells. However, some points concerning the paper by McKee and Lawrence [22] should be considered. Firstly cells to be irradiated were resuspended at a concentration of approximately 10^8 cells/ml, at which a marked hypoxia was obtained [29]. Second, the cells were oxygenated by air rather than oxygen and, as the authors themselves stated, their results were obtained under conditions of low rather than high anoxia. Therefore, it cannot be excluded that these points might indicate the absence of any obvious differences between OER values for wild-type and mutant cells. In a work [30] the genetic control of repair of radiation damage produced under euoxic and anoxic conditions in diploid yeast *S. cerevisiae* was investigated. The OER was shown to be reduced in the *rad*2, *rad*9, *rad*18, *rad*50, *rad*51, and *rad*57 mutants indicating that these genes could be associated with the repair of gamma radiation damage produced under hypoxic conditions. Yeast cells have a number of other genes controlling their radiosensitivity and ability to recover from radiation damage. It would be of interest to estimate oxygen effect for a number of haploid and homozygous diploid yeast cells. Yeast cells are perhaps less important for direct predictions of biological effects on mammalian cells, but similarities and differences among their response may help to understand the molecular mechanisms of the effects in mammalian cells. It is also expected that such an investigation may be informative in regard to protecting mechanisms involved.

The following yeast cells were used in this study: *S. cerevisiae* – haploid (S288C) and diploid (XS800) strains of wild-type and their haploid (*rad*52, strain g160/2b) and diploid (*rad*52/*rad*52, strain XS1898) radiosensitive mutants. Before irradiation, cells were incubated for 3–5 days at 30 °C to the stationary phase. Cell suspensions were saturated with oxygen-free argon or oxygen by passing the gas over the irradiation vessels. After gassing, the irradiation vessels were sealed and the cultures were irradiated by an electron beam from a 25 MeV pulsed linear accelerator. The average dose rate was 200 Gy/min. For dosimetry the Super Fricke method (10^{-2} M $FeSO_4$, 0.8 N H_2SO_4, O_2 saturated) was used in the position of cell suspension. At the end of treatment, the samples were further diluted suitably and plated on nutrient agar to assess colony-forming ability.

Figure 2.5 shows gamma-radiation survival response curves of haploid and diploid wild-type strain of *S. cerevisiae*. Curves 1 were obtained for cells in equilibrium with air, curves 2 – for cells in equilibrium with argon. It is obvious that cell response was dependent on the gaseous environment during irradiation. As usual, homozygous diploid cells of the wild-type were much more radioresistant than their haploid cells.

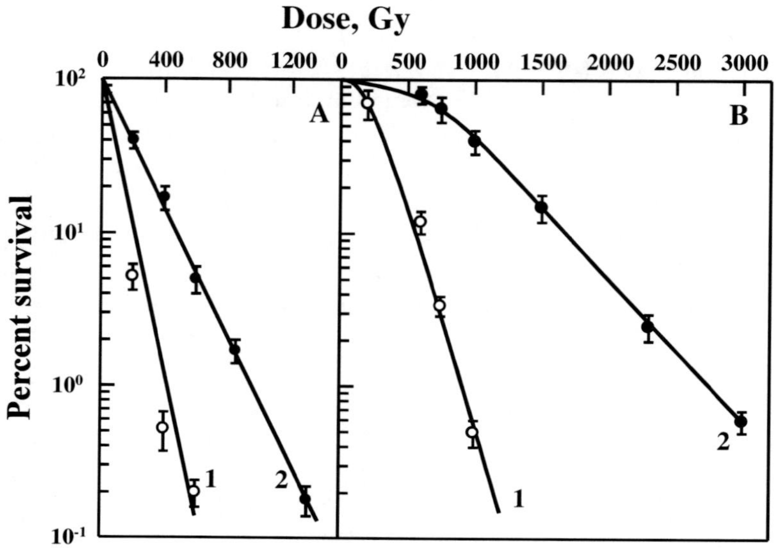

Figure 2.5. The survival curves of haploid (strain S288C, A) and diploid (strain XS800, B) yeast cells of *S. cerevisiae* exposed to γ-rays with (curves 1) and without (curves 2) oxygen. The standard error of the mean of 5 replicate experiments is shown.

The oxygen radiation sensitization is normally "dose-modifying", i.e., dose-effect curves obtained after exposure with or without oxygen can be superimposed by changing the dose scale by an appropriate factor which has been termed "oxygen enhancement ratio" (OER). In this study, the OER is defined as the ratio of the mean lethal dose D_0 estimated at argon equilibrium to the D_0 at oxygen condition. Induced in Table 2.4 are the D_0 and OER values calculated on the basis of experimental survival curves of all yeast strains analyzed. From these data it is evident that the haploid cells of wild-type *S. cerevisiae* are much more sensitive to ionizing radiation than diploid cells. It is now explained by the ability of diploid cells, in contrast to haploids, to repair double-strands breaks induced in DNA [20]. It can also be seen that the efficiency of anoxic protection was less expressed for haploid yeast cells of wild-type (OER = 2.2) as compared with that for the wild-type diploid cells (OER = 2.9). It means that the diploid-specific recovery can take a considerable part in the mechanism of the anoxic radiation protection. To test this hypothesis, similar experiments were carried out with couples of haploid (*rad*52) and diploid (*rad*52/*rad*52) strains, in which diploid cells are defective in the diploid-specific recovery. Survival curves for haploid and diploid *rad*52 mutant cells are presented in Figure 2.6. Radiobiological parameters of these strains are also included in Table 2.4.

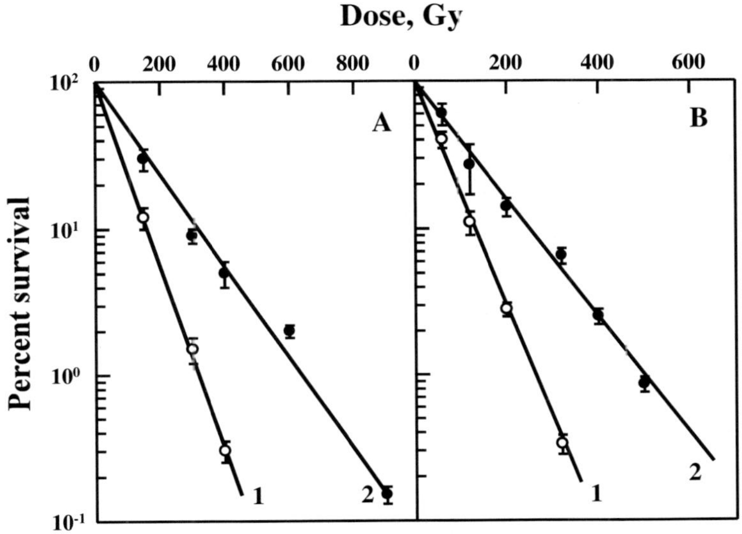

Figure 2.6. The survival curves of haploid (strain g160/2b, A) and diploid (strain XS1898, B) yeast cells of *S. cerevisiae* exposed to γ-rays with (curves 1) and without (curves 2) oxygen. The standard error of the mean of 3 replicate experiments is shown.

Table 2.4. The efficiency oxygen effect in DNA-proficient and DNA–deficient yeast cells of *S. cerevisiae*

Strain, genotype	D_0, Gy with oxygen	D_0, Gy with argon	OER
S288C, RAD	117 ± 6	262 ± 15	2.2 ± 0.2
XS800, RAD/RAD	180 ± 10	515 ± 25	2.9 ± 0.3
g160/2b, *rad*52	82 ± 5	150 ± 7	1.8 ± 0.2
XS1898, *rad52/rad*52	65 ± 5	115 ± 5	1.8 ± 0.1

These mutant cells have an affected pathway involving a mechanism related to recombination between homologous chromosomes in diploid cells. Due to this fact, the radiosensitivity of haploid and diploid cells were not differed markedly from each other The main points which emerge from the data presented in Table 2.4 for this two couple of strains are: (i) oxygen effect is evident in cells of all these strains; (ii) both haploid and diploid cells in *rad*52 state show the identical OER values.

The most interesting information, obtained from the survival curves obtained, pertains to the oxygen effect. While oxygen effect is evident in cells of all strains, the magnitudes of these effects differ significantly. The OER's range from 1.8 to 2.2 for DNA-deficient cells incapable to recover DNA double-strand breaks and was 2.9 for wild-type diploid cells. These results can be discussed in terms of the possibility that mechanism of the oxygen effect may involve the capacity of cells to recover radiation damages. The recovery may be mediated by recombination-like mechanism, for which two homologous DNA duplexes are required. A serious argument supporting this idea can be found in Figure 2.7, where the correlation of D_0

anoxic value with OER for various yeast strains is depicted. Data in Figure 2.7 are based on the results obtained here and published by us previously [25; 26; 28].

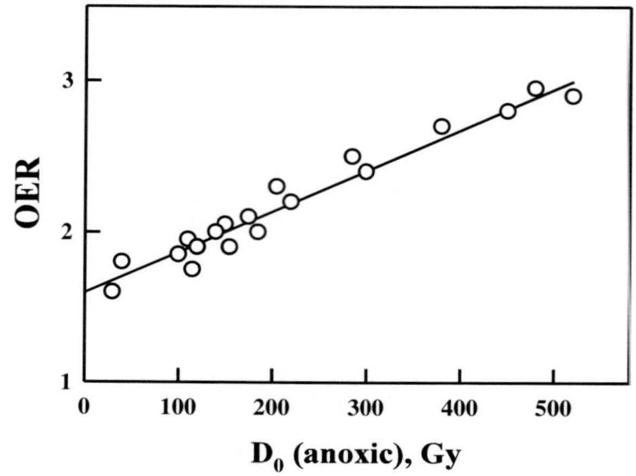

Figure 2.7. The relation of D_0 (anoxic) values with OER for various yeast strains.

Almost a linear relationship is found between these values. The correlation coefficient is 0.96 and the regression line is OER = 1.6 + 0.003· D_0 (anoxic). This means that more radioresistant cells, which are presumably capable of recovery in a great extent, show the greater OER values. Thus, it may be concluded that the oxygen effect may be related with a difference in the amounts of damage sustained under oxygen and argon condition, but also with a difference in reparability of radiation damage inflicted under these conditions.

Further information on this area will be presented in the next section, where the survival curves of the different yeast strains will be adduced after simultaneous irradiation of cells under of oxic, hypoxic and hypoxic + chemical sensitizers conditions.

2.3. Radiosensitization of Hypoxic Cells by Chemical Compounds in Wild-Type and Radiosensitive Mutants of Yeast

It is of interest to compare the effectiveness of hypoxic radiosensitizers for yeast strains differing in their ability to recover from radiation damage. In this section the oxygen effect and radiosensitization of hypoxic wild-type and radiosensitive mutants of yeast cells by metronidazole and misonidazole are compared. Metronidazole (synthesized by Polfa) was used at a concentration of 0.01 M which gives maximal sensitization and no toxicity under the condition used. Misonidazole (Ro 07-0582) was synthesized by Roche Products Ltd. It was tested at 0.01 M concentration; concentrations in excess of 0.01 M were toxic. The compounds were added to the cell suspensions immediately before gassing because they need to be present in the cells at the time of irradiation. In some cases, including all cases where anoxic sensitization was demonstrated, the compounds were also tested in parallel experiments in which oxygen was present throughout.

Radiation survival response curves of several haploid and diploid mutants and the parent wild-type strains irradiated in oxic (open circles), hypoxic (open triangles) and hypoxic + metronidazole conditions are presented in Figures 2.8 and 2.9, respectively. For haploid cells, exponential survival curves were obtained. Table 2.5 gives the mean lethal dose (D_0), OER and enhancement ratio by metronidazole (ER) calculated as follows: OER = D_0 (argon)/ D_0 (oxygen) and ER = D_0(argon)/D_0(argon + drug). These and the following Figures and Tables include the averaged results of our previous publications [25; 26; 28]. The errors involved in the OER have been calculated from interexperimental variations and included in this and other tables. As can be seen from the data, radiosensitization by oxygen is evident in cells of all strains: the OER is 2.2 for wild-type cells and varied over a narrow range from 1.7 up to 2.0 in various mutant haploid cells. Metronidazole was not effective for all haploid strains investigated.

Figure 2.9 exhibits the survival curves of wild-type diploid yeast cells and different mutants homozygous with respect to various *rad* loci. Radiobiological parameters of these curves are listed in Table 2.6. While radiosensitization by oxygen is evident in all strains, its magnitude differs significantly in wild-type and mutant cells. It is well known that diploid yeast cells of the wild-type are capable of LHR. As was shown above, some diploid homozygous *rad* mutant cells (*rad*53, *rad*55, *rad*56, and *rad*57) were capable of LHR, while other mutant cells (*rad*50, *rad*51, *rad*52, and *rad*54) were not. As can be seen from the data, the OER values ranged from about 2.0–3.0 in diploid cells capable of LHR and from 1.6–2.0 in strains incapable of LHR. The metronidazole was found to sensitize those diploid strains which show an OER more than 2.2. Other strains, with lower OER, including all *rad* diploid mutants incapable of LHR, show no sensitizing effect in tile presence of metronidazole. The radiosensitization was approximately dose-modifying [28]. The metronidazole sensitization increases the slope of the survival curve of anoxic, but not of well-oxygenated cells. Table 2.6 shows that metronidazole sensitizes only hypoxic cells of definite genotype with an ER between about 1.1 and 1.3. These values are in fair agreement with reported values for yeast cells [19] as well as for mammalian cells in culture and other *in vivo* systems.

The present experimental results may suggest that radiosensitization by metronidazole is strongly related to oxygen sensitization. Sapora et al. [32] and Fielden et al. [14] have established an empirical relationship (Eq. 2.1), relating the enhancement ratio produced by sensitizer and oxygen-enhancement ratio of a bacterium. The limiting value of α was 0.75. In our case, for wild-type diploid yeast cells α is about 0.47. If this relationship holds for yeast cells and metronidazole, we should expect that metronidazole would not sensitize any *rad* mutant strain whose OER was less than 1/0.47 = 2.1. This prediction is completely fulfilled (Tables 2.5 and 2.6). It seems of interest to make similar experiments with a sensitizer which gives larger enhancement ratios, approaching the higher OER.

Misonidazole is a typical representative of effective electron-affinic compounds. We used the compound synthesized by Roche Products Ltd. (1-(2-nitroimidazol-1-yl)-3-methoxypropan-2-ol, Ro 07-0582) and tested it mainly for hypoxic cells; only in some experiments it was also present in a cell suspension in which oxygen was present throughout.

Table 2.5. The oxygen effect and radiosensitizing effect of metronidazole (0.01 M) in haploid yeast cells of *S. cerevisiae* of various genotypes

Strain	Mutant repair loci	D_0, Gy oxygen	D_0, Gy argon	D_0, Gy argon + drug	OER	ER
S288C	RAD	96.0	215.0	215.0	2.2 ± 0.07	1.0
g218/4d	*rad*50	52.0	96.5	96.5	1.8 ± 0.09	1.0
g218/7a	*rad*51	61.0	114.0	114.0	1.9 ± 0.09	1.0
gl60/2b	*rad*52	71.0	142.0	142.0	2.0 + 0.13	1.0
gl51/12c	*rad*53	75.0	137.0	137.0	1.8 ± 0.07	1.0
g218–3b	*rad*54	80.0	160.0	160.0	2.0 ± 0.10	1.0
X3465/3d	*rad*55	62.0	126.0	126.0	2.0 ± 0.11	1.0
g151/10d	*rad*56	96.0	163.0	163.0	1.7 ± 0.06	1.0
g246/6b	*rad*57	71.0	125.0	125.0	1.8 ± 0.07	1.0

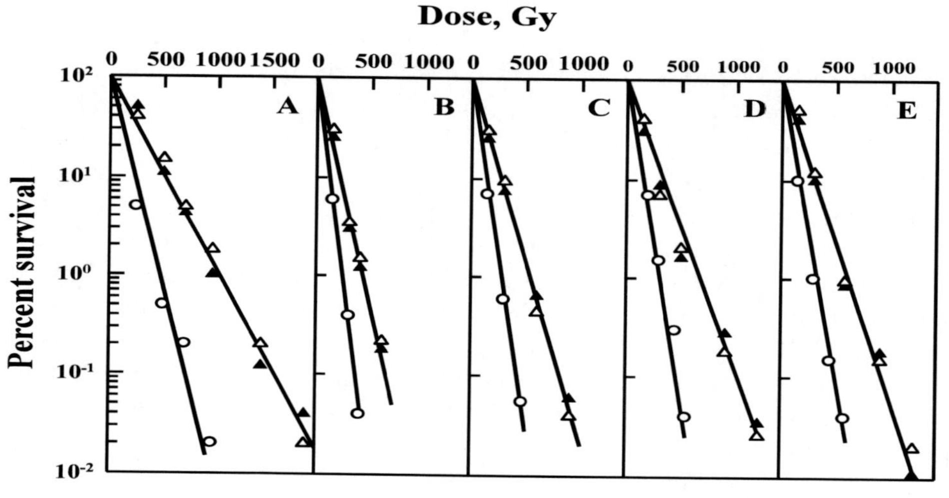

Figure 2.8. Survival curves of haploid yeast cells of *S. cerevisiae* of various genotypes irradiated in oxygenated (circles) and hypoxic (triangles) conditions, with (closed symbols) and without (open symbols) of 0.01 M metronidazole. A, S288C (RAD); B, g218/4d (*rad*50) C, g218/7a (*rad*51); D, gl60/2b (*rad*52); E, gl51/12c (*rad*53).

We determined the dependence of ER by misonidazole as a function of molar concentration for the XS800 strain (Figure 2.10). It is evident that the effectiveness of radiosensitization was increased with drug concentration. The highest concentration tested was 0.01 M, which is almost the solubility limit at 20 °C. A maximal enhancement ratio of 2.0 was observed at this concentration, which gave somewhat less sensitization than that produced by oxygen with these cells (OER = 2.9). The ER obtained in the present series of experiments had a higher value compared with that for metronidazole under comparable conditions. The maximal α of 0.70 was estimated from these results and Eq. (2.1). Using this value and Eq. (2.1), we calculated the expected values of ERs for the other strains tested and compared them with the experimentally observed data.

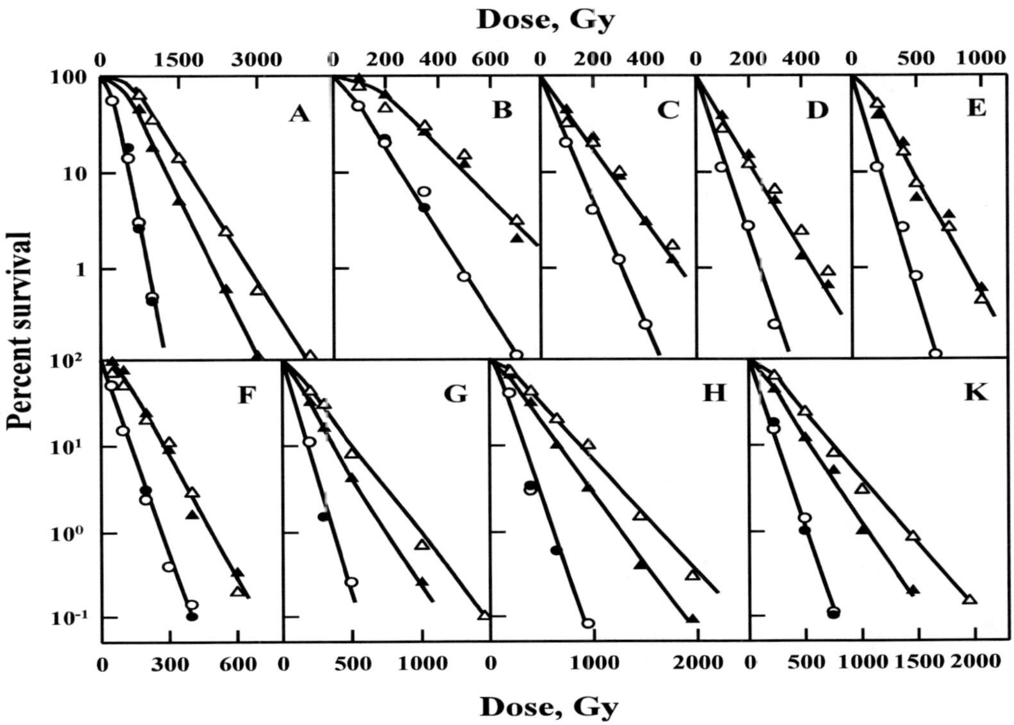

Figure 2.9. Survival curves of diploid yeast cells of *S. cerevisiae* of various genotypes irradiated in oxygenated (circles) and hypoxic (triangles) conditions, with (closed symbols) and without (open symbols) of 0.01 M metronidazole: A, XS800 (RAD/RAD); B, XS1806 (*rad*50/*rad*50); C, XS806 (*rad*51/*rad*51); D, XS1898 (*rad*52/*rad*52); E, XS1889 (*rad*53/*rad*53); F, L9 (*rad*54/*rad*54); G, XS1935 (*rad*55/*rad*55); H, XS1943 (*rad*56/*rad*56); K, XS1988 (*rad*57/*rad*57).

Table 2.6. The oxygen effect and radiosensitizing effect of metronidazole (0.01 M) in diploid yeast cells of *S. cerevisiae* of various genotypes

Strain	Mutant repair loci	D_0, Gy oxygen	D_0, Gy argon	D_0, Gy argon + drug	OER	ER
XS800	RAD/RAD	167	488	372	2.9 ± 0.08	1.3
XS1806	*rad*50/*rad*50	80	152	152	1.9 ± 0.09	1.0
XS806	*rad*51/ *rad*51	64	120	120	1.9 ± 0.10	1.0
XS1898	*rad*52/ *rad*52	55	103	103	1.9 + 0.07	1.0
XS1889	*rad*53/ *rad*53	87	177	177	2.0 ± 0.09	1.0
XS1935	*rad*55/ *rad*55	88	201	178	2.4 ± 0.11	1.3
XS1943	*rad*56/ *rad*56	127	305	240	2.4 ± 0.13	1.3
XS1988	*rad*57/ *rad*57	110	285	215	2.3 ± 0.14	1.3
211	RAD/RAD	191	493	455	2.5 ± 0.14	1.1
L5	RAD/RAD	167	448	343	2.7 ± 0.12	1.3
L6	RAD/RAD	165	476	395	2.9 ± 0.09	1.2
L9	*rad*54/*rad*54	57	110	110	1.9 + 0.09	1.0
T1	RAD/RAD	185	520	410	2.8 ± 0.13	1.3
T3	*rad*54/*rad*54	24	44	44	1.8 ± 0.10	1.0
T4	*rad*2*rad*54/*rad*2*rad*54	22	36	36	1.6 ± 0.12	1.0

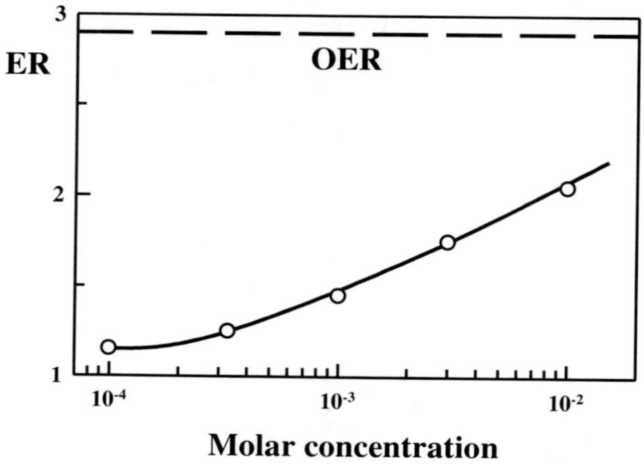

Figure 2.10. The dependence of the enhancement ratio (*ER*) of misonidazole on its concentration; data obtained for diploid *S. cerevisiae* yeast cells (strain XS800).

Table 2.7. Radiobiological parameters of haploid yeast cells *S. cerevisiae* of various genotypes irradiated in oxygen, argon and argon + drug conditions, oxygen enhancement ratio (OER), and enhancement ratio (ER) by 0.01 M misonidazole obtained in experiments and theoretically predicted

Strain	Mutant repair loci	D_0, Gy oxygen	D_0,Gy argon	D_0,Gy argon + drug	OER	ER_{exp}	ER_{theory}
S288C	RAD	117	260	170	2.2 ± 0.07	1.5	1.5
g218/7a	*rad*51	70	130	110	1.9 ± 0.06	1.2	1.3
gl60/2b	*rad*52	80	149	112	1.9 ± 0.10	1.3	1.3
g151/12c	*rad*53	71	130	106	1.8 ± 0.08	1.2	1.3
g151/10d	*rad*56	66	114	93	1.7 ± 0.06	1.2	1.2
g246/6b	*rad*57	75	130	110	1.7 ± 0.07	1.2	1.2

Radiation survival curves of various haploid mutants and the parent wild-type strain irradiated in oxic and hypoxic conditions, both in the absence and in the presence of 0.01 M misonidazole, are presented in Figure 2.11. Table 2.7 gives the mean lethal dose D_0, OER, and ER, determined experimentally and predicted by Eq. (2.1) if $\alpha = 0.70$. These data show that all haploid cells were sensitized by misonidazole, the experimental and predicted values of the ER being close to each other.

Figure 2.12 shows survival curves of diploid yeast cells with various *rad* mutations exposed to ionizing radiation under the conditions investigated. The most sensitive strains showed exponential survival curves, and survival curves of other strains displayed sigmoid form.

Table 2.7 summarizes the radiobiological parameters of these curves. Table 2.8 gives the mean lethal dose D_0, OER, and ER, determined experimentally and predicted by Eq. (2.1) if $\alpha = 0.70$. These data show that all diploid cells were sensitized by misonidazole, the experimental and predicted values of the ER being close to each other.

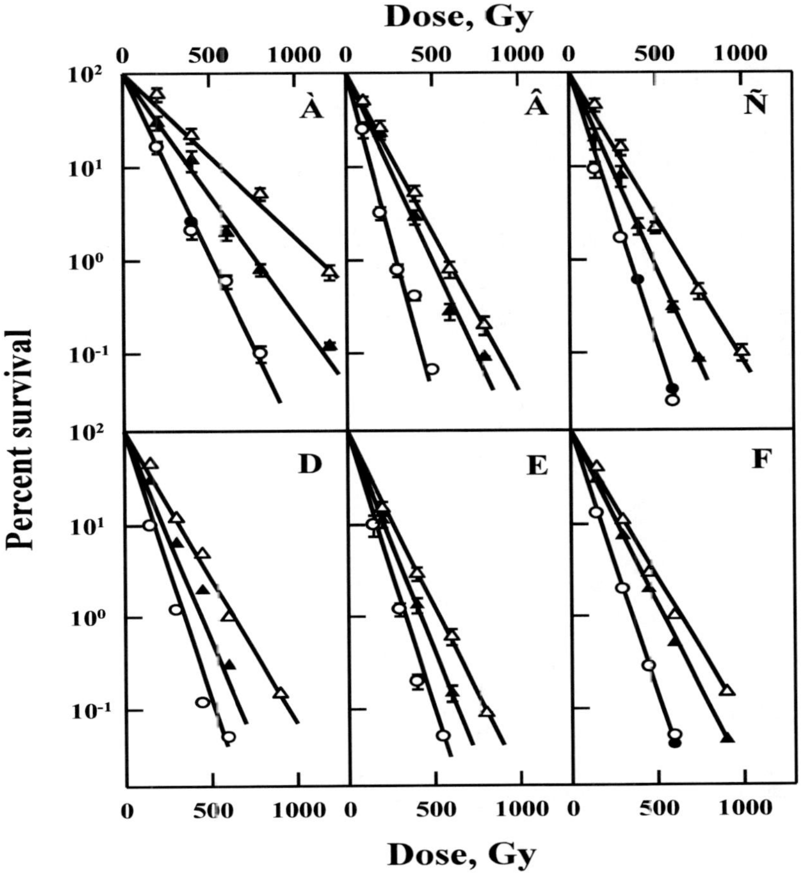

Figure 2.11. Survival curves of haploid yeast *S. cerevisiae* of various strains irradiated in oxygen (circles) and argon (triangles) conditions with (closed symbols) and without (open symbols) 0.01 M misonidazole. A, S288C (RAD); B, g218/7a (*rad*51); C, g160/2b (*rad*52); D, g151/12c (*rad*53) E, g151/10d (*rad*56); F, g246/6b (*rad*57).

Table 2.7. Radiobiological parameters of haploid yeast cells *S. cerevisiae* of various genotypes irradiated in oxygen, argon and argon + drug conditions, oxygen enhancement ratio (OER), and enhancement ratio (ER) by 0.01 M misonidazole obtained in experiments and theoretically predicted

Strain	Mutant repair loci	D_0, Gy oxygen	D_0,Gy argon	D_0,Gy argon + drug	OER	ER_{exp}	ER_{theory}
S288C	RAD	117	260	170	2.2 ± 0.07	1.5	1.5
g218/7a	*rad*51	70	130	110	1.9 ± 0.06	1.2	1.3
gl60/2b	*rad*52	80	149	112	1.9 ± 0.10	1.3	1.3
g151/12c	*rad*53	71	130	106	1.8 ± 0.08	1.2	1.3
g151/10d	*rad*56	66	114	93	1.7 ± 0.06	1.2	1.2
g246/6b	*rad*57	75	130	110	1.7 ± 0.07	1.2	1.2

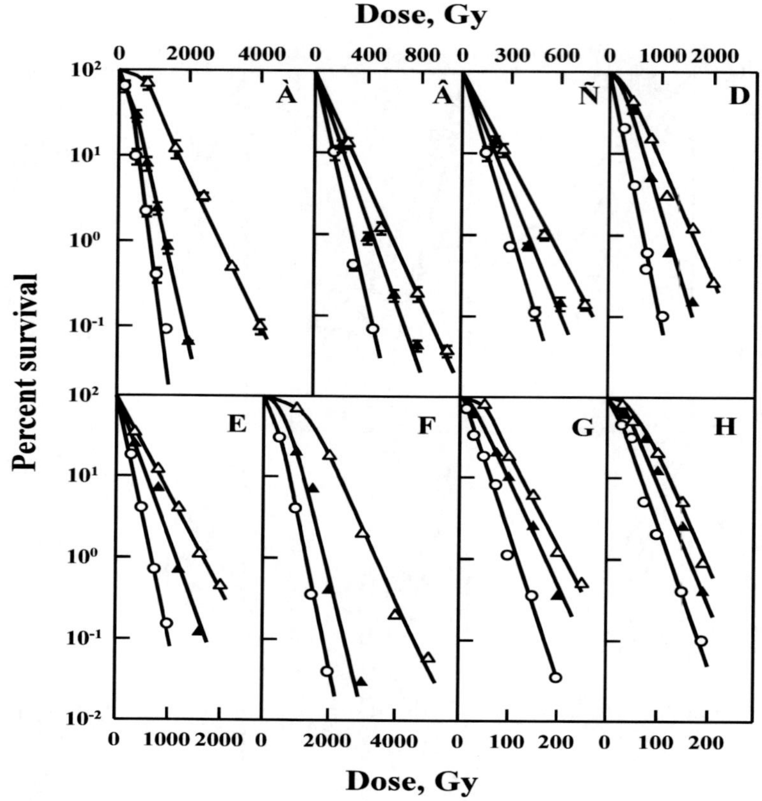

Figure 2.12. Survival curves of diploid yeast cells of *S. cerevisiae* of various genotypes irradiated in oxygenated (circles) and hypoxic (triangles) conditions, with (closed symbols) and without (open symbols) of 0.01 M misonidazole: A, XS800 (RAD/RAD); B, XS806 (*rad*51/*rad*51); C, XS1898 (*rad*52/*rad*52); D, XS1943 (*rad*56/*rad*56); E, XS1988 (*rad*57/*rad*57); F, T1 (RAD/RAD); G, T3 (*rad*54/*rad*54); H, T4 (*rad*2*rad*54/*rad*2*rad*54).

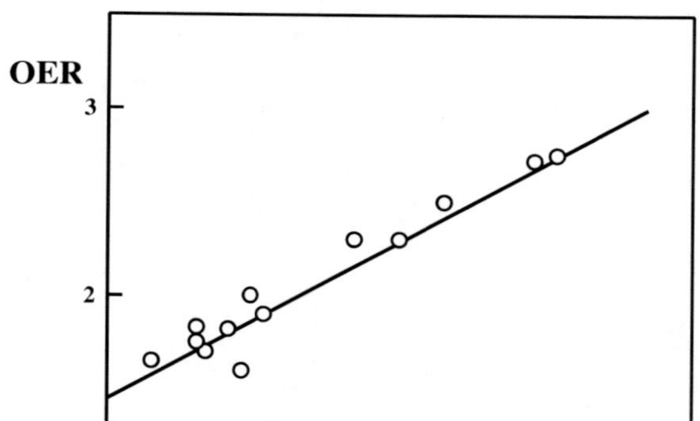

Figure 2.13. The dependence of the enhancement ratio (*ER*) against the oxygen enhancement ratio (*OER*) for the corresponding yeast strain of *S. cerevisiae*.

Figure 2.13 gives the ER produced by 0.01 M misonidazole plotted against the corresponding OER for all strains. The straight line is $ER = \alpha \cdot OER$, where $\alpha = 0.70$. The experimental points are not greatly dispersed around the curve.

We studied the LHR of some yeast strains capable of recovery and irradiated under various conditions. By comparing the survival curves obtained after immediate and delayed plating [26], we estimated the LHR quantitatively by the irreversible component K. Table 2.9 shows the K values estimated by the ratio of the corresponding D_0. It is evident that the magnitude of the LHR was the smallest (K is the greatest) after anoxic irradiation without drug, and was almost the same after oxic and under argon plus 0.01 M misonidazole irradiation.

As a whole, the effect of hypoxic sensitization by metronidazole and misonidazole was studied for a number of yeast cells *S. cerevisiae*. The effectiveness of sensitization was proportional to the oxygen-enhancement ratio. Both compounds sensitized cells to ionizing radiation by increasing the slope of the survival curve and did not affect the shoulder. There is general agreement that the electron-affinic sensitizers mimic the action of oxygen at the level of DNA [34]. Although the mechanisms responsible for the sensitizing effect are not completely understood, misonidazole is thought to act by the inhibition of enzyme-induced recovery processes or by sensitization of a vital macromolecule resulting in the greater expression of potentially lethal damage. The data presented here show that the radiosensitizing effect can be observed in cells independent of their ability to repair radiation damage to DNA (cell radiosensitivity and the LHR served as indicators of repair activity). Moreover, the results presented here reveal that the sensitizer increases the magnitude of the LHR compared with that in hypoxic conditions. In any case, it can be concluded that whichever model is ultimately shown to be correct, the results presented in this section do not contradict the results obtained with bacterial and mammalian cells and, thus, show that the basis mechanism of hypoxic cell sensitization may be similar in different cell systems.

Table 2.8. Radiobiological parameters of diploid yeast cells *S. cerevisiae* irradiated in oxygen, argon and argon + drug conditions, oxygen enhancement ratio (OER) and enhancement ratio (ER) by 0.01 M misonidazole

Strain	Mutant repair loci	D_0, Gy oxygen	D_0, Gy argon	D_0, Gy argon + drug	OER	ER_{exp}	ER_{theory}
XS800	RAD/RAD	178	510	255	2.9 ± 0.08	2.0	2.0
XS806	*rad*51/ *rad*51	64	120	92	1.9 ± 0.05	1.3	1.3
XS1898	*rad*52/ *rad*52	60	110	87	1.8 + 0.07	1.3	1.3
XS1943	*rad*56/ *rad*56	137	345	206	2.5 ± 0.13	1.7	1.8
XS1988	*rad*57/ *rad*57	140	350	215	2.3 ± 0.14	1.6	1.6
T1	RAD/RAD	183	522	271	2.9 ± 0.11	1.9	2.0
T3	*rad*54/*rad*54	24	43	36	1.8 ± 0.10	1.2	1.3
T4	$\frac{rad2rad54}{rad2rad54}$	23	37	34	1.6 ± 0.12	1.1	1.1

Table 2.9. Irreversible component K of yeast cells *S. cerevisiae* irradiated with ionizing radiation at various conditions

Strain	Mutant repair loci	Oxygen	Argon	Argon+ 0.01 M misonidazole
XS800	RAD/RAD	0.33 ± 0.02	0.73 ± 0.07	0.38 ± 0.03
XS1943	*rad56/rad56*	0.35 ± 0.04	0.64 ± 0.05	0.36 ± 0.03
XS1988	*rad57/rad57*	0.32 ± 0.02	0.47 ± 0.04	0.31 ± 0.02
T1	RAD/RAD	0.37 ± 0.03	0.79 ± 0.05	0.42 ± 0.04

2.4. Radiosensitivity Modification of γ-Irradiated Haplont Yeast Cells

In the preceding sections it was shown that the recovery processes are involved in the mechanism of radiosensitivity modification by radioprotective compounds, oxygen and hypoxic radiosensitizers. It might be expected that the efficiency of radiosensitivity modification of haplont yeast cells incapable of the diploid specific recovery, should not depend on the ploidy and its magnitude should be close to that obtained for *Saccharomyces* haploid yeast and radiosensitive diploid mutants lost their ability to recover. Therefore, it is of interest from theoretical point of view to study the radioprotective effect of hypoxia and several chemical compounds as well as hypoxic radiosensitizers for haplont yeast.

The following haplont yeast cells were used in this study: haploid and diploid strains of *P. pinus* (MH4 and MH4-D, respectively) and *P. guilliermondii* (NRRLY-2075 and NRRLY-2075x2076, respectively). In experiments with radioprotectors cells were irradiated at stationary phase of growth with γ-rays (^{60}Co, 10 Gy/min) at room temperature without chemical compounds and with various concentrations of cysteine and cysteamine. Both compounds (Sigma) were placed to cell suspension (10^6 cells/ml) for 30–40 minutes prior to irradiation. Details of methods were presented above.

Radiation survival-response curves of haploid and diploid *yeast P. pinus* and *P. guilliermondii* irradiated without radioprotector and with various concentrations of cysteine (Figure 2.14) and cysteamine (Figure 2.15) are presented. Experimental points in these and following figures have errors of 2%, that is approximately the size of symbols.

We estimated the efficiency of radioprotection by the dose modifying factor (DMF) defined as the ratio of the mean lethal doses D_o of the survival curves obtained with and without radioprotectors. This value was estimated for the initial exponential part of survival curves for *P. pinus* yeast cells and for the final linear portion of dose-effect curves for *P. guilliermondii*. Radiobiological parameters of the data presented in Figures 2.14 and 2.15 are listed in Table 2.10. It can be seen that cysteine does not have a protective effect both for haploid and diploid cells of *P. pinus* whereas the efficiency of cysteamine for *P. pinus* strains was small (DMF = 1.3) and not dependent on the ploidy. The radioprotective effect of cysteine for *P. guilliermondii* yeast cells was larger, DMF being equal about of 2.0 independently of drug concentration and cell ploidy. Similar effect was observed for cysteamine: DMF values were 1.9 and 1.8 for the haploid and diploid cells, respectively, at 0.01 M solution of cysteamine. Further increase in the concentration of cysteamine (0.03 and

0.05 M) resulted in cell death in control population. Curiously, despite the cell death in non-irradiated yeast population due to toxic effect of cysteamine (Figures 2.15C and 2.15D), its radioprotective properties at high doses were identical to those observed in non-toxic molarity.

The main conclusion that can be judged from these findings is that radioprotective action of both drugs is much smaller for haplont yeast cells than that for wild-type strain of S. cerevisiae (DMF = 2.3–2.9) and similar to those described above (section 2.1) for radiosensitive *rad* mutants of *S. cerevisiae*. Since haplont yeasts are not able to diploid-specific recovery, these data confirm the conclusion that the mechanism of radioprotective action of chemical agents may involve cellular recovery process.

In previous sections it was shown that in mechanisms of radioprotective action, including anoxic protection, a distinction should be made between the competition for free radical intermediates and protection by repair of vital macromolecules in the cell. Such a distinction between the two mechanisms has just been confirmed in experiments with the action of radioprotectors on haplont-yeast, whose responses to ionizing radiation was similar to those observed for radiosensitive *rad* mutants of *S. cerevisiae*. Therefore it is of interest to study oxygen effect and effect of hypoixic sensitizers for haplont yeast cells.

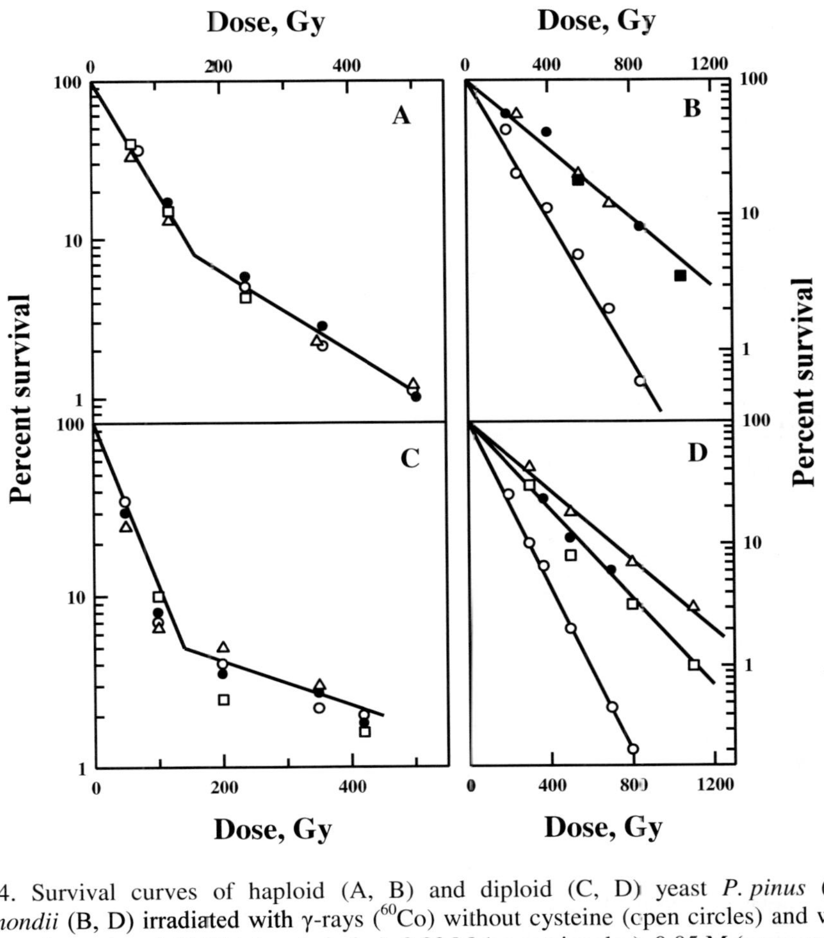

Figure 2.14. Survival curves of haploid (A, B) and diploid (C, D) yeast *P. pinus* (A, C) and *P. guilliermondii* (B, D) irradiated with γ-rays (^{60}Co) without cysteine (open circles) and with various concentration of cysteine: 0.01 M (closed circles), 0.03 M (open triangles), 0.05 M (open squares).

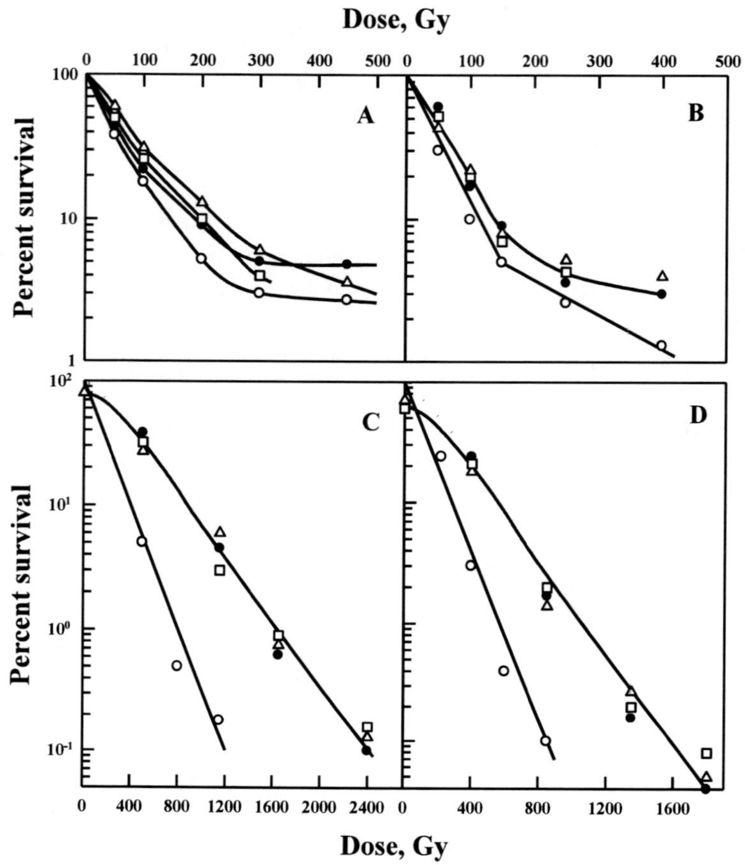

Figure 2.15. Survival curves of haploid (A, C) and diploid (B, D) yeast *P. pinus* (A, B) and *P. guilliermondii* (C, D) irradiated with γ-rays (^{60}Co) without cysteamine (open circles) and with various concentration of cysteamine: 0.01 M (closed circles), 0.03 M (open triangles), 0.05 M (open squares).

Table 2.10. The effectiveness of radioprotective action of cysteine and cysteamine for yeast *P. pinus* and *P. guilliermondii* irradiated with γ-rays of ^{60}Co

Yeast	Ploidy	D_o, Gy without protector	0.01 M		0.03 M		0.05 M	
			D_o, Gy	DMF	D_o, Gy	DMF	D_o, Gy	DMF
Cysteine								
P. pinus	Haploid	62	62	1.0	62	1.0	62	1.0
	Diploid	49	49	1.0	49	1.0	49	1.0
P. guilliermondii	Haploid	171	329	1.9	329	1.9	329	1.9
	Diploid	129	227	1.8	294	2.3	227	1.8
Cysteamine								
P. pinus	Haploid	65	76	1.2	91	1.4	82	1.3
	Diploid	49	63	1.3	63	1.3	63	1.3
P. guilliermondii	Haploid	171	330	1.9	330	1.9	330	1.9
	Diploid	133	241	1.8	241	1.81	241	1.8

The survival curves of haploid (strain NRRLY-2076, A) and diploid (strain NRRLY 2075 x 2076, B) yeast cells of *P. guilliermobdii* exposed to γ-rays with and without oxygen are depicted in Figure 2.16. OER values were 1.8 and 1.6 for haploid and diploid cells, i.e., it does not dependent on ploidy and is similar to that observed for *rad* mutants of *S. cerevisiae.*

The next series of experiments was performed to evaluate the sensitizing action of electron-affinic compounds. Metronidazole (Polfa) and mizonidazole (Roche Products Ltd) was added to the cell suspension (10^6 cells/ml) prior to its saturation with gas. Each experiment was repeated 3–5 times. The effectiveness of radiosensitivity modification was evaluated quantitatively by the calculation of OER and ER. These parameters are estimated by the ratio of corresponding values of D_0. For yeasts *P. pinus* D_0 values were assessed only for the initial parts of survival curves. Figure 2.17 exhibits the survival curves of haploid and diploid yeast strains of *P. pinus* and *P. guilliermondii* yeast cells irradiated in the presence of oxygen or argon, as well as in hypoxic conditions plus of 0.01 M metronidazole or mizonidazole. Radiobiological characteristics of these survival curves are summarized in Table 2.11.

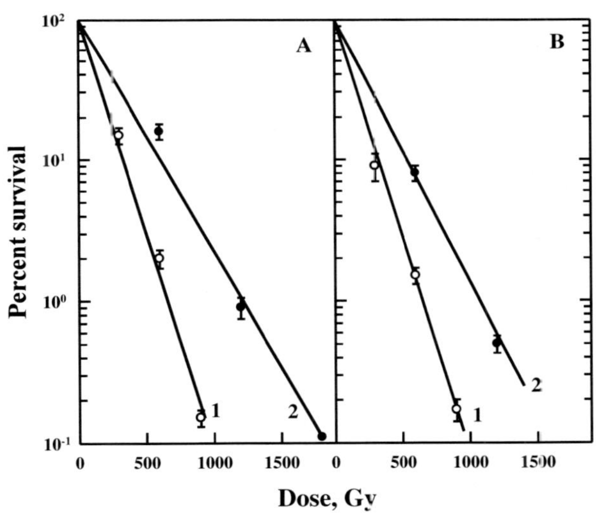

Figure 2.16. The survival curves of haploid (strain NRRLY-2076, A) and diploid (strain NRRLY 2075 x 2076, B) yeast cells of *P. guilliermobdii* exposed to γ-rays with (curves 1) and without (curves 2) oxygen. The standard error of the mean of 3 replicate experiments is shown.

From these data it is evident that radiosensitivity modification by oxygen was observed for all strains (OER = 1.8–2.2), being independent of cell ploidy. These values correspond to those inherent for many *rad* mutants and wild-type haploid rather than diploid *Saccharomyces* yeast cells capable of diploid-specific recovery. It can be also seen that metronidazole did not increase the radiosensitivity of hypoxic cells of all strains studied, like many *Saccharomyces rad* mutant cells. More effective hypoxic sensitizer misonidazole increased the radiosensitivity of all strains (ER = 1.20–1.54). Here again, the effect was independent on cell ploidy and ER values were closed to that observed for many *Saccharomyces rad* mutant cells. In other words, radiobiological responses of haplont yeast are reminiscent to those typical for haploid cells of wild-type and both for haploid and diploid yeasts *S. cerevisiae* with artificial mutations that block the recovery of damage inflicted by ionizing radiation. These results are also consistent with the opinion that diploid-specific recovery is mild or completely absent in haplont yeast cells.

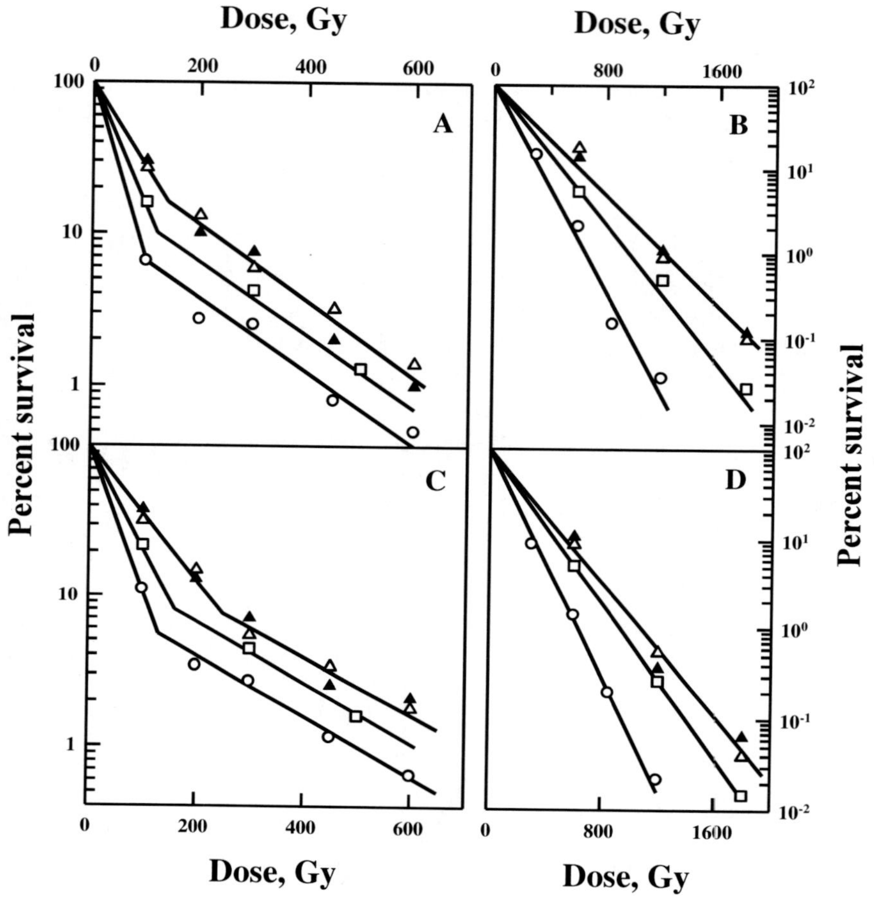

Figure 2.17. Survival curves of haploid (A, B) and diploid (C, D) yeast *P. pinus* (A, C) and *P. guilliermondii* (B, D) irradiated with electrons from 25 MeV accelerator in the presence of oxygen (open circles), argon (open triangles), argon + 0.01 M metronidazole (closed triangles), argon + 0.01 M misonidazole (open squares).

A quantitative study of the effects of some radiosensitizers upon the survival of bacterial cells of wild-type and radiosensitive mutants indicated that ER is directly proportional to OER. As was shown in the preceding section, this relationship is suitable for yeast cells and the coefficient of proportionality $\alpha = 0.47$ for metronidazole and $\alpha = 0.70$ for misonidazole. Using these values and the empirical expression (Eq. 2.1.), we predicted the expected value of ER for haplont yeasts. The results are also inserted in Table 2.11. It can be seen that the experimental and theoretical values of ER close to each other for both compounds. This means that mechanism of electron-affinic sensitizer action mimics oxygen effect on the DNA level. Taking into account these new results and the data presented in the preceding sections, it can be concluded that the mechanism of the hypoxic sensitizers may involve cellular recovery process mediated by a recombination-like mechanism, for which the diploid state is required. Thus, the mechanism of the radioprotector action, oxygen and hypoxic sensitizer effect may be realized on the primary radiation damage production as well as on cellular biochemical processes including recovery from radiation damage.

Table 2.11. Modification of radiosensitivity by oxygen (OER) and hypoxic sensitizers (ER) for yeast *P. pinus* and *P. guilliermondii* irradiated with electrons from 25 MeV accelerator in the presence of oxygen, argon or argon + *sensitizer*

Strain	D_o, Gy O_2	D_o, Gy Ar	OER	Ar + metronidazole			Ar + misonoidazole		
				D_o, Gy	ER exp.	ER theory	D_o, Gy	ER exp.	ER theory
P. pinus									
MN4	35	77	2.2	77	1.0	1.03	52	1.48	1.54
MN4-D	45	97	2.1	97	1.0	1.00	65	1.49	1.47
P. guilliermondii									
NRRLY-2075	148	261	1.8	261	1.0	1.0	216	1.21	1.24
NRRLY-2075x2076	139	244	1.8	244	1.0	1.0	203	1.28	1.20

2.5. Participation of Recovery Systems in Cell Radiosensitivity Modification

A possible mechanism of oxygen effect in cells has been proposed by Eidus and Korystov (1984). The mechanism involves two subsequent processes: 1) a physicochemical modification of primary damage of a target by oxygen and thiols, and 2) enzymatic repair of the lesions modified by endogenous thiols. The first process occurs during irradiation; it is dependent on the thiol/oxygen ratio in the cell nucleus and determines constant K; the second process occurs in the postirradiation period; it is dependent on repair efficiency and determines constant m. Both these constants have been suggested by Alper and Howard-Flanders (1956). In accordance with these authors, the empirical relationship between the increase in radiosensitivity (S) and oxygen concentration (O_2) in the medium is

$$\frac{S(O_2)}{S(N_2)} = \frac{m[O_2]+K}{[O_2]+K}, \tag{2.2}$$

where $S(N_2)$ and $S(O_2)$ represent the radiosensitivity at the partial pressure zero and at oxygen concentration $[O_2]$, respectively, m and K are constant. It is obvious that Eq. (2) reflects the competition of the damaging effect of oxygen with the effect of another agent, which somehow reduces the injury. Depending on the assumed nature of the competitive agents, different hypothesis for the oxygen effect mechanism were advanced.

Kiefer [18] suggested that repair processes take part in the mechanism of oxygen effect. Such repair is thought to be a fast physico-chemical reaction occurring in the course of irradiation, and does not explain the dependence of oxygen effect on postradiation conditions and on enzymatic repair processes. Besides, the already known data of thiol participation in the realization of the oxygen effect were not discussed.

Oxygen effect was explained [2] by the competition of oxygen with thiols in their interaction with radiation-induced macromolecules radicals: these donors of hydrogen atoms restitute the primary lesion of cell target, whereas oxygen fixes them (lethal damages LD). These suggestions were supported by the fact that at the irradiation of phages, DNA and some enzymes oxygen effect was observed only after addition of SH-substances:

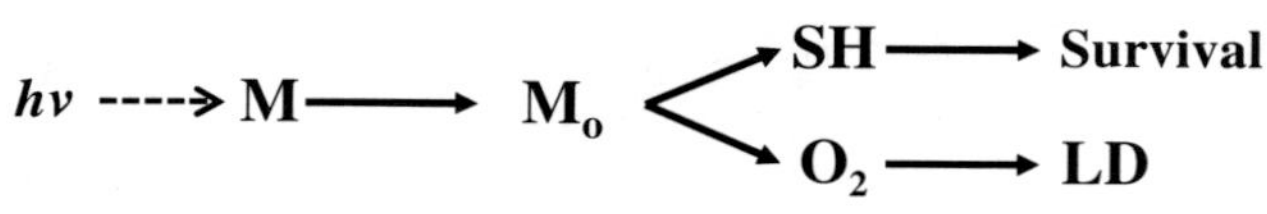

where M and M_O are the intact and hit cell targets. According to this scheme the complete repair of target is the result of pure physico-chemical processes.

It was shown [2] that OER depends on postradiation conditions and on the efficiency of enzymatic repair (ER) processes. Especially illustrative were experiments with bacterial and yeast mutants deficient in their DNA repair capacity. According to Dertinger and Jung [12] the primary lesion, M_O, is also repairable, but enzymatic means, in contrast with the previous scheme:

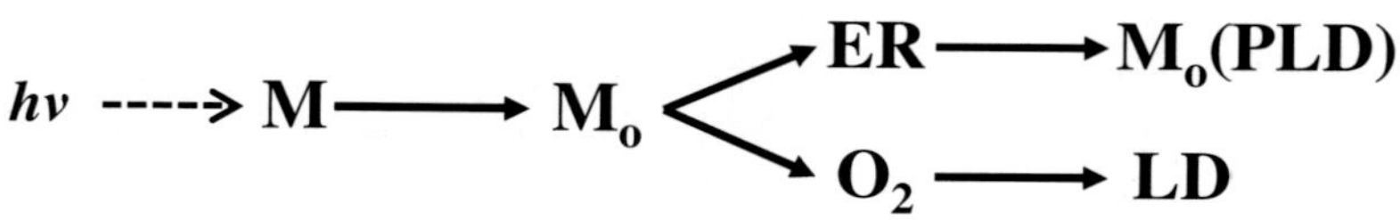

It is proposed here that in the absence of the repair the lesion M_O is lethal both with and without oxygen. But it remains unclear how the protracted repair processes can compete with the rapid action of oxygen. This shortcoming was overcome by Chapman et al. [9] by combining the two previous schemes:

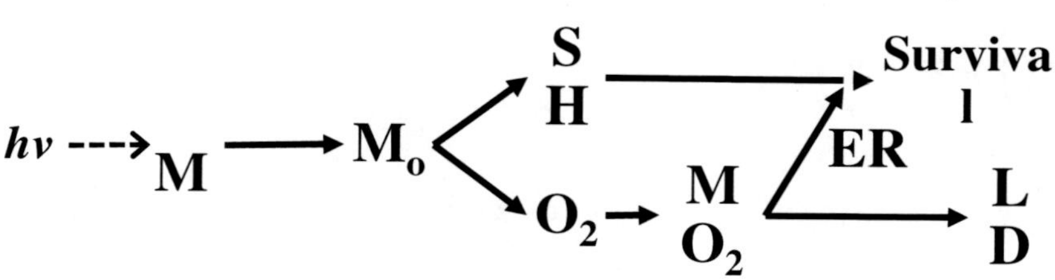

This scheme includes into the oxygen mechanism the restitution of the primary lesions by SH-groups and enzymatic repair of the fixed lesions. In other words, oxygen induces not irreversible but potentially lethal damage (MO_2) which can be enzymatic repaired.

However, this scheme does not explain the decrease of oxygen effect in the cells deficient in repair ability.

Eidus and Korystov [13]) suggested more complex scheme to explain all the above mentioned peculiarities of oxygen effect:

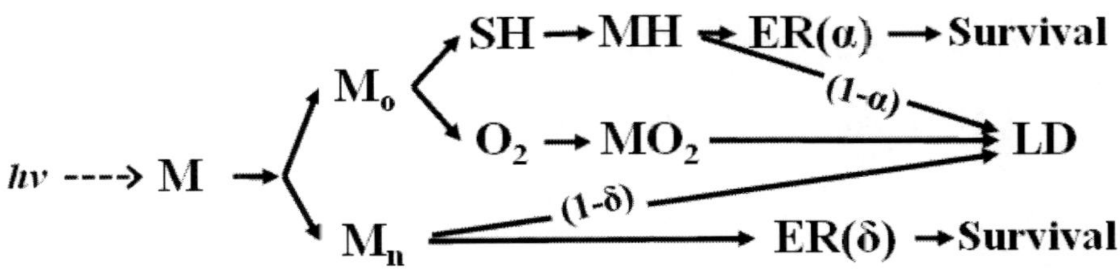

According to this scheme, two types of radiation damage can be produced in the cell target, one of which is sensitive (M_0) and another insensitive (M_n) to oxygen. Both these damage are lethal without further modification. Oxygen sensitive damage M_0 after interaction with protector SH transferred into MH state that can be either irreversible or enzymatic repaired with probability α. But if the repair ability decreased or totally absent (radiosensitive mutants) the damage MH are lethal. After interaction with oxygen M_0 damage become completely irreversible MO_2. Another type of primary damage M_n can be also either irreversible or enzymatic repaired with probability δ. Some contradiction of this scheme was also demonstrated [24]. Quantitative estimations have shown that in accordance with this scheme the ratio b of O_2-dependent and O_2-independent lesions should be more 2 and under another way of calculation it should be less than ⅓. To eliminate this contradiction, we supposed [24] that MO_2 lesion can be also repaired with some probability γ. Then we came to the following scheme.

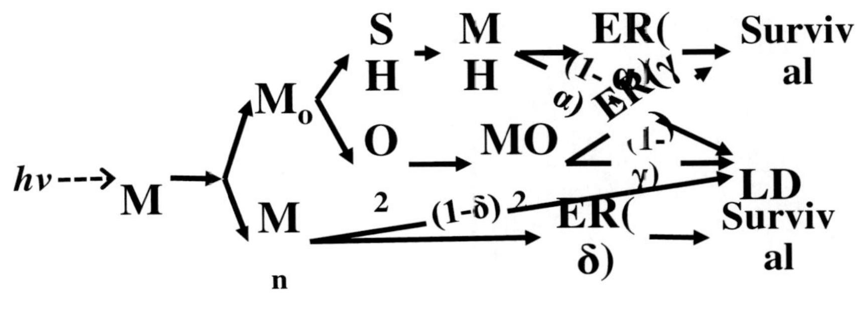

In equation presented below we used indexes r and s for resistance and sensitive strains, respectively. It was shown [8, 9] that in aerobic condition the fixation of O_2-dependent lesion is responsible for 82% of lethal lesions. Then we assume that the ratio b =

4.5. It was shown that $\alpha = \delta_r = 0.97$. Omitting details of mathematical calculations published in our work earlier [24], we were able to deduce equations for estimation of OER and probabilities δ and γ of enzymatic repair of M_n and MO_2 lesions:

$$OER_r = \frac{(1-\delta_r)+b(1-\gamma_r)}{(1-\delta_r)(b+1)}, \tag{2.3}$$

$$OER_s = \frac{(1-\delta_s)+b(1-\gamma_s)}{(1-\delta_s)(b+1)}, \tag{2.4}$$

$$\gamma_r = 1-(1-\delta_r)\left[\frac{m_r(b+1)-1}{b}\right], \tag{2.5}$$

$$\delta_s = 1-(1-\delta_r)\frac{D_{0r}(Ar)}{D_{0s}(Ar)}, \tag{2.6}$$

$$\gamma_s = 1-\frac{(1-\delta_r)}{b}\left[\frac{D_{0r}(O_2)}{D_{0s}(O_2)}OER_r(b+1)-\frac{D_{0r}(Ar)}{D_{0s}(Ar)}\right]. \tag{2.7}$$

As was shown above, ER_{max} by misonidazole may be expressed as $ER_{max} = 0.75 \cdot OER$. Then we can write

$$ER_{max} = 0.75\frac{(1-\delta)+b(1-\gamma)}{(1-\delta)(b+1)}. \tag{2.8}$$

We employed these equations to calculate the probabilities δ and γ of enzymatic repair of lesions M_n and MO_2 in the dependence of the mean lethal doses D_0 (Figure 2.18) for a number of yeast strains differing in their ability to recover radiation damage. Experimental values of OER(γ), $D_{0r}(Ar)$, $D_{0s}(Ar)$, $D_{0r}(O_2)$, $D_{0s}(O_2)$ for different yeast strains have been presented above. The results are collected in Table 2.12.

Table 2.12. The probabilities δ and γ of enzymatically repair of lesions M_n and MO_2 for various strains of yeast *S. cerevisiae*

Strain	Repair mutant loci	δ	γ
T1	RAD/RAD	0.97	0.90
T3	*rad*54/*rad*54	0.59	0.29
T4	*rad*2*rad*54/*rad*2*rad*54	0.56	0.26
XS800	RAD/RAD	0.97	0.90
XS1806	*rad*50/*rad*50	0.90	0.78
XS806	*rad*51/*rad*51	0.87	0.72
XS1898	*rad*52/*rad*52	0.84	0.67
XS1889	*rad*53/*rad*53	0.92	0.81
L9	*rad*54/*rad*54	0.85	0.69
XS1935	*rad*55/*rad*55	0.91	0.79
XS1943	*rad*56/*rad*56	0.95	0.86
XS1988	*rad*57/*rad*57	0.94	0.84

It is evident that the probability δ of enzymatic repair of M_n lesions changes for most strains within 0.84–0.97, being the smallest (0.56–0.59) for mutant *rad*54 and double mutant *rad*2*rad*54/*rad*2*rad*54. For all strains studied, the probability γ of enzymatic repair of MO_2 lesions was smaller than that for M_n lesions.

To illustrate the further usefulness of the scheme discussed, we applied it to describe the dependence of ER by misonidazole on its concentration $[C]$. It is quite reasonable to suppose that such a dependence should be similar to that described for oxygen effect [1, 2, 24], i.e.,

$$ER = \frac{ER_{max}[C] + k}{[C] + k}, \tag{2.9}$$

where k is equal to concentration of electron affinic compounds when cell radiosensitivity is middle between the greatest value in the presence of sensitizer and without it. Taking into account the equation for ER_{max}, we obtain

$$ER = \frac{0.75\dfrac{(1-\delta)+b(1-\gamma)}{(1-\delta)(b+1)}[C] + k}{[C] + k}. \tag{2.10}$$

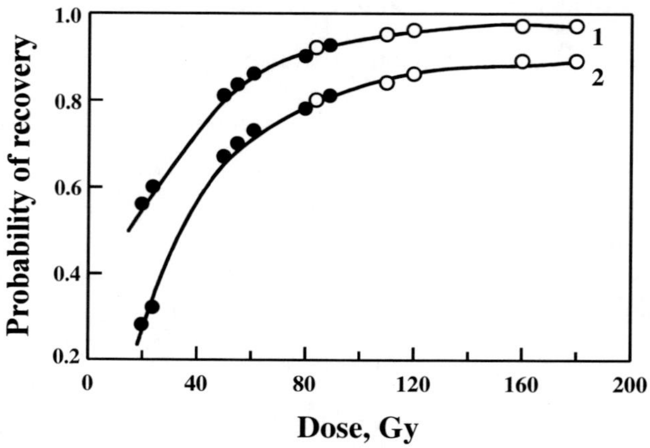

Figure 2.18. The dependence of the probabilities δ (curve 1) and γ (curve 2) of the enzymatic repair of lesions M_n and MO_2 on the mean lethal doses D_0 for yeast strains capable and incapable of the LHR (open and closed circles, respectively).

The constant k for misonidazole was shown to be of 10^{-3} M [25]. Using this value and the latter equation, we calculated the theoretically expected dependence of ER on the concentration of misonisazole $[C]$ for yeast *S. cerevisiae* (strain XS800). The findings are presented in Figure 2.19. One can see a good coincidence of theoretical predicted (solid line) and experimental data presented above (open circles).

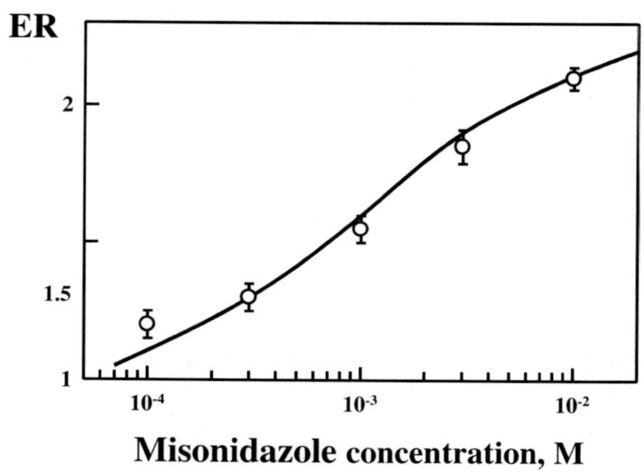

Figure 2.19. Theoretically predicted (solid line) and experimentally obtained (circles) dependence of the enhancement ratio (*ER*) of misonidazole on its concentration; data obtained for diploid *S. cerevisiae* yeast (strain XS800).

Eidus and Korystov [13] tried to interpret the molecular nature of some type of damage presented in the scheme of oxygen effect for bacterial *E. coli*. It is now generally accepted that DNA is the target (M) for reproductive cell death. They suggested that MO_2 damage can be a single strand breaks and alkali labile sites. Symbol M_n denotes DNA lesions whose yield does not depend on oxygen tension. These include lesions such as thymine glycols, γ-endonuclease sensitive DNA sites, pyrimidine dimers. The multiple DNA injuries produced by ion clusters from densely ionizing radiation can be also considered as M_n type of damage. These lesions induced at the physico-chemical step of oxygen effect realization. If the primary lesions in the cell do not undergo enzymatic repair, the OER for survival should be determined totally by the physico-chemical step.

Hence, strains deficient in reparation systems should display lower OER values that were confirmed by the vast data presented in this Chapter for yeast cells. It is difficult to identify the molecular nature of damage presented in the schemes for eukaryotic cells. It is plausible to affirm that potentially lethal damage denoted in the scheme by symbol γ may represents the elimination of DSB known to restore in mammalian and yeast cells [20]. As a whole, the idea that cell radiosensitivity modification can be closely linked with the occurrence of recovery processes is of fundamental importance for inactivation of various cells, in which DNA is the primary radiation sensitive structure.

CONCLUSION

Results presented in this chapter reveal the correlation between the radioprotective efficiency of cysteamine, cystamine, cysteine and mexamine with cell-repair capacity of yeast cells: those strains which were most resistant, i.e., were more "capable of recovery", had the highest radioprotection, while those strains which were most sensitive, i.e., were deficient in "repair capacity", had significantly less radioprotection values. It follows that the explanation of radioprotection by scavenging of free radicals is incomplete and specific repair mechanisms should be involved. Differential radioprotection was not expressed for wild-type and *rad* mutant haploid yeast cells. On this basis, it may be concluded that radioprotection may involve cellular recovery processes, which may be mediated by a recombination-like mechanism. Saeki et al. [31] showed the good correlation between the ability of induction of the mitotic recombination, budding resistance and the ability of the diploid-specific recovery, defined as the ratio of radiation doses for diploid and haploid cells producing 10% survival. In some of their strains, *rad*50, *rad*53, *rad*56 and *rad*57, the induction of mitotic recombination was slightly less than that of the wild-type, while in another group, *rad*51, *rad*52 and *rad*54, induction of both mitotic non-reciprocal gene conversion and reciprocal recombination were greatly reduced. Our data indicate that *rad* mutant groups classified by the diploid-specific recovery correspond fairly well to the groups classified by the protection ability. Indeed, the DMF values for *rad* mutants of the first group fell into a range from 1.9 to 2.2 and, for another group, from 1.3 to 1.5. Results of the previous chapter revealed similar correlation of the RBE of densely ionizing radiation with diploid yeast cell recovery. Taking these data as a whole, one can conclude that the diploid-specific recovery may be responsible for various modification of yeast cell radiosensitivity.

There are several indications that oxygen operates in more than one way in affecting the radiation sensitivity of the cell. Alper [1] was first who put forward a model based on two types of primary events with different intrinsic OER, designated type N and O. Both forms of damage may have an oxygen effect although magnitudes may differ, the OER for the N type being quite small. It would follow from the formal statement of the model that, if a family of strains differs in their capability to repair one type of damage, there would be a linear relationship between OER and D_0 [1]. The predicted relationship holds qualitatively for a group of *E. coli* strains in which the effect of type O damage hypothetically remained constant [1]. Such a relationship was also obtained here for a group of yeast strains. According to the Alper model, the intercept of OER on the axis $D_0 = 0$ (Figure 2.7) gives an approximate OER value for recoverable damage (type N). It is interesting that these values (OER = 1.6) are almost identical for bacterial and yeast cells.

A biphasic response to radiation as a function of oxygen concentrations has been found in bacterial spores and mammalian cells [23]. In the latter paper, the OER versus oxygen concentration curve displays two plateaux corresponding to the OER of 1.9 (the oxygen concentration ranges from 0.4 to 1.5 μM) and 3.1 (the oxygen concentration more than 280 μM). This information might be also interpreted as an indication that oxygen sensitizes anoxic cells by more than a single process.

The two-site hypothesis was suggested to explain a two-component oxygen effect observed by a rapid-mix technique [33]. In this paper the oxygen effect for mammalian cells was resolved into two time-dependent components. For oxygen added to cells 3–5 ms before irradiation, the OER was 1.7 and was independent of oxygen concentrations, while for oxygen added to cells 5–40 ms before irradiation, the OER increases up to 2.9 and was dependent upon oxygen concentration. These data were interpreted as indicating that there are two regions in the cell where oxygen interacts with molecular damage to increase the over-all biological effect.

It is curious, that in our investigation the oxygen effect varied over the same range from 1.6 for the most sensitive mutant to 2.9 for the wild-type diploid yeast cells. It is also possible that the response which we observed was a manifestation of a two-type radiation lesion, as suggested by Alper [1] or of a two-component oxygen effect which was observed by Shenoy et al. [33] and by Millar et al. [23]. These possibilities need not be mutually exclusive, and it is quite possible that they are interrelated or are a consequence of the same two primary radiation mechanisms or targets.

Diploid and other polyploid yeast cells of *S. cerevisiae* are known to be significantly steadier to damage of environmental factors, including ionizing radiation, in comparison with that of haploid cells. It is generally interpreted by the ability of diploid cells, in contrasts to haloids, to recover from damage inflicted by ionizing radiation. *S. cerevisiae* yeasts are diplont – they exist as diploid in the natural conditions. Haplont yeast species exist in nature with one set of chromosomes and their diploid cells were composed artificially in laboratory conditions. As was demonstrated above, a common feature of haplont yeast cell *of P. pinus* and *P. guilliermondii* is that they exhibit a haploid state as a normal vegetative state in natural conditions. It was shown that the transition from the haploid to the diploid state in these species is not accompanied by increased radioresistance, and diploid strains were unable to perform liquid-holding recovery. The absence of diploid-specific recovery in diploid haplont strains was also supported by the fact that the RBE of α-particles was almost identical for haploid and the corresponding diploid strains being much smaller than that observed for typical wild-type diploid strains capable of diploid-specific recovery. For haplont yeast cells (*P. pinus* and *P. guilliermondii*), incapable of the diploid-specific recovery, radiosensitivity modification by radioprotective drugs, oxygen and hypoxic sensitizers was independent on ploidy and quantitatively remind that observed for Saccharomyces haploid yeast and radiosensitive diploid *rad* mutants lost their diploid-specific recovery.

To explain the action of radiobiological modifiers, several authors had suggested hypothetical reaction schemes for the events leading to the formation of injuries in vital macromolecules within the cell [13, 18]. If the cell radioresistance to ionizing radiation may, to some extent, serve as a measure of cell sensitivity to other environmental factor inflicted genetic damage, one may suppose that only the yeasts capable of diploid cell recovery will have a selective advantage. On the contrary, for yeasts defective in repair systems, haploid cells may possess a selective advantage. It is not excluded that owing to this fact, the vegetative phase in natural conditions is presented by diploid cells in *S. cerevisiae* and by haploid cells in *P. pinus* and *P. guilliermondii.* It follows that the development of repair systems, providing more full recovery of diploid cells from genetic damages in comparison with haploid cells, may have served as a prerequisite for the transition, during evolution processes, from haplont to diplont cells. The results suggest that haplont yeast may have evolved to diplont yeast via the development of a specific repair system conferring specific resistance in the diploid state.

The main conclusion that can be judged on the basis of the analysis of kinetic schemes of the radiation damage formation is the real participation of the enzymatic repair in the mechanism of the radioprotective action, oxygen hypoxic sensitizers effect. Moreover, both schemes suggested by Eidus and Korystov [13] and us clearly demonstrate that the effectiveness of radiation protectors, oxygen and hypoxic sensitizers should be reduced in strains with the reduced repair ability. According to these schemes thiol restitutes the oxygen-sensitive DNA radicals while oxygen converts them into stable one. The dependence of the OER value on enzymatic repair is found to be due to the change of the contribution of oxygen-sensitive and oxygen resistant lesions to cell death. The proposed scheme fits well the experimental data which could not be explained by the previous oxygen effect schemes. As was mentioned above, the predicted results from these schemes have been confirmed by many authors in experiments with bacteria. Our extensive data with chemical protectors, oxygen and hypoxic sensitizers acting on irradiated yeast cells of different genotypes are also supported this idea for eukaryotic cells.

In this Chapter modification of cell radiosensitivity was expressed mathematically in terms of known experimental values of OER for different yeast cell strains. It was shown that the parameters used permit to describe the dependence of the efficiency of radioprotectors and radiosensitizers upon yeast cell genotype. The data indicate the participation of recovery processes in the expression of the oxygen effect in yeast cells and are an additional argument in favor of the plausibility of these schemes.

Shortly, our new findings described in this Chapter can be formulated as follows.

1. The correlation between the effectiveness of radioprotective action of various chemical compounds and cell repair capacity was demonstrated for diploid yeast cells. Such a correlation was not expressed for wild-type and *rad* mutant haploid yeast cells. It was concluded that the radioprotective action may involve cellular recovery processes, which may be mediated by a recombination-like mechanism, for which the diploid state is required.

2. Radiosensitization of hypoxic yeast cells by oxygen was found to be dependent on yeast cell genotype. Yeast strains which were deficient in their ability to repair radiation damage have lower values of OER than their more radioresistant parents and radiosensitive mutants which showed LHR. The relationship between the OER and cell radioresistance was almost linear. This observation was expressed in great extent for diploid than for haploid yeast cells.
3. The effect of hypoxic sensitization by metronidazole and misonodazole was studied for a number of haploid and diploid yeast cells. Both compounds sensitized cells to ionizing radiation by increasing the slope of the survival curve and did not affect the shoulder. The effectiveness of sensitization was proportional to the oxygen-enhancement ratio and was expressed in great extent for diploid than for haploid yeast cells.
4. Radioprotective and radiosensitizing action of chemical drugs was much smaller for all haplont yeast cells than for wild-type strain of *S. cerevisiae* and similar to those obtained for radiosensitive *rad* mutants. The result suggests that haplont yeast may have evolved to diplont yeast via the development of a diploid specific repair which may be mediated by a recombination-like mechanism, for which the diploid state is required.
5. As a whole, mechanism of the radioprotective action, oxygen and hypoxic sensitizers effect may be realized on the primary radiation damage production as well as on cellular biochemical processes including the diploid-specific recovery from damage inflicting by ionizing radiation.
6. The dependence of cell radiosensitivity modification on the diploid-specific recovery was explained by the suggested kinetic scheme of radiation damage development which take into account cell ability to enzymatic repair.

REFERENCES

[1] Alper, T. Mechanisms of lethal radiation damage to cells. In: Ebert HG. editor. *Second Symposium on Microdosimetry*. Brussels: Commission of the European Communities, 1970; 5-36.

[2] Alper, T. *Cellular Radiobiology*. Cambridge University Press, 1979.

[3] Antoku, S. Chemical protection of bacteria and cuhured mammalian cells by sulphur-containing compounds. *J Radiat Res.*, 1975; 16, 28-36.

[4] Aptikaeva, GF; Ganassi, EE; Zaichkina, SI et al. Relationship between chemical protection and reparation of radiation-induced damages to chromosomes. *Radiobiol.*, 1976; 16, 35-38.

[5] Averbeck, D; Ebert, M. Genetic control of sensitization by oxygen involving repair systems in Saccharomyces. *Int. J. Radiat. Biol.*, 1972; 21, 493-502.

[6] Bacq, ZM. *Chemical Protection Against Ionizing Radiation*. Springfield: Charles C Thomas Publisher, 1965.

[7] Bresler, SE; Noskin, LA; Stepanova, IM et al. Mechanism of radioprotecting action of chemical compounds on Escherichia coli cells. *Mol. Gen. Genet.*, 1978; 163, 75-85.

[8] Chapman, JD; Reuvers, AP; Borsa, J et al. Chemical radioprotection and radiosensitization of mammalian cells growing in vitro. *Radiat Res.*, 1973; 56(2), 291-306.

[9] Chapman, JD; Dugle, DL; Reuvers, AP et al. Letter: Studies on the radiosensitizing effect of oxygen in Chinese hamster cells. *Int. J. Radiat. Biol.*, 1974; 26(4), 383-389.

[10] Davies, DR. Absence of an oxygen effect and the genetically determined repair processes. Nature, 1967; 215(5103), 829-832.

[11] Deering, RA; Smith, MS; Thomson, BK et al. Gamma-ray-resistant and -sensitive strains of slime mold (Dictyostelium discoideum). *Radiat. Res.*, 1970; 43(3), 711-728.

[12] Dertinger, H; Jung, H. *Molecular Radiation Biology, The Action of Ionizing Radiation on Elementary Biological Objects*. Berlin, Heidelberg, New York: Springer-Verlag, 1970.

[13] Eidus, LKh; Korystov, YuN. *The Oxygen in Radiobiology*. M: Energoatomizdat; 1984.

[14] Fielden, EM; Sapora, O; Loverock, PS. The effect of electron affinic sensitizers Ro 07-0582 and Ro 03-6156 on the survival of several *E. coli* mutants. *Int. J. Radiat. Biol.*, 1978; 31, 41-45.

[15] Hernadi, F; Karl, Cs; Nagy, Zs. The role of repair in the decreased radiosensitivity of *E. coli* cells induced by asynchronous macromolecular synthesis. *Stud. Biophys.*, 1969; 18, 71-76.

[16] Hollaender, A. *Radiation Protecuon and Recovery*. New York Oxford London Paris: Pergamon Press, 1960.

[17] Johansen, I; Gulbrandsen, R; Pettersen, R. Effectiveness of oxygen in promoting x-ray-induced single-strand breaks in circular phage lambda DNA and killing of radiation-sensitive mutants of *Escherichia coli*. *Radiat. Res.*, 1974; 58(3), 384-397.

[18] Kiefer, J. The oxygen effect: physical, chemical and biological aspects. Booz J, Ebert HG, Smith BGR editors. *Fifth Symposium on Microdosimetry*. Luxemburg: Commission of the European Communities, 1976; 849-868.

[19] Krishnan, D; Singh, DR; Madhvanath, UA. Possible combination on hypoxic cell sensitizer with an oxic protection: implication for radiotherapy. *Br. J. Cancer*, 1978; 37, 1026-1032.

[20] Luchnik, AN; Glaser, VM; Shestakov, SV. Repair of DNA double-strand breaks requires two homologous DNA duplexes. *Mol. Biol. Rep.*, 1977; 3, 437-442.

[21] Maisin JR. Radioprotective action, toxicity and mechanisms of action of sulphydryl radioprotectors. *Stud. Biophys.*, 1975; 53, 121-124.

[22] McKee, RM; Lawrence, CW. Genetic analysis of γ-ray mutagenesis in yeast. III. Double-mutant strains. *Mutat. Res.*, 1980; 70, 37-48.

[23] Millar, BC; Fielden, EM; Steele, JJ. A biphasic radiation survival response of mammalian cells to molecular oxygen. *Int. J. Radiat. Biol.*, 1979; 36, 177-180.

[24] Neklyudov, AG; Komarov VP; Petin VG. Quantitative estimation of radiosensitivity modification. *Radiobiology*, 1987; 27(1), 46-51.

[25] Petin, VG. Influence of misonidazole on survival of wild-type and radiosensitive mutants of yeast. *Mutat. Res.*, 1983; 108, 121-131.

[26] Petin, VG. *Genetic Control of Cell Radiosensitivity Modification*. Moscow: Energoatomizdat, 1987.

[27] Petin, VG; Matrenina, VL. Radioprotection action of chemical compounds on gamma-irradiated yeast cells of various genotypes. *Mol. Gen. Genet.*, 1981; 183, 152-157.

[28] Petin, VG; Ryabchenko, VI. A study of oxygen effect and radiosensitization of hypoxic cells by metronidazole in wild-type and radiosensitive mutants of yeast. *Int. J. Radiat. Biol.*, 1982; 42, 491-500.

[29] Pohlit, W. Radiation sensitivity and repair capacity depent on cell concentration? *Biophysik*, 1973; 10, 33-37.

[30] Reddy, NM; Rao, BS. Genetic control of repair of radiation damage produced under euoxic and anoxic conditions in diploid yeast *Saccharomyces cerevisiae*. *Radiat. Environ. Biophys.*, 1981; 19, 187-196.

[31] Saeki, T; Machida, I; Nakai, S. Genetic control of diploid recovery after gamma-irradiation in the yeast *Saccharomyces cerevisiae*. *Mutat. Res.*, 1980; 73, 251-265.

[32] Sapora, O; Fielden, EM; Loverock, PS. A comparative study on the effect of two classes of radiosensitizer on the survival of *E. coli* B and K12 mutants. *Radiat. Res.*, 1977; 69, 293-305.

[33] Shenoy, MA; Asquith, JC; Adams, GE et al. Time resolved oxygen effects in irradiated bacteria and mammalian cells: a rapid-mix study. *Radiat. Res.*, 1975; 62, 498-512.

[34] Sugahara, T. *Modification of Radiosensitivity in Cancer Treatment*. Edition. Tokyo: Academic Press, Inc., 1984.

[35] Sverdlov, AG; Bresler, SE; Noskin LA et al. DNA repair and chemical protection against irradiation in *E. coli*. *Stud Biophys.*, 1978; 69, 225-232.

[36] Thomson, JF. *Radiation Protection in Mammals*. New York: Reinhold Publishing Co, 1962.

[37] Yonei, S. Modification of radiation effects on *E. coli B/r* and a radiosensitive mutant B_{S-1} by membrane-binding drugs. *Int. J. Radiat. Biol.*, 1979; 36, 547-551.

Chapter 3

QUANTITATIVE REGULARITIES OF COMBINED ACTION OF VARIOUS AGENTS

ABSTRACT

Definitions of interactions of physical and chemical agents in this work are presented. A method of the synergistic enhancement ratio estimation is proposed. As an example of additive interaction, experimental results of combined action of laser (λ = 10.6 microns) and ionizing radiation (^{60}Co γ-rays) are offered. Some selected examples of antagonistic interaction are analyzed. Among them, an unusual effect of triggered cell response after combined application of α-particles and UV light to yeast cells of various ploidy is given. It was shown that UV light mitigated the lethal effect of densely ionizing radiation. The effect of oscillating changes in cell survival was registered. This antagonistic interaction of α-particles and UV light was independent of the sequence in which the two factors were applied and noted both in haploid and diploid yeast cells of the wild-type. It is concluded that radiation-induced repair system may have significance for therapeutic radiology. Another example of antagonistic interaction is photoreactivation after combined action of ionizing radiation and visible light. Some new experimental data of this phenomenon are given. These data are in line with early suggested idea that the photoreactivable damage induced by ionizing radiation might be UV-like damage produced by Čerenkov light emitted by charged particles when their velocity exceeds the phase velocity of light in the medium. A simple mathematical model for biological estimation of UV light dose concomitant to ionizing radiation was suggested. This approach was applied to determine the dependence of equivalent UV light dose accompanied 100 Gy of ionizing radiation on the energy of sparsely ionizing radiation and on the volume of the exposed cell suspension. On the basis of the results obtained, two important issues are discussed: a possible synergistic interaction of the damage produced by ionizations and excitations and a probable participation of UV component of ionizing radiation in the mechanism of different low-dose responses: adaptive reaction, radiation hormesis, bystander effect.

INTRODUCTION

There are many examples of interactions between ionizing radiation and other toxic and non-toxic physical or chemical agents. Biological responses obtained range from strong antagonism to strong synergism. In view of the many different agents that may be involved in combined exposure with radiation and complexity of the possible interactions, it is worth to give some insight to the definitions existing in this area. Relatively many different terms are employed in the literature to determine character of the combined effects of various environmental factors and their interaction [4; 17; 26; 59]. These definitions, arguments, discussion, mathematical calculations and conclusions may be applied to any environmental hazards. This Chapter will give briefly a quantitative insight into the interaction processes.

For concreteness sake, it is of interest to consider examples of combined interaction of ionizing radiation with heat, which is often used for sterilization and radiation therapy. To estimate quantitatively the sensitization action of hyperthermia, the thermal enhancement ratio is frequently employed, for example, [62]. This parameter defined as the ratio of dose (or exposure duration) required to give a specified level of damage induced by ionizing radiation alone to the dose required to give the same level of damage after combined action with heat. However, it does not reflect the kind of interaction (whether it was antagonistic, independent or synergistic). In some cases, for practical and theoretical reasons, it is of importance to provide information concerning the interaction. To calculate the synergistic effect, we used the synergistic enhancement ratio defined as the ratio of the exposure time expected for an independent effect of both modalities to that observed from the experimental survival curve for the combined action of hyperthermia and ionizing radiation [17; 46]. One of the tasks of this Chapter was to compare both these parameters and stress the advantage of the synergistic enhancement ratio. The problem of synergy is worth of our particular attention and will be considered in the following Chapters.

Independent interaction of agents does not bring new ideas for our understanding of the biological effects of ionizing radiation and other agents, so that in the following presentation we will use it for quantitative comparison with the real observed results to evaluate the type of interaction. Antagonistic kind of interaction is of great interest from fundamental and practical view of radiobiology, radiation therapy, radioecology and medicine. In the preceding Chapter we have considered a number of new results related with the radioprotective compounds. Several selected examples of additive and antagonistic interaction will be described in this Chapter.

Some studies have indicated an antagonistic interaction of ultraviolet light and ionizing radiation [8], i.e., protective effect of UV light against the lethal effects of ionizing radiation was observed. Similar effect was demonstrated for protozoa and yeast cells [1]. In particular, the survival of haploid yeast cells irradiated in exponential phase of growth with α-particles increased after the action of UV light. It is curiously that the change in survival was dependent on the dose of UV-light when the final curve after combined actions characterized by unusual shapes [1]. To explain this observation, the authors introduced the hypothesis of triggered system of recovery which resulted in an unusual dose-effect curve: cell survival changed in oscillating way with one or several maxima and minima before it began to decrease monotonically. Because of unusual and unexpected effect of "triggered" or "induced" repair system, it would be of interest to examine the effect of the combined application of α-particles and UV light to yeast cells of various genotypes exposed at stationary phase of growth.

The phenomenon of photoreactivation of UV-induced damage has been known for a long time and is observed in various biological objects illuminated in postirradiation period by visible light. Several investigators have made efforts to obtain photoreactivation of X-ray-induced damage but the effect was found to be negligible. It is a well-known fact that the charged particles emit Čerenkov emission when their velocity exceeds the phase velocity of light in a medium [19]. This emission includes a wide spectra of visible light and UV light; its biological significance was demonstrated by the photoreactivation of UV-hypersensitive bacterial cells irradiated with ionizing radiation [32; 35; 36; 40; 41; 54-56; 66]. It would be of interest to confirm the role of photereactivable damage after irradiation of yeast cells, the simplest model of eukaryotic cells. The proportion of damage susceptible to photoreactivation was shown to be dependent on the energy of sparsely ionizing radiation [41; 56] and the volume of irradiated suspension [35; 36]. Furthermore, it was demonstrated that photoreactivation after ionizing radiation exhibits some properties identical to those after UV-irradiation [36; 56; 60; 65; 66]. The data provide evidence that similar types of photoreactivable damage can be produced by UV and ionizing radiation, lending support to the proposed biologically important role of excitation of DNA and indicate that the biological effect of ionizing radiation was not totally due to the ionization of molecules, and excitation of molecules due to Čerenkov emission might be important in some cases. It is of interest to get an additional arguments supporting this idea.

In particular, it was shown that the extent and the rate of photoreactivation in *E. coli* following ionizing radiation decreased when compared with UV light [40]. These effects may be proceeded through the impairment of the process of photoreactivation itself, i.e., through the decrease in the probability of recovery, or through changes of the repairable and irreversible portions of the damage inflicted. In the latter case, the probability of recovery may either decrease or stay unchanged. However, the data distinguishing these possibilities are lacking in the literature.

We also intend to develop a mathematical approach to estimate the relative contribution of ionization and excitation to the lethal effect of ionizing radiation making use of an elementary approximation for the first step. Some predictions from the model will be considered. This model will be applied to estimate UV light dose concomitant to ionizing radiation of high energy.

Thus, in this Chapter we are going to implement the next tasks.

1. To analyze quantitative estimations of interaction after combined action of various agents.
2. To consider several examples of additive and antagonistic interaction.
3. To present new experimental data for photoreactivation after combined action of ionizing radiation and visible light.
4. To evaluate parameters of cell photoreactivation after separate exposure to UV light and different ionizing radiations.
5. To develop a mathematical model for determination of UV light dose concomitant to ionizing radiation of high energy.

3.1. Quantitative Estimation of Interaction after Combined Actions

There are many examples of interactions between ionizing radiation and other toxic and non-toxic physical or chemical agents. Biological responses obtained range from strong antagonism to strong synergism. The most effective simultaneous exposure to ionizing radiation and another applied factor may result in four possible modes of interaction: (i) no interaction: the observed effect is based on the most toxic agent; (ii) additivity: the effects produced by separate application of each agents are simply summed up; this mode can be subdivided on the additive and independent interaction; (iii) antagonism: the effect is less than the expected additivity; (iv) synergism: the effect is great than expected additivity.

In accordance with the definition, independent interaction means that the probability of the ultimate effect is determined by the product of the probabilities of the effects produced by each agent applied separately. Then for the probability of cell to be survival we can write

$$S = S_1 \cdot S_2, \tag{3.1}$$

where S is the probability of cell survival after simultaneous or successive action of two agents, S_1 and S_2 are the probabilities, for instance, cell survival after exposure to each agent separately. Taking the logarithm, we have

$$\begin{aligned} \ln S &= \ln S_1 + \ln S_2 \\ \lg S &= \lg S_1 + \lg S_2 . \end{aligned} \tag{3.2}$$

This means that the independent action of two agents is characterized by the summation of logarithms of cell survival probability after exposure to each agent separately. According Poisson's statistics [13; 24], the number of damage produced by any harmful agent may be estimated by

$$N = -\ln S. \tag{3.3}$$

Eq. (3.2) suggests that the total numbers of damage for independent interaction of two agents is determined by the simple summation of damage produced by each of the relevant factors. This type of interaction is classified as an additive and the biological response is also sometimes referred to as an additive. Note that the interaction of damage especially after successive application of two agents can be subdivided into additive and independent summation of the effects produced.

Antagonistic interaction is observed when any physical or chemical factor results in a decrease of the biological effectiveness of ionizing radiation (or another agent) and the biological response is called as antagonistic. In this case, the damaging effect of ionizing radiation and another modifying agent is less than that expected form Eqs. (3.1) and (3.2) and the agent can be used to reduce the consequences of radiation injury. It is said about radioprotective action when chemical agent reduces the biological effectiveness of ionizing radiation. The term antagonistic interaction has never been used to classify radioprotective action, although in fact a real antagonistic interaction takes place in this case too.

Synergism occurs when the agent used for radiation effect modification enhances the action of ionizing radiation, i.e., cell survival after the combined action is less than expected for independent interaction. In this case, the final effect of combined application of two agents is due to synergistic interaction of damage produced by each of the applied factors.

To determine whether the interactions are additive or synergistic, data are usually analyzed according to a definition recommended by the International Commission on Radiation Units [17], and their definition has since been used by many authors [3; 45]. An independent interaction (additivity) is determined by SF_C = SF(HT) x SF(RT) and the synergy is observed if SF_C < SF(HT) x SF(RT), where SF(HT) and SF(RT) stand for the surviving fractions after treatments with hyperthermia and radiation applied alone, respectively. SF_C stands for the surviving fraction after the simultaneous treatment of both modalities.

Before clearing up of the basic set of experiments and definitions used in our findings with synergistic interaction, briefly mention materials and methods used to obtain the data presented in Figure 3.1 The following yeast strains were used in the experiments: *S. cerevisiae* (two diploid strains, XS800 and T1), *Endomyces magnusii* (diploid, strain K7). Yeast cells were incubated before irradiation for 3–5 days at 30 °C on a complete nutrient agar layer to a stationary phase. The repair-proficient bacterial strain *Escherichia coli* B/r was grown in YEP broth (yeast extract: 1%, glucose: 1%, peptone: 1%, NaCl: 1%) at 37 °C during 18 h of incubation. Homogeneous cell populations were treated at stationary phase of growth. Final suspensions contained 10^6 and 10^8 cells per ml for yeast and bacterial cells.

The ^{60}Co γ-ray source was a Gammacell 220 (Atomic Energy of Canada LTD). The γ-ray dose rate, estimated by ferrous sulphate dosimetry and by a Siemens ionization chamber, was 10 Gy/min. For UV exposure, cells were irradiated with germicidal lamps that emitted predominantly UV light of wavelength 254 nm at fluence rate of 0.25 W/m^2. The fluence rate was measured using a calibrated General Electrical germicidal meter. To avoid photoreactivation, UV exposure, dilution and other procedures were performed under red ambient light, while post-irradiation incubation was carried out under dark conditions. A continuous mode of sonication was accomplished by a 20 kHz ultrasonic unit (Fisher sonic dismembrator Model 300). The diameter of the ultrasonic transducer was 15 mm. The ultrasonic intensity was 0.05 W/cm^2 which were measured by the calorimetric method. For sonication, 0.1 ml of cell suspension (10^8 cells/ml) at room temperature was put into 9.9 ml of sterile water which was placed into a metal vessel constructed together with the transducer. To determine the effect induced by ultrasound alone, the absorbed ultrasound heat was completely removed by cooling with water.

Hyperthermia was given in a water bath where a desired temperature ±0.1 °C was maintained. For the simultaneous action of hyperthermia and other agents, the time interval between the introduction of the cells into the pre-heated water and the beginning of exposure was about 0.1–0.3 min, which was significantly less than the total treatment time. At the end of the treatment, the samples were rapidly cooled to room temperature. Therefore, the durations of physical agent treatment and heat exposure were identical.

After treatments, a known number of cells were plated so that 150–200 colonies per dish would form by the surviving yeast cells after 5 days of incubation at 30 °C and 50–100 colonies would form by the survival bacterial cells after 24 h of incubation at 37 °C. All experimental series were repeated 3–5 times. The details were described elsewhere [50].

Figure 3.1 exhibits some examples of the basic experimental data used in this investigation. Here, survival curves were obtained for diploid yeast cells of different species and bacterial cells after heat treatment alone (curves 1) or after exposure with another inactivating agent applied alone at room temperature (curves 2). Curves 3 represent the theoretical survival curves that would be expected by the simultaneous combined action of both modalities if their inactivation actions were independently added. In other words, the expected probability of surviving (surviving fraction) after combined action was equal to the product of surviving fractions observed after each agent applied separately (Eq. 3.1). Curves 4 shows the real survival curves obtained experimentally after the simultaneous action of heat and another inactivating agent. These four types of survival curves depicted in Figure 3.1 have been obtained in the following Chapters for every temperature and intensity of ionizing or ultraviolet radiations, ultrasound, chemical compounds and cell kind investigated. They were not always presented but were used to evaluate the effectiveness of the interaction. Figure 3.2 exhibits similar survival curves obtained for haploid yeast cells after simultaneous action of ionizing radiation and high temperatures.

To estimate quantitatively the sensitization action of hyperthermia, one can employ, for example [62], the thermal enhancement ratio (TER) defined as the ratio t_3 / t_1 or D_3 / D_1 (Figures 3.1 and 3.2), i.e., by the ratio of dose (or exposure duration) required to give a specified level of damage induced by the applied inactivating agent without heat to the dose required to give the same level of damage produced by the same agent used together with heat:

$$TER = t_3 / t_1 = D_3 / D_1. \tag{3.4}$$

This ratio indicates an increase of cell sensitivity by high temperature. However, it does not reflect the kind of interaction (whether it was antagonistic, independent or synergistic). To calculate the synergistic effect we used the synergistic enhancement ratio (k) defined as the ratio of the exposure time (or dose) expected for an independent effect of both modalities to that observed from the experimental survival curve for the simultaneous action of hyperthermia and another inactivating factor [46]:

$$k = t_2 / t_1 = D_2 / D_1. \tag{3.5}$$

It is worth to note that such a definition does not contradict to that discussed by others [6; 61]. This parameter was calculated at 10% survival for sigmoid survival curves (Figures 3.1A, B, C). For purely exponential survival curves, k does not depend on the surviving level (Figures 3.1D and 3.2). Sometimes in order to estimate the synergistic enhancement ratio, we used the ratio of the corresponding mean lethal dose D_0.

Both these parameters are plotted in Figure 3.3 against the exposure temperature for the simultaneous action of hyperthermia with various inactivating agents and cell systems. The remarkable feature of Figure 3.3 is that the thermal enhancement ratio (curves 1) increases with increasing exposure temperature, while the synergistic enhancement ratio (curves 2) increases at first, then reaches a maximum, which is followed by a decrease. This pattern was observed independently of the agent (ionizing radiation, UV light, ultrasound, and zinc sulphate) and the cell line (yeast or bacterial cells) used in this investigation.

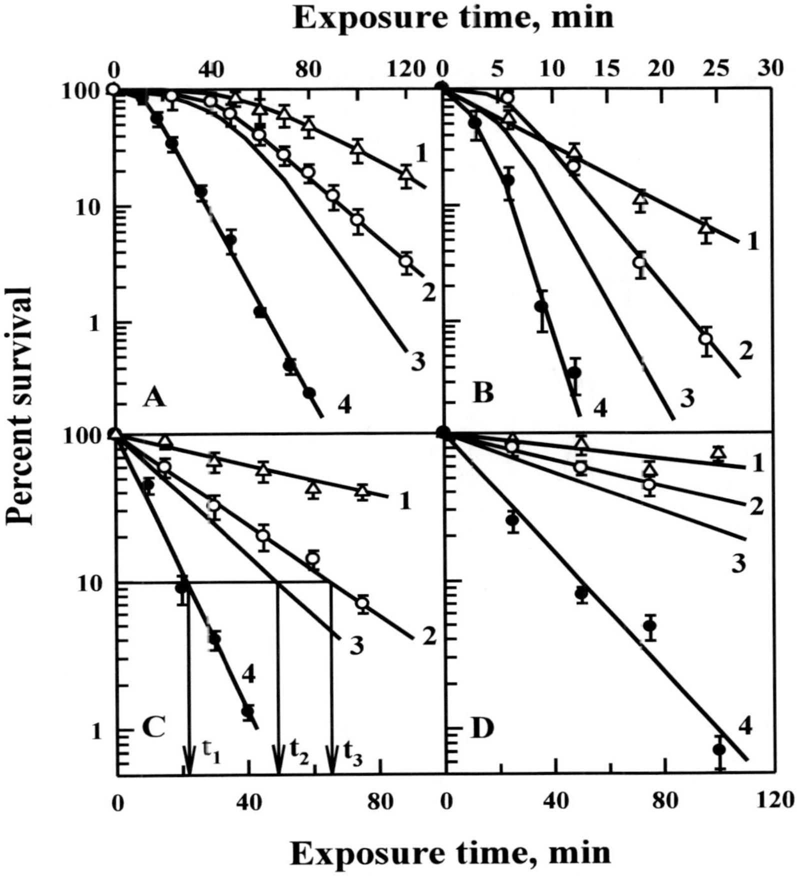

Figure 3.1. Survival curves of diploid yeast cells (A: *Endomyces magnusii* K7; B, C: *S. cerevisiae*, strain XS800) and D: bacterial *Escherichia coli* B/r cells exposed to various agents. Curves 1, heat treatment alone (A, 40 °C; B, 52,5 °C; C, 48 °C; D, 40 °C); curves 2, another inactivating agent applied alone at room temperature (A, γ-ray of ^{60}Co, 10 Gy/min; B, 254 nm UV light, 0.25 W/m^2; C, 20 kHz ultrasound, 0.05 W/cm^2, D, 0.01 M zinc sulphate); curves 3, calculated curve for the independent action of the agents used; curves 4, experimental surviving curves after the simultaneous action of the agents used. Error bars show inter-experimental standard errors.

Enhancement of biological action of ionizing radiation can be subdivided into two classes depending on the properties of the second agent. For example, if any factor combined with ionizing radiation also causes the observed biological effect, there is a real synergism or synergistic interaction. If the second agent is not able to produce biological effect but enhance the effect of ionizing radiation, such a biological action is called sensitization and the agent could be considered as a sensitizer. The problem of sensitization and the development of new type of sensitizers are relevant to radiation therapy of tumors and methods of sterilization. Oxygen is a classic and universal sensitizer in radiobiology. Some novel directions in study of oxygen effect and radiosensitizing action of chemical compounds have been presented in the preceding Chapter. The sensitization can be considered as a particular case of synergy.

Indeed, the phenomena of sensitization and synergy are close to each other in a number of common properties. First, they both result in an increase the final radiobiological effects. Secondly, in both cases after the combined application with ionizing radiation the cell capability to recover from radiation damage is gradually decreased with an increase in the effectiveness of the agent applied. Thirdly, the efficiency of synergistic and sensitizing action seems to decrease with an increase in LET of radiation [11; 44; 52; 53].

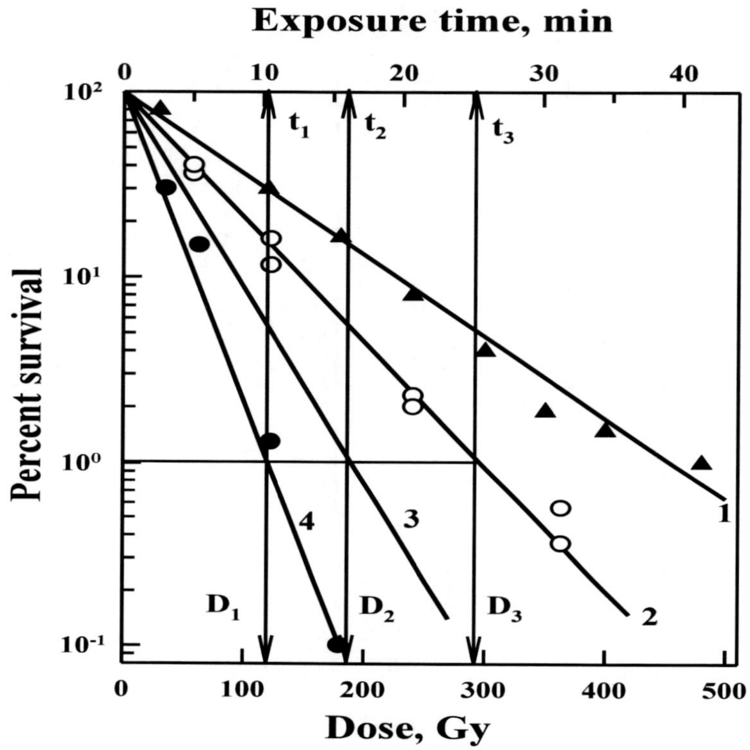

Figure 3.2. Survival curves for haploid yeast *Zygosaccharomyces bailii* K17: 1, cells were subjected only to hyperthermia (45 °C); 2, irradiation with ^{60}Co γ-rays (10 Gy/min) at room temperature; 3, theoretically expected curve for additive interaction of ionizing radiation and heat; 4, experimental curve obtained after simultaneous treatment of these agents.

There is an objective uncertainty in the determination of interaction of environmental factors, which was most clearly manifested in the sequential application of two agents. Suppose, for example, we used a sequential application of a number of doses of ionizing radiation and a single dose of the modifying agent. If the survival curve after action of ionizing radiation alone has a simple exponential shape (Figure 3.4A), the kind of interaction with another agent can be easily determined. Assume that the single dose of agent applied in combination with ionizing radiation decreases cell survival to 4%. Then additive curve should be also exponential starts at 4% survival and has a slope that is identical to the initial curve. If the real experimental survival curve after sequential application of these agents is located above the curve 2, we deal with antagonistic interaction, if it is lower than expected additive curve, the character of interaction is synergistic.

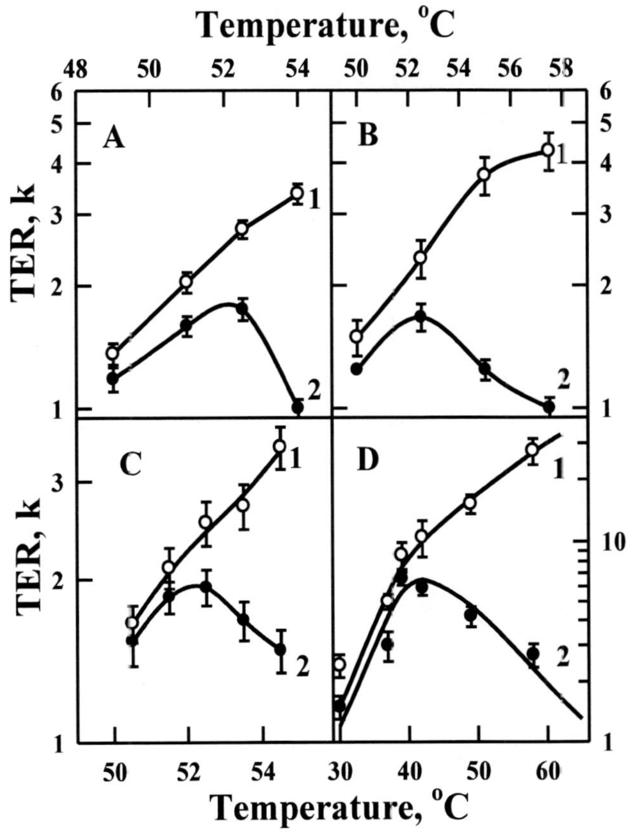

Figure 3.3. The dependence of the thermal enhancement ratio TER (curves 1) and the synergistic enhancement ratio k (curves 2) upon exposure temperature after the simultaneous action of high temperature and another inactivating agent (A, 25 MeV electron radiation, 25 Gy/min; B, 254 nm UV light, 0.25 W/m^2; C, 20 kHz ultrasound, 0.05 W/cm^2; D, 0.01 M zinc sulphate) on *S. cerevisiae* diploid yeast cells, strain XS800 (A, B, C) and *E. coli* bacterial cells, strain B/r (D). Error bars show inter-experimental standard errors.

Such an univocal certainty would not exist for sigmoid dose-effect curve (Figure 3.4B, curve 1) which is characterized by the presence of a "shoulder' at low doses, when the detected effect increases with dose is relatively slow due to the formation and/or repair of sublethal damage [16; 24; 42; 67]. In this case there are several options for summation of the effects produced by each agent. Let the mechanism of action and radiosensitive target be totally different for ionizing radiation and the agent used for modification of radiation response. Suppose that cell survival after exposure to this agent is 10%. Then the expected survival curve after combined action would repeat the shape of the initial survival curve, but originating from 10% survival (Figure 3.4B, curve 2).

If the real experimental survival curve after sequential application of these agents is located above the curve 2, we deal with antagonistic interaction, if it is lower, the character of interaction is synergistic. Note that such a summation of biological effects was recommended to be called as an independent [17]. The second hypothetically possible, but also unlikely situation: mechanisms of action and target are identical to the applied agent and ionizing radiation. Then the additive survival curve (Figure 3.4B, curve 3) should repeat the shape of the initial survival curve starting from 10% survival. Just this case was recommended to be called as an additive [17]. If the experimental survival curve after sequential application of these agents is located above the curve 3, we deal with antagonistic interaction, if it is lower than expected additive curve, the kind of interaction is synergistic.

However, as a rule, the mechanisms of action and the pertinent targets responsible for detectable biological effect are not known exactly, there is some uncertainty (shaded region in Figure 3.4B, the envelope of additivity) within which the type of interaction could not be determined. To avoid this uncertainty in the synergistic enhancement ratio calculation, we compared the survival curves obtained after successive combined action of two agents with the additive (according to the ICRU definition) dose-effect curve. In this case, there may be some underestimation of the synergistic interaction effectiveness, but the kind of the interaction would be determined exactly. In the next section we will present some selected examples of additive and antagonistic interaction.

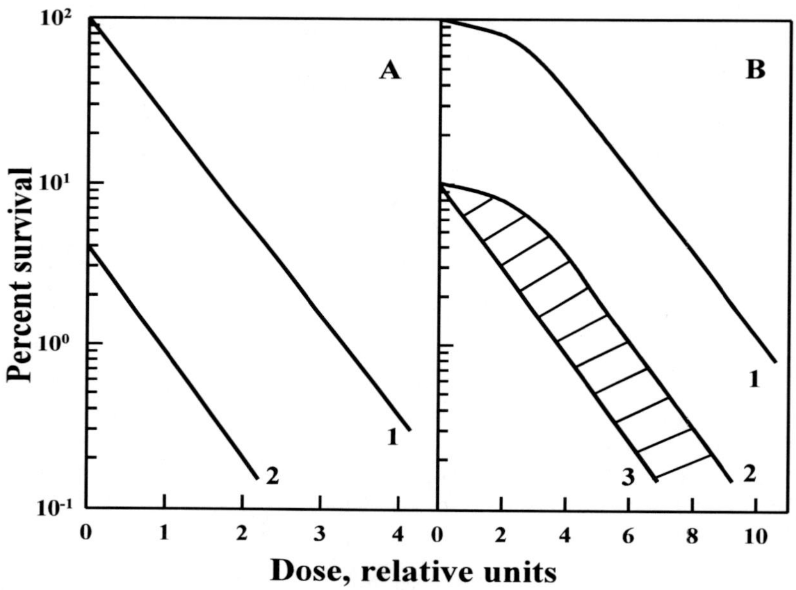

Figure 3.4. To the determination of additivity, antagonism and synergy after the sequential application of two inactivating agents.

3.2. Additive and Antagonistic Interaction: Selective Examples

Organisms in their natural environment are always simultaneously exposed to a number of physical and chemical agents such as ionizing radiation, UV light, nonionizing electromagnetic radiations of various wavelengths, magnetic and electrical natural and man made fields as well as enormous line of chemical pollutants. The number of toxic substances in the environment and workplace is extremely large and increases every year. An interaction between two agents may be considered as additive or antagonistic when the damaging effect is equal or smaller than the sum of the two separate responses. In many cases, the biological effects produced by harmful agents are independent of each other and simply summarized. Just that very case is specified in safety standards of various countries. Antagonistic interaction is extremely infrequent in the natural conditions and it is seldom taken into account in practice. Synergistic interaction happens very often but for its manifestation some specific conditions are required that will be considered in detail in the following Chapters. In this section, some selected examples of additive and antagonistic interaction of various acting agents will be presented.

Irradiation by coherent electromagnetic radiation was carried out with CO_2 laser in which the pulse power changed due to the passage of radiation through the plastic film. The pulse duration (100 ns) was constant and its repetition frequency was 3 pulses per a minute. Carbon dioxide laser emits in the infrared region at a wavelength of 10.6 microns. The simultaneous irradiation and power measurement was due to the spilling of the beam into two orthogonal ones. Irradiation with ionizing radiation was carried out with ^{60}Co γ-rays at dose rate of 15 Gy/min.

In the experiments, the following isogenic polyploid series of *S. cerevisiae* yeast strains was used: 211-1 (haploid), 211 (diploid), 400 (tetraploid), 600 (hexaploid) obtained from W. Laskowski, who described their radiosensitivity [23], and three strains of bacteria *E. coli*: B, B/r and B_{s-1}. Yeast strains of different ploidy contained different number of chromosome sets and therefore differed in their sensitivity to ionizing radiation. Bacterial strains were also different in their sensitivity both to ionizing radiation and especially to UV light because of defects in reparation systems [13]. Cells were irradiated in stationary phase of growth. Because of low penetration ability, a monolayer of cells was exposed to CO_2 laser irradiation: 0.02 ml of cell suspensions were placed onto the surface of non-nutrient agar and before irradiation the water was evaporated from this droplet. Immediately after irradiation, samples were diluted to the appropriate concentration with distilled water and plated on nutrient agar in Petri dishes for the assay of colony-forming ability. Colonies were scored after 5 days of yeast incubation at 30 °C and after 24 h bacterial growth at 37 (all three strains) or 42 °C (strain B).

Figure 3.5 exhibits the dependence of cell survival on the intensity of laser radiation for yeast strains with different numbers of chromosome sets that were irradiated with a single pulse (curve 1) and with a set of pulses of the same intensity (0.2 J/cm^2). It is evident that for both modes of exposure the cell sensitivity did not depend on cell ploidy and the form of dose-effect curve was sigmoid. It should be noted that radiosensitivity of these strains to low- and high-LET radiation at 10% survival differed in 4 and 3 times, respectively [23]. The most radiosensitive was haploid strain and the most resistant – diploid. The sensitivity of these strains to hyperthermia was differed in 2 times and direct proportional to cell size, haploid cells being more resistant and then cell thermal sensitivity increased with further increase in the number of chromosome sets [44]. It is evident that cell sensitivity to laser irradiation did not depend on the number of chromosome sets or the cell size (Figure 3.5). Such a unification of cell sensitivity is unique and has not been observed for yeast cells after any form of inactivating factors. At the same time, a typical radiobiological response can be noted: single laser pulses of high intensity were much more effective in contrast with a series of low intensity pulses (curves 1 and 2, Figure 3.5).

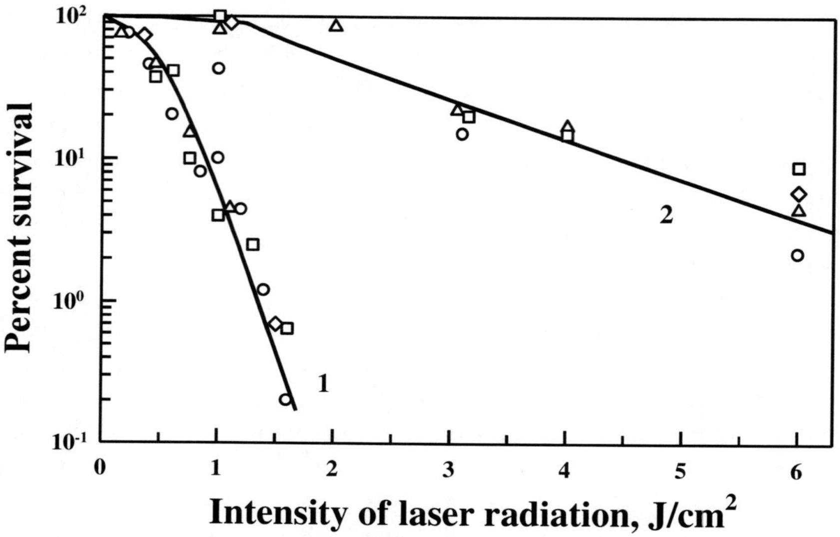

Figure 3.5. The dependence of cell survival on the intensity of laser radiation for yeast strains with different numbers of chromosome sets: haploid (circles), diploid (triangles), tetraploid (squares), hexaploid (rhombs). Cells were irradiated with a single pulse (curve 1) and with a series of pulses (curve 2).

It would be of interest to obtain analogous dose-effect curves for prokaryotic cells. The dependence of cell bacterial survival on the intensity of laser radiation for E. coli of wild-type (strain B) and its radioresistant (strain B/r) and radiosensitive (strain Bs-1) mutants. The results are presented in Figure 3.6. It can be seen that the dose-effect survival curves after exposure to laser radiation are identical for all of these strains. It is curiously that the sensitivity of these strains was identical after exposure to high-LET ionizing radiation [13; 39]. It was proved that this effect might be explained by the production of more severe irreparable damage associated with the excessive absorbed energy.

We may apply similar arguments to interpret the data obtained with laser irradiation. It is quite plausible that because of inhomogeneous distribution of laser energy density a part of cells absorbed more energy than it is required for cell killing ("severe" damage) while survival cells absorbed significantly less energy. This hypothesis is supported by some other observations. In particular, microscopic analysis showed that after laser irradiation cells killed without any postirradiation division. Such form of cell inactivation is known to encounter only after giant doses of ionizing radiation decreasing cell survival below 0.001%. Besides, both yeast and bacterial cells were not able to recover from damage inflicted by laser radiation. Diploid and other polyploid yeast cell survival after exposure to laser irradiation, in contrast to irradiation with ionizing radiation, was not increased after their holding in non-nutrient condition during postirradiation period. It is known that survival of bacterial cell exposed to ionizing radiation increased after their cultivation during positirradiation period at 42 rather than at 37 C due to their ability to repair radiation damage. Such kind of reparation was not observed after laser irradiation (Figure 3.6). The colonies produced by surviving cells after laser irradiation differed from control neither in the rate of formation nor in the morphology patterns. This means that surviving cells does not contain any sublesions produced by laser irradiation. Then it can be expected that preliminary laser irradiation should not affect the cell sensitivity to the subsequent exposure to ionizing radiation. Figure 3.7 shows the results of experiments that confirm this conclusion. It is obvious that interaction of these agents is strictly additive.

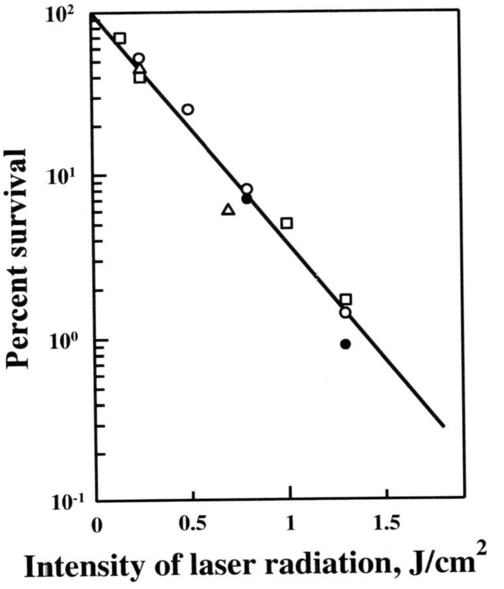

Figure 3.6. The dependence of cell bacterial survival on the intensity of laser radiation for strains *E. coli*: B/r (squares), B (cultivated during postirradiation period at 37, open circles, or 42 °C, closed circles) and B_{s-1} (triangles).

Some studies have indicated an antagonistic interaction of ultraviolet light and ionizing radiation [8], i.e., protective effect of UV light against the lethal effects of ionizing radiation can be observed for some doses. Similar effect was demonstrated for protozoa and yeast cells [1]. In particular, the survival of haploid yeast cells irradiated in exponential phase of growth with α-particles increased after the action of UV light. It is curiously that the change in survival was dependent on the dose of UV-light in an unusual way: cell survival changed in oscillating way with one or several maxima and minima before it began to decrease monotonically [1]. To explain this observation, the authors introduced the hypothesis of triggered system of recovery.

Because of unusual and unexpected effect of "triggered" or "induced" repair system [1], we examined the effect of the combined application of α-particles and UV light to haploid and diploid yeast cells of various genotypes exposed at stationary phase of growth. The following yeast strains of *S. cerevisiae* have been used in this investigation: haploid and homozygous diploid of wild-type (S288C, RAD, and XS800, RAD/RAD) and their radiosensitive mutants (XS774-4d, *rad*51, and XS806, *rad*51/*rad*51) as well as UV-sensitive mutant (XS1924, *rad*18/*rad*18). For UV exposure, cells were irradiated with germicidal lamp that emitted predominantly UV light 254 nm at fluence rate of 0.36 W/m^2 for strain XS1924 and 2 W/m^2 for other strains. Exposure to α-particles, the source of ^{239}Pu (17.4 Gy/min), was carried out.

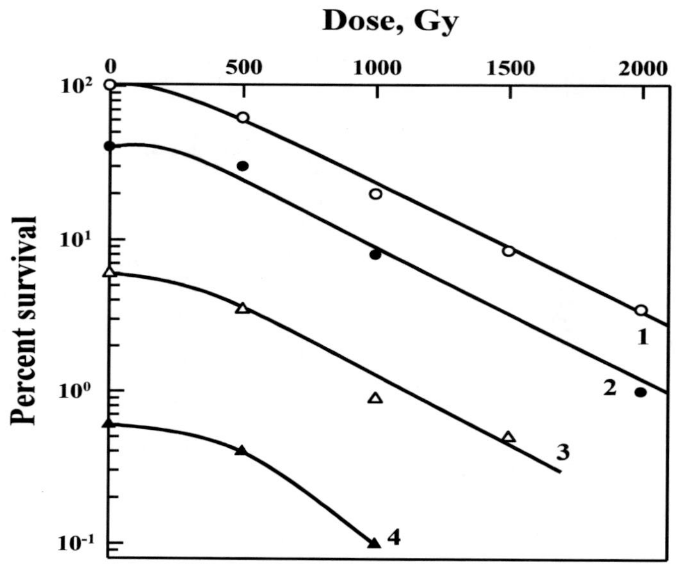

Figure 3.7. Survival curves of *S. cerevisiae* yeast cells (diploid strain 5a3bα) after exposure to γ-rays alone (curve 1) and after preliminary laser exposure decreasing cell survival to 40% (curve 2), 7% (curve 3), and 0.6% (curve 4) and the subsequent irradiation with γ-rays of ^{60}Co (15 Gy/min).

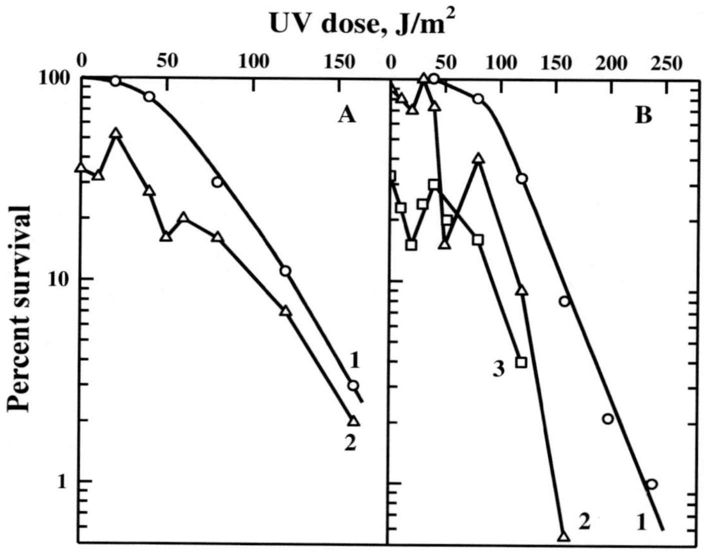

Figure 3.8. The dose-effect curves of haploid (A, S288C) and diploid (B, XS800) *S. cerevisiae* yeast cells irradiated with UV light alone (curves 1) and previously exposed to α-particles at dose of 17.4 Gy (A, curve 2), 52.2 and 104.4 Gy (B, curves 2 and 3, respectively) and the subsequent irradiation with various doses of UV light.

Figure 3.8 shows the results of these experiments. The time interval between the influencing factors was less than 10–15 minutes. Cell viability was assessed by the scoring of visible macrocolonies after 5 days of growth on solid standard nutrient media at 30 °C. It is evident that α-particles mitigate the lethal effect of UV light radiation. With combined irradiation, the survival curves showed an unusual shape with one or two "peaks". The protective effect of α-particles on UV-irradiated yeast cells was independent of the sequence in which the two factors were applied and noted both in haploid and diploid yeast cells of the wild-type. It should be stressed that the oscillatory dose-response curves for both strains were observed at doses of UV light, which do not lead to a significant decrease in cell viability. Almost additive effect of high doses of UV light and ionizing radiation was expressed for haploid strain incapable of the DNA double strand break recovery (Figure 3.8A). For diploid yeast cells possessing such a capability, the synergistic interaction of damages inflicted by α-particles and high doses of UV light was noted as can be judged by the final slope of the corresponding curves (Figure 3.8B). Similar to experiments described by Calkins [1], in our investigation the location and magnitude of the maxima of dose-response curves for the combined effect of high-LET radiation and UV light can vary significantly from one experiment to another, but the existence of antagonistic interaction of α-particles with low doses of UV light for wild-type yeast cells irrespective of their ploidy reproduced regularly.

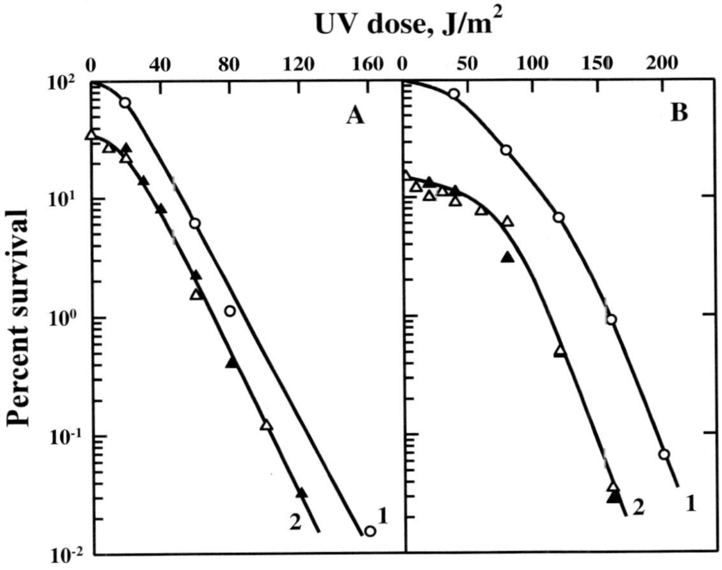

Figure 3.9. The dose-effect curves of haploid (A, XS774-4d, *rad*51) and diploid (B, XS806, *rad*51/*rad*51) *S. cerevisiae* yeast cells irradiated with UV light alone (curves 1) and exposed to α-particles at dose of 17.4 Gy (A, curve 2), and 26.1 Gy (B, curve 2) and further irradiated with various doses of UV light. The sequence α + UV is designated with open triangles and the reverse sequence – with closed triangles.

The results of experiments with yeast mutants sensitive to ionizing radiation are depicted in Figure 3.9. Haploid cells with mutation in RAD51 gene (*rad*51) were more sensitive to UV light than their homozygous diploid cells (*rad*51/*rad*51). It is seen that in this case the small dose of UV light does not lead to reactivation of the damage caused by α-particles, i.e., interaction of densely ionizing radiation with UV light was rigorously additive. Induction of gene conversion and reciprocal recombination are greatly reduced in *rad*51 cells than that of the wild-type cells [58]. It can be assumed on this basis that the trigger recovery can be related to these molecular processes.

Figure 3.10 shows the results obtained with UV-sensitive mutant (*rad*18/*rad*18). Cells were exposed to UV light alone (curve 1) and in combination with exposure to α-particles in different order of application (curve 2). It can be seen that the RAD18 gene mutation resulted in the disappearance of the shoulder of UV survival curve and in an increase of its slope compared with those of the parent wild-type strain (curve 1, Figure 3.8B). The combined effects of (α + UV) or (UV + α) did not lead to oscillations of the survival curve observed for wild-type cells (curves 2 and 3, Figure 3.8B). Nevertheless at low doses of UV light some antagonistic interaction was obtained due to the appearance of a shoulder at the survival curve. After high doses of UV light interaction was synergistic: the slope of the survival after combined action was significantly greater than after irradiation with UV light alone.

Taking our data with those published in literature [1] as a whole, one can conclude that the protective effect of UV light against both low- and high-LET ionizing radiation is observed for protozoa, yeast cells of various ploidy and mammalian cells.

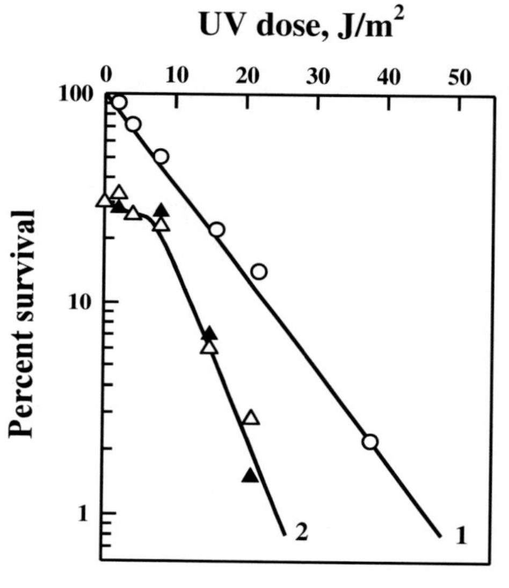

Figure 3.10. The dose-effect curves of diploid (XS1924, *rad*18/*rad*18) *S. cerevisiae* yeast cells irradiated with UV light alone (curve 1) and exposed to α-particles at dose of 55 Gy and further irradiated with various doses of UV light (curve 2). The sequence α + UV is designated with open triangles and the reverse sequence – with closed triangles.

3.3. Photoreactivation after Combined Action of Ionizing Radiation and Visible Light

For a long time it had been considered that significance of excitation was negligible and the biological effect of ionizing radiation is totally due to ionizations. The discovery of the phenomenon of photoreactivation after exposure of some bacterial and yeast cells to ionizing radiation showed that a part of damage produced by ionizing radiation is UV-like and due to molecule excitation [40; 55; 56; 65]. Such UV-like damage can be detected only by the photoreactivation in the UV hypersensitive cells. There are indications that the photoreactivable part of damage is similar whether induced by ionizing radiations or UV light and this damage was thought to arise from production of an excited state [36; 66]. Several lines of experimental data and theoretical considerations are in favor of the idea that the photoreactivable damage induced by ionizing radiation might be UV-like damage produced by Čerenkov light emitted by charged particles when their velocity exceeds the phase velocity of the light in the medium [32-35; 38; 48; 54; 60]. Last years interest in the phenomenon of photoreactivation resumed. There appeared some new evidence that the ability of cells to photoreactivation may reduce solar injury, prevents or reduces the risk of tumors induced by ultraviolet light. The decrease in the effectiveness of UV and γ-disinfection of liquids may be also related with photoreactivation of microorganisms [5; 14].

In this section, we deal with our new data concerning the dependence of photoreactivation of bacterial and yeast cells both on the energy of sparsely ionizing radiation and the volume of the irradiated cell suspension. Bacteria *E. coli* (strain B_{s-1}), yeast *S. cerevisiae* (strain T4) and *P. pinus* (strain MN4-D) were used for experiments. The culture conditions, irradiation and dosimetry of sparsely and densely ionizing radiation and the illumination with photoreactivating light have been described elsewhere [40; 41]. The X-ray sources were machines RUM-7 and RUM-17 operated at 40 kV and 150–200 kV. Pulsed X-ray beam was produced by electron accelerator (25 MeV). Gamma irradiation was performed using a ^{60}Co source (Gammacell 220, Atomic Energy of Canada, Ltd.) and a ^{137}Cs source ('Kolos'). Ferrous sulfate dosimetry, ionization chambers, and LiF dosimeters were used for measuring dose rates. For photoreactivation, a part of irradiated suspension was illuminated for 20 minutes by a visible light. All other manipulations were performed under dark or red ambient light. Colony-forming ability was assessed by plating of irradiated cells on nutrient agar and counting macrocolonies after 24 hours incubation of bacterial cells at 37 °C and 3 days of yeast cell growth at 30 °C. The magnitude of the photoreactivation was evaluated by the dose-modifying factor (DMF) which was defined as the ratios of radiation doses producing the same effect in survival curves with and without photoreactivation. The data obtained for *E. coli* are presented in Figure 3.11. As can see, the photoreactivation phenomenon takes place after exposure to sparsely ionizing radiation (X-rays of different energy and γ-rays) beginning from a certain energy. The DMF is increased from 1.00 for low voltage X-rays up to 1.6 for high-energy X-rays. A statistically significant effect was observed for ^{137}Cs-irradiation (670 keV).

The cells of *E. coli* K-12 (*uvr*A*rec*A) and *E. coli* B_{s-1} (*uvr*A*lexr*A) for which the photoreractivation phenomenon was demonstrated are hypersensitive to UV-irradiation. It is of interest that biochemical investigations have shown that 100 Gy of ^{137}Cs γ-radiation is equivalent to 0.07 J/m^2 of 254 nm-radiations in the production of pyrimidine dimmers that can be photoreactivated [66]. Since the photoreactivation effect of visible light was observed after exposure to low-LET ionizing radiations in strains, which are defective in some major dark repair systems, one may suppose that photoreactivable damage is produced in all cells irradiated but it is not manifested for cells capable of recovery from this damage due to thymine dimmer excision and postreplication repair.

A simple mathematical model was suggested [47] to estimate the relative contribution of excitations and ionizations in lethal effect of ionizing radiation. The proposed model provides an explanation for the difficulties encountered in observing photoreactivation after ionizing radiation in eukaryotic cells. In particular, yeast cells are characterized by a considerably smaller amount of photoreactivation after UV light exposure (DMF = 2.5) in comparison with bacterial cells (DMF = 5–7). The cited model predicts that for equal contribution of excitations to the lethal effect of yeast and bacterial cells the extent of photoreactivation after ionizing radiation will also be much smaller for eukaryotic than for prokaryotic cells.

If the photoreactivated damages are mainly due to Čerenkov light, the increase in both volume of irradiated suspension and energy of ionizing radiation would favor the greater yield of such damages since the yield of Čerenkov light is known to be increased with increasing parameters mentioned [19]. This, indeed, was found to be the case for bacterial cells [35; 41]. The oxygen effect is known to be exerted predominantly on the damages inflicted by ionization events rather than excitation ones, therefore hypoxic conditions allows to increase the total dose of ionizing radiation and results in the relatively larger contribution of Čerenkov light to the lethal effect.

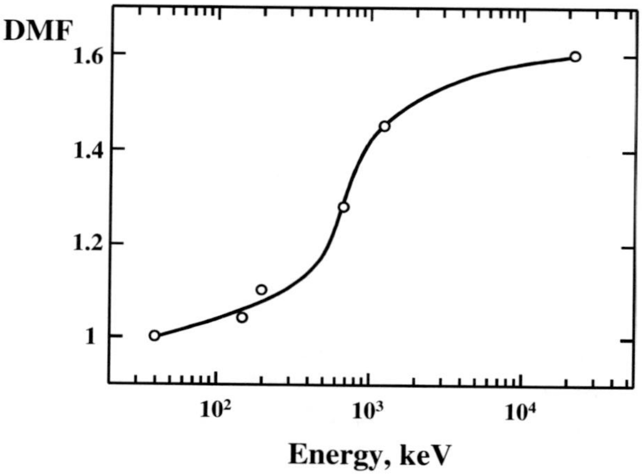

Figure 3.11. The dependence of the magnitude of photoreactivation (DMF) of *E. coli* B_{s-1} on the energy of low-LET ionizing radiation.

Hence, to observe photoreactivation after ionizing radiation exposure in eukaryotic cells, one needs optimum conditions for the induction of photoreactivable damage: a large volume of irradiated suspensions, a high energy of sparsely ionizing radiation and hypoxia. A lot of strains of wild-type and various rad-mutants of yeast cells were tested in these experiments. However, an appreciable photoreactivation was observed only for two diploid yeast strains which were the most UV-sensitive among the other investigated. One of these was a double mutant (*rad2rad54/rad2rad54*) of *S. cerevisiae*, strain T4, defective in both excision and double strand break repair, and a wild-type strain (MN4-D) of *P. pinus*.

To verify the dependence of photoreactivation on the volume of exposed suspension, a 10 and a 1,000-ml yeast cell suspensions were irradiated. To check the dependence of the effect on energy of ionizing radiation, the cells were exposed to 25 MeV photons from pulsed electron accelerator (dose rate 50 Gy/min) and a ^{60}Co γ-ray (dose rate 10 Gy/min). To test the dependence of photoreactivation on the degree of hypoxia, we have used two different concentration of yeast suspension (10^6 and 10^7 cells/ml). It is known that at high cell concentrations the oxygen consumption by maintenance energy metabolism of cell leads to anoxia under certain conditions. It was shown [51] that irradiation (without stirring) of 3 ml cell suspension in small cylindrical plastic dishes (3.2 cm in diameter) results in hypoxic or anoxic conditions for the cell concentration about of $3\cdot 10^7$ and $6\cdot 10^7$ cells/ml, respectively. In our experiments (without stirring) the area of contact of yeast suspension with air was smaller (2 and 5 cm^2) and the volume of irradiated suspension was larger (10 and 1,000 ml). It causes a partial hypoxia at the cell concentration about of 10^7 cells/ml. Concentrations in excess of 10^7 cells/ml result in increasing of optic density and therefore in decreasing of the value of photoreactivation [36].

The magnitude of photoreactivation (DMF) for yeast cells are listed in Table 3.1 for yeast *P. pinus* (strain MN4-D) and *S. cerevisiae* (strain T4) exposed to ionizing radiation at various conditions. The errors involved in the DMF-values have been calculated from inter-experimental variations. The results revealed the highest photoreactivation was observed for both strains at the following optimum conditions: 25 MeV X-rays, 1,000 ml suspension and cell concentration of 10^7 cells/ml.

Table 3.1. The dose modifying factor (DMF) after photoreactivation of yeast cells exposed to ionizing radiations at various experimental conditions

Strain	Irradiation	Experimental conditions			
		10^6 cells/ml		10^7 cells/ml	
		10 ml	1,000 ml	10 ml	1,000 ml
Saacharomyces cerevisiae, T4	^{60}Co γ-rays	1.00 ± 0.03	1.04 ± 0.03	1.01 ± 0.02	1.07 ± 0.04
	25 MeV X-rays	1.02 ± 0.04	1.08 ± 0.03	1.05 ± 0.03	1.15 ± 0.03
Pichia pinus MN4-D	^{60}Co γ-rays	1.01 ± 0.03	1.04 ± 0.03	0.99 ± 0.03	1.10 ± 0.05
	25 MeV X-rays	1.02 ± 0.02	1.12 ± 0.04	1.07 ± 0.04	1.31 ± 0.04

The presented results demonstrates that postirradiation illumination by visible light of yeast cells of certain genotypes exposed to ionizing radiation leads to an increase of the cell survival; the increment over the controls is not very large, but consistent throughout replica experiments, and highly significant for the optimum condition producing the maximum yield of Čerenkov light. From the foregoing, it would appear that the effect of photoreactivation of yeast cells exposed to ionizing radiation may depend on the energy of sparsely ionizing radiation as well as on the volume of irradiated suspensions and hypoxic condition during irradiation. Hence, these data support the viewpoint that Čerenkov emission is mainly responsible for the induction of photoreactivable damage in both bacterial and yeast cells irradiated with high energy sparsely ionizing radiation. The hypothesis that Čerenkov light produces the photoreactivable damage in cells irradiated with ionizing radiation has been advanced by several authors [32; 35; 36; 54]. Some other observations are in favor of it. If the hypothesis was correct, then no photoreactivable damage should be produced by densely ionizing slow particles. Precisely this has been observed experimentally [41]. Another confirmation of the hypothesis comes from the fact that the survival fraction of ^{60}Co γ-irradiated *E. coli* B_{s-1} decreases whereas the photoreactivation increases with increasing volume of irradiated suspension in the range from 0.5–500 ml [35]. Our calculations [47] have shown that the excitation contribution in these experiments increased from 12% up to 50% with increasing suspension volume. Further direct confirmation comes from experiments measuring the effect of photoreactivation in bacterial cells exposed to ^{137}Cs γ-rays when using glass or quartz irradiation vessels (0.8 ml) with or without 800 ml of distilled water in the surrounding vessel [36]. The surrounding water results in an increased inactivation and also in an increase in the amount of photoreactivation, however, only for quartz and not for glass tubes. It was inferred that these effects were due to UV radiation produced by Čerenkov radiation in the surrounding water [36]. For these conditions, the highest excitation contribution in lethal effect of ionizing radiation was found to be of 56% [47].

Michael et al. [32] have also shown that a large part of the photoreactivation that can be achieved after ionizing radiation involves repair of damage induced by UV Čerenkov light. This was demonstrated by irradiation of suspensions either with electrons and their concomitant Čerenkov light, or with a similar flux of net Čerenkov light that was produced by stopping the electrons in high purity fused silica placed immediately in front of the suspension; in the latter case the dose-modifying factor of photoreactivable light was about 5, as was estimated from their published graphical data [32]. It worth noting that this value obtained for net Čerenkov light are within the above mentioned range (DMF = 5–7) observed for the dose-modifying effect of photoreactivatable light after UV-exposure alone.

It should be noted that our experiments with photoreactivation after ionizing radiation described above were performed under conditions where photoreactivation is not always realized at optimal conditions: the intensity of the visible light used for photoreactivation, duration of exposure, the temperature at which the photoreactivation took place. These factors may lead to an underestimation of photoreactivation after ionizing radiation [21; 22]. The data presented in Figure 3.12 have been obtained recently. Irradiation with ^{60}Co γ-rays was occurred at dose rate 27.7 Gy/min. Pulsed X-ray beam formed by 18 MeV electron accelerator (SL-20, Phillips) was utilized at an average dose rate 2.9 Gy/min. For dosimetry measurements, calibrated ionization chambers (Siemens and Keithleg) have been used. For photoreactivation, bacterial cells irradiated in the dark condition with ionizing radiation were illuminated with visible light (5 W/m^2) at 37 °C for 30 min.

It is evident that the value of photoreactivation of bacterial cells irradiated with ^{60}Co γ-rays at 500 ml is larger (DMF = 1.9) compared to the irradiation at 0.5 ml (DMF = 1.3). A comparison of curves 1 and 2 shows that the optimization of photoreactivation resulted in an increase in the effect observed. A similar series of experiments was carried out with 18 MeV X-rays. Comparing curves 2 and 3 one can draw a conclusion that the magnitude of cell photoreactivation is significantly larger after irradiation with high-energy of ionizing radiation for all volumes of cell suspension. The dependence of the magnitude of photoreactivation on the volume of irradiated cell suspension and energy of ionizing radiation can be explained by the fact that electron track is increased with the increase in both the energy of ionizing radiation and the irradiated volume of the suspension and, consequently, UV light flux, which is a part of Čerenkov radiation accompanying the action of ionizing radiation, is also increased. These data are in agreement with the idea that the photoreactivable damage induced by ionizing radiation might be UV-like damage produced by Čerenkov light emitted by charged particles when their velocity exceeds the phase velocity of light in the medium.

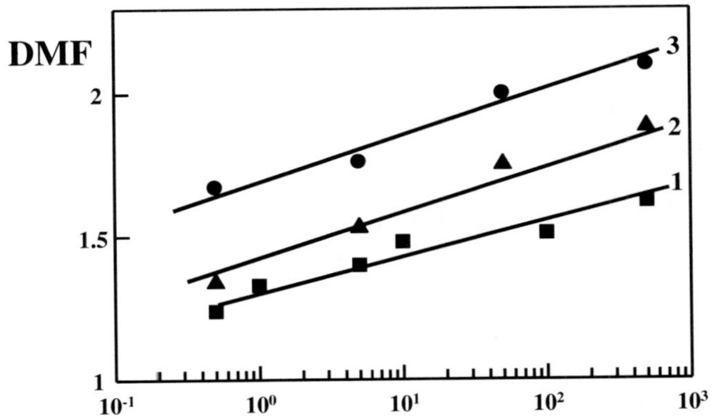

Figure 3.12. The dependence of the magnitude of photoreactivation (DMF) of *E. coli* B_{s-1} on the volume of irradiated cell suspension and energy of low-LET ionizing radiation. Curve 1, recalculated data after cell irradiation with ^{60}Co γ-rays [35]; curve 2, results of our new experiments with ^{60}Co γ-rays performed at the optimal conditions of photoreactivation; curve 3, cells were exposed to X-ray from 18 MeV electron accelerator.

3.4. Parameters of Bacterial Photoreactivation after Exposure to UV Light and Ionizing Radiation

The extent and the rate of bacterial and yeast cell photoreactivation were shown to be modified by a number of chemical and physical factors [35; 48; 55; 56]. After exposure to ionizing radiation, the extent of photoreactivation was found to increase with the energy of damage-inducing ionizing radiation in both the bacterial [36; 41; 56;] and yeast cells [48]. But in none of the cited papers the probability of photoreactivation and the irreversible portion of radiation damage have been evaluated quantitatively. The decrease in both the extent and the rate of recovery may proceed through the impairment of the process of photoreactivation itself, i.e., through the decrease in the probability of recovery, or through the change in the repairable and irreversible portions of the damage produced. In the latter case, the probability of recovery may either decrease or stay unchanged. However, the data distinguishing these possibilities are lacking in the literature. Here we will compare quantitatively the parameters of photoreactivation of a UV hypersensitive strain of *E. coli* B_{s-1} irradiated with UV light and ionizing radiations of different energies. On this basis one may elucidate whether the change in the extent and the rate of recovery is proceeding through the impairment of the recovery process itself or through the alteration of the repairable and irreversible portions of the damage inflicted. We will also evaluate the influence of the different physical and chemical factors, which modify the cell radiosensitivity, on the parameters of photoreactivation kinetics of the bacterial cells exposed to ionizing radiation.

Figure 3.13A shows the survival curves for the bacterial cells irradiated with graded doses of UV light, pulsed X-ray beam, γ-ray of ^{60}Co and ^{137}Cs. The shape of the survival curve was purely exponential after UV light irradiation and there was a little deviation from the exponential shape after irradiation with ionizing radiations. Figure 3.13B exhibits the photoreactivation kinetics for the bacterial cells exposed to the same kinds of radiation. The recovery kinetics were studied at approximately equal initial levels of survival. The number of viable cells increased as a function of time, reached a plateau after about 15 min. Here again we used an optimal conditions for photoreactivation mentioned above. The present data suggest that both the rate and the extent of photorecovery decreased in the following order of the inactivating agents: UV light, pulsed X-ray beam, γ-ray of ^{60}Co and γ-ray of ^{137}Cs. The rate of photoreactivation can be judged by the tangent of angle (the slope ratio) of curves describing photorecovery kinetics. It is also apparent that this rate gradually decreased with increasing duration of photorecovery. But the question whether or not this decrease is due to the increase in the portion of the irreversible damage or to the impairment of photorecovery itself, i.e., to the change in the probability of recovery or to both these effects is still unclear.

Using the results presented in Figure 3.13, we estimated (Figure 3.14A) the decrease in the relative part of the unrecovered radiation damage, $K(t)$ (Eq. 1.11), as a function of the recovery time for the cells exposed to UV light, pulsed X-ray beam, γ-ray of ^{60}Co and ^{137}Cs. These data demonstrate that $K(t)$ decreased, reaching a plateau after 15–20 min of photoreactivation. It also appears from these data that the ultimate quantities of $K(t)$, i.e., the values of the irreversible component of radiation damage, $K(plateau)$ (Eq. 2.12), increased in the following order: UV light, pulsed X-ray beam, γ-ray of ^{60}Co and γ-ray of ^{137}Cs (Table 3.2).

The results show the expected trend in terms of the proportion of irreversible damage as a function of photon energy with the highest proportion occurring with the lowest photon energy. This is in agreement with previous results that indicate that the proportion of UV-like damage reduces in such an order [41; 48]. The experimental data presented make it possible to calculate the relative part of the reparable damage $A(t)$, defined by Eq. (1.14). The outcome is shown in Figure 3.14B. It is evident that this function decreases exponentially with the photorecovery time and does not depend on the quality of the radiation. Using Eq. (1.14) and the results shown in Figure 3.14B, we obtained the recovery constant $\beta = 0.37$ min^{-1} for all the four kinds of radiation used here (Table 3.2). This means that about 37% of the residual reparable damage is recovered every 1 min independently of radiation quality.

Table 3.2. Radiobiological parameters of the photoreactivation of the *E. coli* B_{s-1} irradiated with UV light (254 nm), 25 MeV photons, γ-ray ^{60}Co and γ-ray ^{137}Cs

Treatment	Irreversible component $K(plateau)$	Recovery constant β, min^{-1}
UV light (254 nm)	0.04 ± 0.01	0.37 ± 0.04
25 MeV photons	0.63 ± 0.03	0.37 ± 0.04
γ-ray ^{60}Co	0.70 ± 0.03	0.37 ± 0.04
γ-ray ^{137}Cs	0.84 ± 0.03	0.37 ± 0.04

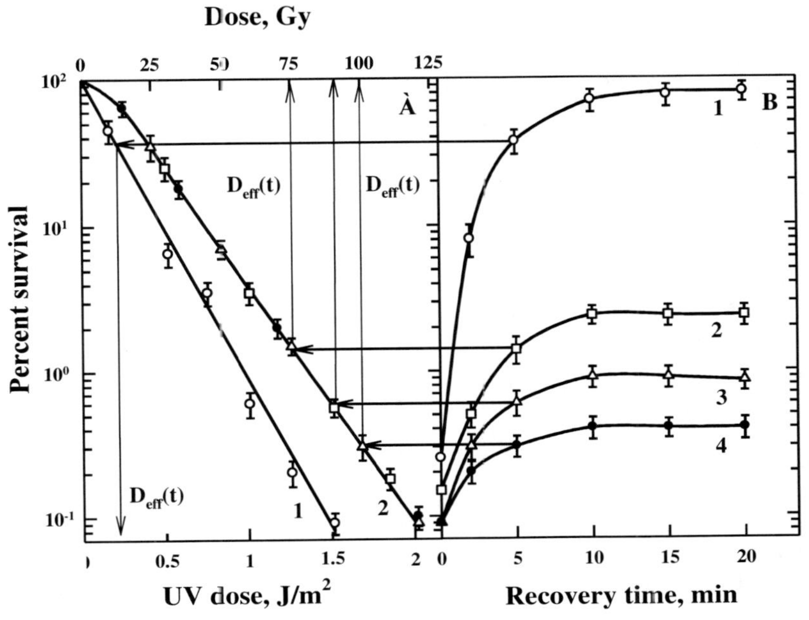

Figure 3.13. Survival curves (A) and photoreactivation kinetics (B) for bacterial cells *E. coli*, strain B_{s-1}. A, cells were exposed to graded doses of UV light (curve 1), γ-ray of ^{137}Cs (curve 2, closed circles), γ-ray of ^{60}Co (curve 2, triangles) and pulsed X-ray beam (curve 2, squares). Cells were plated on nutrient agar immediately after irradiation. B, cells were exposed to UV light (curve 1, circles), pulsed X-ray (curve 2, squares), γ-ray of ^{60}Co (curve 3, triangles) and γ-ray of ^{137}Cs (curve 4, closed circles) and were plated on nutrient agar after various photoreactivation times. The arrows indicate the examples of the effective dose, $D_{eff}(t)$, the determination after 5 min of photoreactivation. Points are the means from at least three independent experiments. Error bars indicate the standard error of the mean. Curves were fitted to the data points by eye.

Thus, the data presented show that the irreversible component and the probability of recovery vary in different ways with the kind of radiation employed in this study. While the probability of photorecovery appears to be identical for the four kinds of radiation tested, the irreversible component of radiation damage was much smaller after UV light exposure as compared with that for ionizing radiation treatment. For ionizing radiation, this component seems to decrease with the increase in photon energy. Hence, it may be concluded that the change in the extent and the rate of photoreactivation may be entirely explained by the formation of irreversible damage rather than by the impairment of recovery process itself. The fact that the probability (not the rate) of photoreactivation is the same for UV light and whatever photon energy was used indicates a commonality of the photoreactivation process both for UV and ionizing radiations.

It would be of interest to analyse treatments that selectively modify the cell sensitivity through the change in the relative yield of damage inflicted by ionization and excitation (UV-like damage). For example, some chemical radioprotectors and radiosensitizers, which are effective for ionizing radiation, are known to exert their effect predominantly on events inflicted by ionizations. On the other hand, illumination with visible light during the post-irradiation period is more effective in modifying UV effects called forth by excitations. It would seem feasible that the increase in the relative yield of excitations due to radioprotective action should result in an increase of the extent of photoreactivation after exposure to ionizing radiation. An inverse situation would be expected after radiosensitizing action of some chemicals. Thus, it would be of interest to evaluate the influence of different physical and chemical factors, which modify the cell radiosensitivity, to the parameters of the photoreactivation kinetics of the bacterial cells exposed to ionizing radiation.

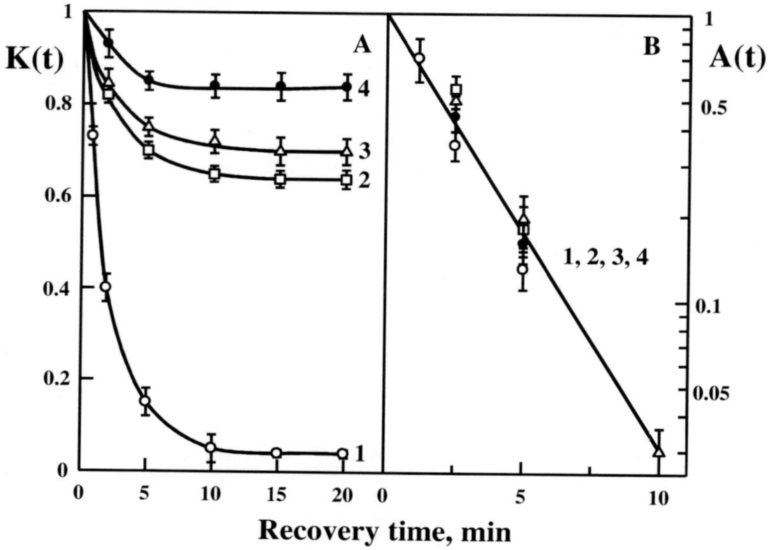

Figure 3.14. Decrease in the portions of the relative part of unrecovered, $K(t)$ (A), and recovered, $A(t)$ (B), radiation damage as a function of photoreactivation time for bacterial cells *E. coli*, strain B_{s-1}. Cells were exposed to UV light (curve 1, circles), pulsed X-ray beam (curve 2, squares), γ-ray of ^{60}Co (curve 3, triangles) and γ-ray of ^{137}Cs (curve 4, closed circles). Cells were plated on nutrient agar after various photoreactivation times. Points are the means from at least three independent experiments. Error bars indicate the standard error of the mean. Curves were fitted to the data points by eye (A) and by the least squares method (B).

Table 3.3. Radiobiological parameters of photoreactivation of *E. coli* B_{s-1} irradiated with 6 MeV photons

Condition of irradiation	Irreversible component K	Recovery constant β, min^{-1}
N_2 + DTT	0.65 ± 0.03	0.20 ± 0.03
O_2 + DTT	0.78 ± 0.03	0.20 ± 0.03
O_2	0.80 ± 0.03	0.11 ± 0.02
N_2	0.82 ± 0.03	0.11 ± 0.02
O_2 + caffeine	0.88 ± 0.04	0.05 ± 0.01
N_2 + caffeine	0.90 ± 0.04	0.05 ± 0.01

Extensive studies of this matter have been published [55; 56]. The authors came to the following three major conclusions: (1) the extent of photoreactivation of ionizing radiation-induced damage was independent of the presence of oxygen during irradiation; (2) the presence of caffeine during the photoreactivation process reduced the rate of the recovery by a factor of about 4. This was comparable with the effect of caffeine, a known inhibitor of photoreactivation enzyme, on the kinetics of the photoreactivation of the UV-induced damage [12]; (3) the presence of the radioprotective agent dithiothreitol (DTT) during irradiation does not suppress the production of the photoreactivatable damage. It is of interest to apply mathematical approach describing the recovery process as a decreasing of the effective dose to their data to evaluate quantitatively the influence of these factors on the irreversible component and the probability of recovery. The output is presented in Figure 3.15. The analysis was based on the work of Redpath and Zabilansky [56].

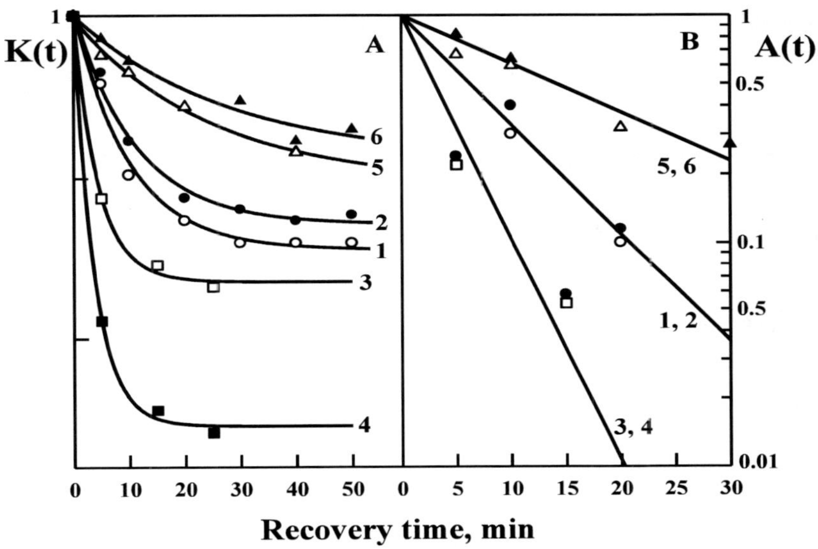

Figure 3.15. Decrease in the portion of the relative part of unrecovered, $K(t)$ (A), and recovered, $A(t)$ (B), radiation damage as a function of photoreactivation time for bacterial cells *E. coli*, strain B_{s-1}. Cells were exposed to 6 MeV photons in the presence (open circles, curve 1) and absence (closed circles, curve 2) of oxygen; in the presence of 10 mM DTT during irradiation under oxic (open squares, curve 3) and anoxic (closed squares, curve 4) conditions; in the presence of 16 mg/ml caffeine during photoreactivation after irradiation under oxic (open triangles, curve 5) and anoxic (closed triangles, curve 6) conditions. Curves were fitted to the data points by eye (A) and by the least squares method (B).

Here, the decrease in the portion of the relative part of the unrecovered, $K(t)$ (Figure 3.15A), and recovered, $A(t)$ (Figure 3.15B), radiation damage as a function of the photoreactivation time for the bacterial cells *E. coli* are depicted. Cells were exposed to 6 MeV photons in the presence and absence of oxygen; in the presence of 10 mM DTT during irradiation under oxic and anoxic conditions; in the presence of 16 mg/ml caffeine during photoreactivation after irradiation under oxic and anoxic conditions. It is of important to note that during these experiments [56] post-irradiation illumination was performed with low intensity light for the following conditions of irradiation: O_2, N_2, O_2 + caffeine, N_2 + caffeine. In contrast, a high intensity of post-irradiation illumination was used for N_2 + DDT and

O_2 + DDT conditions. Thus, the conditions of photorecovery were quite different for the two series of experiments which undoubtedly resulted in the different rates and probabilities of photorecovery. Therefore, it is not possible to derive any conclusions based on the comparison between these two series of experiments. However, comparisons within each of these series are valid. The parameters of the photoreactivation calculated on the basis of the curves presented in Figure 3.15 are listed in Table 3.3.

The following conclusions can be drawn from these data.

(1) Both the extent and the rate of photoreactivation of ionizing radiation-induced damage was dependent of the presence of oxygen during irradiation; it was most noticeably seen for the (O_2 + DTT) and (N_2 + DTT) conditions of irradiation. However, the probability of the photoreactivation of ionizing radiation-induced damage was independent of the presence of oxygen during irradiation for all the versions of the experiment.

(2) The presence of caffeine during the photoreactivation affected both the parameters of photoreactivation. It resulted in the decrease in both the rate and the probability of photoreactivation. The proportion of irreversible damage was higher in the presence than absence of caffeine and this makes sense in that caffeine is known to sensitize cells to the effects of ionizing radiation-induced damage. If ionizing radiation really produces a part of UV-like lesions, this would be consistent with the data that caffeine inhibits photoreactivation of UV-induced damage by competing with the photoreactivation enzyme for its substrate [12].

(3) The presence of the radioprotective agent (DTT) during irradiation does not suppress the production of the photoreactivatable damage. Moreover, the portion of the irreversible damage decreased while photoreactivatable damage increased in the presence of the radioprotective agent DTT during irradiation. The effect was more clearly revealed in the absence of oxygen during irradiation. Both of these facts may be associated with a decrease in the relative yield of lesions induced by ionizations and a corresponding increase in the part of UV-like lesions inflicted by excitations due to the protective action of the chemical agent and anoxia. This is in agreement with the known effects of this thiol acting as a radioprotector through free radical scavenging. This also suggests that the photoreactivatable damage was not induced by free radicals, the main conclusion of the original article [56].

3.5. Estimation of UV Light Dose Concomitant to Irradiation with Ionizing Radiation

In the previous sections of this Chapter our experimental data and the results published by others concerning the phenomenon of photoreactivation after cell exposure to ionizing radiation have been summarized and it was concluded that the UV-like damage are produced by Čerenkov light emitted by charged particles when their velocity exceeds the phase velocity of the light in the medium. Therefore it is of interest to estimate a UV light dose concomitant to irradiation with ionizing radiation. We will describe a simple biological method determining the UV light dose concomitant to ionizing radiation exposure and present the results of calculations of the dependency of this dose of equivalent UV light accompanied to 100 Gy of ionizing radiation on the energy of ionizing radiation and on the volume of the exposed cell suspension.

The mathematical model that will be used here has been partly published [47; 49]. So, it is worth mentioning in short the main points of the model. The survival curves of UV-hypersensitive bacterial and yeast cells exposed to ionizing radiation can be simply presented as

$$S = e^{-kD}, \tag{3.6}$$

where S is the cell survival fraction, D the irradiation dose and k is the overall cell sensitivity, defined as the inverse of the mean lethal dose ($k = 1/D_0$). It is reasonable to assume that excitation and ionization events act stochastically and independently of each other for sparsely ionizing radiation. Then Eq. (3.6) can be rewritten as follows:

$$S = e^{-(k_1+k_2)D}, \tag{3.7}$$

where k_1 and k_2 are the cell sensitivities to ionizations and excitations, respectively. We can normalize the overall radiosensitivity and consider that

$$k = k_1 + k_2 = 1. \tag{3.8}$$

Then k_1 and k_2 represent the relative contribution of ionization and excitation to cell inactivation. It is not difficult to choose some factors which can modify cell radiosensitivity through a selective influence to the consequences produced by ionization and excitation. For example, the oxygen effect is known to be predominantly exerted on cell inactivation produced by ionization. On the contrary, illumination by visible light during a postirradiation period is more effective in modifying the effects caused by excitation.

Let us suppose that the initial survival curve (Eq. 3.6) was obtained under oxic and dark conditions. Then the survival under hypoxia will be given by

$$S = e^{\frac{-kD}{F_1}} = e^{-\left(\frac{k_1}{f_1}+k_2\right)D}, \tag{3.9}$$

where F_1 is the DMF of the experimental survival curve obtained under hypoxic and dark conditions and f_1 is the DMF for the ionization component only. If the initial survival curve (Eq. 3.6) was obtained under an oxic condition during irradiation and the photoreactivating light was used after irradiation, the corresponding survival curve may be described by

$$S = e^{\frac{-kD}{F_2}} = e^{-\left(k_1+\frac{k_2}{f_2}\right)D}, \tag{3.10}$$

where F_2 is the DMF of the experimental survival curve obtained under oxic and light conditions and f_2 is the DMF for the excitation component only.

Taking into account Eqs. (3.8)–(3.10), we attain the following set of equations:

$$\begin{cases} \dfrac{k_1}{f_1} + k_2 = \dfrac{1}{F_1} \\ k_1 + k_2 = 1 \\ k_1 + \dfrac{k_2}{f_2} = \dfrac{1}{F_2} \end{cases}, \tag{3.11}$$

where k_1, k_2, f_1, and f_2 are unknown values, while F_1 and F_2 are known from experiments. The DMF for the excitation component (f_2 = 4.5) was estimated from the experiments on photoreactivation of cells exposed to UV light (254 nm) only. Then the solution for the set in Eq. (3.11) yields

$$k_1 = \frac{f_2 - F_2}{F_2(f_2 - 1)} \tag{3.12}$$

$$k_2 = \left(1 - \frac{1}{F_2}\right) \cdot \left(1 + \frac{1}{f_2 - 1}\right). \tag{3.13}$$

Figure 3.16 shows a typical example of the basic experimental data observed in this study. Figure 3.16A exhibits the survival curves for the bacterial cells irradiated with graded doses of pulsed X-rays (25 MeV) and plated on nutrient media without and with post irradiation photoreactivation. The results revealed that postirradiation illumination by visible light of bacterial cells exposed to ionizing radiation leads to an increase of cell survival; the increment over the control cells is consistent throughout replica experiments and the dose modifying factor F_2 = 1.60.

Using Eqs. (3.12) and (3.13), we obtained that values of k_1 = 0.57 and k_2 = 0.43. This means that the contribution of the ionizations to the total lethal effect induced by pulsed X-rays is 57%, while the contribution of the excitations is 43%. Subdividing the initial survival curve after ionizing radiation action (Figure 3.16B, curve 1) into two components caused by ionizations (57%, Figure 3.16B, curve 2) and excitations (43%, Figure 3.16B, curve 3), i.e., mainly by UV light, we found that for the dose of 100 Gy the value of the cell survival for the UV component was 7%. The survival for the action of 254 nm UV light only [21] corresponds to the UV dose of 0.56 J/m^2.

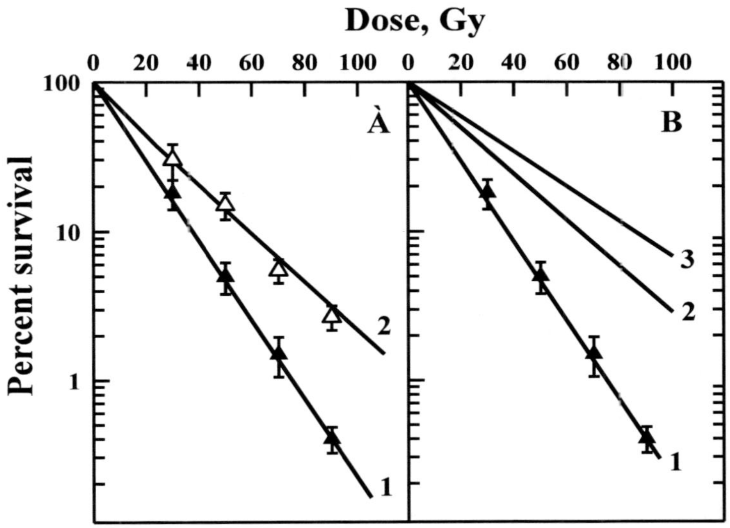

Figure 3.16. Survival of *E. coli* B_{s-1} cells after pulsed X-irradiation (25 MeV): A, without (curve 1) and with (curve 2) photoreactivation and B, cell inactivation induced by both ionizations and UV light damage without photoreactivation (curve 1), only by ionizations (curve 2) and excitations (curve 3) after pulsed X-irradiation (25 MeV).

Thus, the irradiation of bacterial cells (strain B_{s-1}) with pulsed X-rays (100 Gy) was simultaneously accompanied by cell exposure to the UV dose of 0.56 J/m^2. It is important to note that for the sake of simplification we compared the whole spectra of the Čerenkov emission with the most effective equivalent fluence for the lethal and genetic effects of 254 nm wavelength of UV light.

Extensive experimental data regarding the dependence of the extent of photoreactivation of bacterial cells of *E. coli* (strain B_{S-1}) on energy of ionizing radiation have been published [41]. Using these data as well as the model presented here and the above described example of the calculations of the UV dose based on the relative contributions of ionizations and excitations to the lethal effect of ionizing radiation, we calculated the dependence of the relative contribution of excitation to the lethal effect (k_2) and dose of the equivalent UV light accompanied to 100 Gy of ionizing radiation on energy of ionizing radiation. The results are presented in Figure 3.17. The excitation contribution to the total lethal effect of ionizing radiation seems to increase from 3% for 0.04 MeV X-rays to 43% for 25 MeV pulsed X-rays while the dose of the equivalent UV light accompanied by 100 Gy of ionizing radiation increases from 0.04 to 0.56 J/m^2.

As was shown above, the survival fraction of the γ-irradiated bacterial cells of *E. coli* (strain B_{S-1}) decreased, whereas the value of the photoreactivation increased with an increase in volume of the irradiated suspension. This effect can be explained by the corresponding increase in the UV light dose and the excitation contribution to the total lethal effect of ionizing radiation. Therefore, it would be of interest to use the above described approach to estimate the dependence of UV light dose concomitant to exposure with ionizing radiation on volume of irradiated cell suspension.

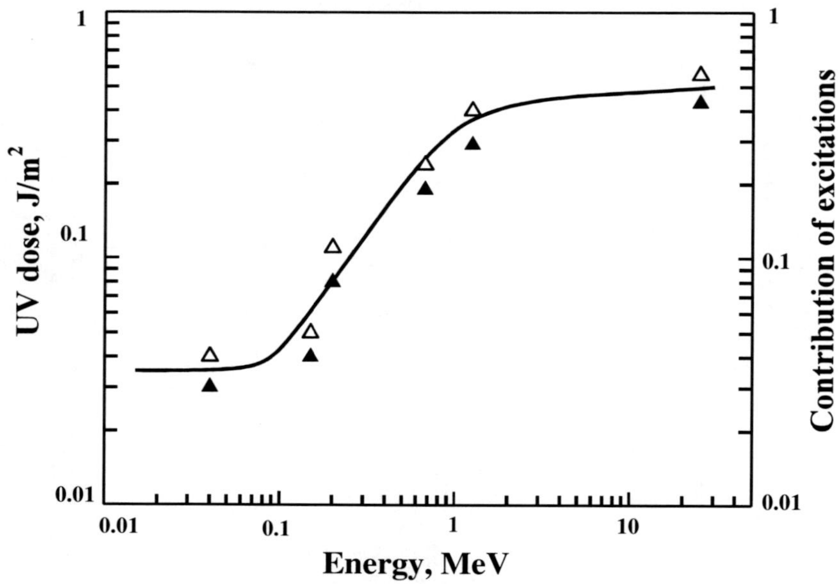

Figure 3.17. UV light dose (open triangles) and the relative contribution of UV damage to the lethal effect of ionizing radiation (closed triangles) as a function of the energy of sparsely ionizing radiation.

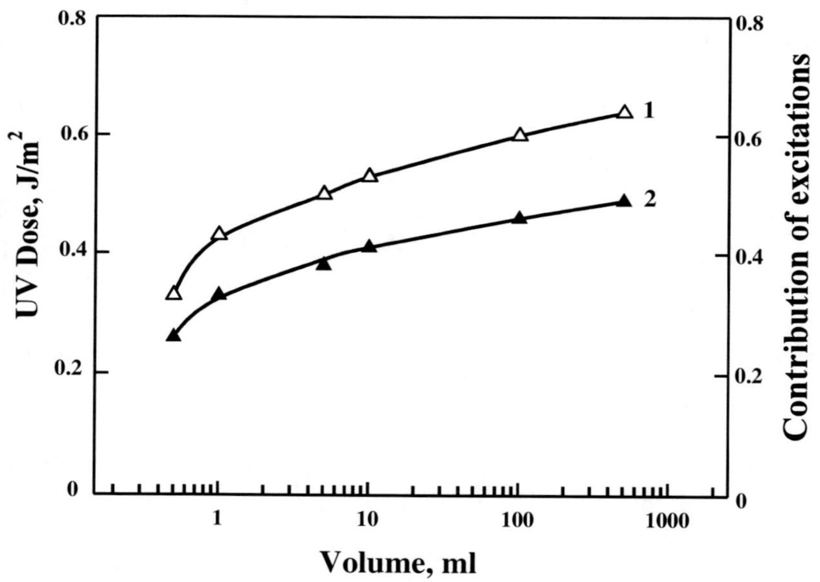

Figure 3.18. UV light dose (open triangles) and the relative contribution of UV damage to the lethal effect of ionizing radiation (closed triangles) as a function of the volume of the exposed cell suspension.

Using the results published before [35], we calculated the dependency of the volume of irradiated suspension of *E. coli* for the relative contribution of excitation in the lethal effect (k_2) and the dose of the equivalent UV light accompanied to 100 Gy of ionizing radiation. The results are depicted in Figure 3.18.

One can see that the excitation contribution to the total lethal effect of ionizing radiation increases from 26% for 0.5 ml to 49% for 500 ml of irradiated suspensions while the dose of the equivalent UV light accompanied by 100 Gy of ionizing radiation increases from 0.33 to 0.64 J/m^2.

CONCLUSION

The suggested synergistic enhancement ratio was shown to have a certain advantage against traditional thermal enhancement ratio used in radiation therapy for thermoradiation treatment. Besides indication the kind of interaction, the remarkable feature of this parameter is that the thermal enhancement ratio increases permanently with increasing exposure temperature, while the synergistic enhancement ratio increases at first and then reaches the highest value, which is followed by a decrease. This pattern was observed independently of the agent (ionizing radiation, UV light, ultrasound, and zinc sulfate) and cell line (yeast or bacterial cells). Noteworthy is the fact that such a dependence of the synergistic effect on the temperature under which the exposure occurred was also obtained for various other biological objects subjected to the combined action of ionizing radiation and heat. This implies that, irrespective of the biological system investigated, the synergistic interaction between hyperthermia and other inactivating agents is observed only within a certain temperature range. This data indicate the possibility to achieve an optimal regimen of thermoradiation action. Hence, one can conclude that for a given intensity of physical factors or concentration of chemical agents there would be a specific temperature that maximizes the synergistic interaction. The problem of synergy is worth of a special attention and will be considered in the following Chapters in detail.

As an example of independent interaction, the results of the successive exposures of yeast and bacterial cells of different strains to coherent electromagnetic (CO_2 laser of high intensity) and ionizing radiations have been presented. Antagonistic kind of interaction is of great interest from fundamental and practical view of radiobiology, radiation therapy, radioecology and medicine. Two important examples of antagonistic interaction have been described in this Chapter.

First, an unusual effect of triggered cell response after combined application of α-particles and UV light to yeast cells of various ploidy is given. Some studies have indicated an antagonistic interaction of ultraviolet light and ionizing radiation [8]. Similar effect was demonstrated for protozoa and yeast cells [1]. In particular, the survival of haploid yeast cells irradiated in exponential phase of growth with α-particles increased after the action of UV light. It is curiously that the change in survival was dependent on the dose of UV light when the final curve after combined actions characterized by unusual shapes [1]: cell survival changed in oscillating way with one or several maxima and minima before it began to decrease monotonically. Because of this unexpected cell response, we continued these investigations with yeast cells of various genotypes exposed at the stationary phase of growth. It was shown that after combined treatment, the protective effect of α-particles on UV-irradiated yeast cells was independent of the sequence in which the two factors were applied and noted both in haploid and diploid yeast cells of the wild-type. The oscillatory dose-response curves were observed at doses of UV light, which do not lead to a significant decrease in cell viability. Almost additive effect of high doses of UV light and ionizing radiation was expressed for haploid strain incapable of the DNA double strand break recovery. For diploid yeast cells possessing such a capability [9; 10], the synergistic interaction of damages inflicted by α-particles and high doses of UV light was observed. The small doses of UV light does not lead to reactivation of the damage caused by α-particles in haploid and diploid yeast cells with mutation in RAD51 gene. In this case the interaction of densely ionizing radiation with UV light was only additive. It can be considered as an indication of the participation of recovery processes in the effect discussed. It is known that the induction of gene conversion and reciprocal recombination are greatly reduced in *rad*51 cells than that of the wild-type cells [58]. One can assume on this basis that the triggered recovery can be related to these molecular processes.

Moreover, these data indicate a possible use of combined application of UV light and ionizing radiation to reduce radiation damage in normal tissue during radiation therapy. The findings published by Calkins and Greenlaw [2] demonstrate such an opportunity. In their work, activated repair of skin was initiated when UV light and β-particles have been applied separately or in combination on human and rabbit skin. The response was shown to be complex and highly dose-dependent. UV reduced radiation injury at certain levels of β-dose while increased injury at other levels. The observations were in accord with the hypothesis that mammalian cells have a "triggered" or "induced" repair system. The activation of the triggered system required a threshold level of radiation dose. The authors [2] concluded that radiation-induced repair system is undoubtedly significant for therapeutic radiology.

The second example of antagonistic interaction consists in the analysis of phenomenon of photoreactivation observed after combined action of ionizing radiation and visible light. Some new experimental data of this phenomenon for bacterial and yeast cells have been presented: the dependence of photereactivation on the energy of low-LET radiation and the volume of irradiated suspension.. These data are in agreement with early suggested hypothesis that the photoreactivable damage induced by ionizing radiation might be UV-like damage produced by Čerenkov light emitted by charged particles when their velocity exceeds the phase velocity of light in the medium. It is concluded that irradiation of both prokaryotic and eukaryotic cells with sparsely ionizing radiation of high-energy results in production not only single- and double-strand breaks, base damage and cross-links between DNA molecules, but also UV-like damage. pyrimidine dimers similar to those induced by UV light, which may be susceptible of photoreactivation. These results show that excitations in addition to ionizations may be considered as important physical interaction processes which can give rise to a variety of physicochemical, biochemical and biological reactions which finally produce observable effects in biological objects.

An additional argument supporting these data was obtained in the quantitative evaluation of the parameters of photoreactivation after exposure to ultraviolet light and ionizing radiation of different energy. A mathematical model describing the process of photoreactivation in terms of a decreasing effective dose was applied to the experimental data to evaluate quantitatively the probability of photoreactivation per unit time and the irreversible component. Both the rate and extent of photoreactivation decreased in the following order of inactivating agents: UV light, pulsed X-ray beam, γ-ray of ^{60}Co and γ-ray of ^{137}Cs. The irreversible component increased with the same order of radiations whereas the probability of photorectivation per unit time was independent of the kind of radiation. Hence, the extent and the rate of photoreractivation following ionizing radiation decreased in comparison with UV light only due to the less number of damage susceptible of photoreactivation. On this basis it is concluded that the nature of photoreactivable damage produced by UV light and ionizing radiation may be identical.

A simple mathematical model for biological estimation of UV light dose concomitant to ionizing radiation was developed. This approach was applied to determine the dependency of equivalent UV light dose accompanied to 100 Gy of ionizing radiation on the energy of sparsely ionizing radiation and on the volume of the exposed cell suspension. It was shown that the relative excitation contribution to the total lethal effect and the value of UV dose for the same dose of ionizing radiation is increased with an increase both in energy of ionizing radiation and volume of the irradiated cell suspension. Both these observations are in agreement with the supposition that the Čerenkov emission is responsible for the production of UV light damage and the phenomenon of photoreacrivation observed after ionizing radiation. In the model presented here we suggested additive summation of damage produced by ionizations and excitations of molecules. However, it is not improbable that they can interact in a synergistic manner. Some data can be considered as an indication of such an opportunity [18; 30].

So, the action of high energy sparsely ionizing radiation is always accompanied by the simultaneous action of UV light. Čerenkov emission can be considered as a means of UV light delivery to deeply located tissues. The estimated UV doses will undoubtedly take place during irradiation of wild-type cells including eukaryotic cells, plants, animals, men *etc*. However, the induced UV damage may be effectively repaired and not displayed. It is not excluded a probable participation of UV component of ionizing radiation in the mechanism of the manifestation of different low-dose responses: adaptive reaction, radiation hormesis, bystander effect that can be observed after extremely low dose of ionizing radiation when charged particles traversed through only a small part of cell population but the whole population is capable of changing its response to the subsequent action of a higher dose of radiation [7; 15; 20; 27-29; 31; 37; 63;]. Since the number of quantum of Čerenkov emission for the radiation dose of 100 Gy (^{60}Co) is about 10^{13} per 1 cm^3 of irradiated water [35], there would be a tremendous number of these quantum even for a low dose of radiation – for example 10^9 quanta per 1 cm^3 for the dose of 1 cGy. As many quantum of UV light are more uniformly distributed in an irradiated volume than ionizations, it must not be ruled out that the low dose of ionizing radiation including the natural background can trigger reparation processes in cells analogous to the SOS repair by the accompanied UV light, thereby increasing the radioresistance of biological objects to a subsequent irradiation.

The relative excitation contribution to the total lethal effect and the value of the UV dose for the same dose of ionizing radiation (100 Gy) were shown to be considerably increased with an increase in energy of ionizing radiation. Second, these parameters were increased with an increase in volume of the irradiated cell suspension. Both these observations are in agreement with the supposition that the Čerenkov emission is responsible for the production of UV light damage and the phenomenon of photoreacrivation observed after ionizing exposure of bacterial and yeast cells hypersensitive to UV light. This hypothesis has been put forward and discussed in various papers [32; 35; 36; 48; 54]. The plateau of the UV dose with an increase in the energy of ionizing radiation is determined by physical properties of the Čerenkov emission [19]. The plateau of the UV dose with an increase in volume of the exposed suspension is determined by the highest penetration of Čerenkov emission photon, which is about 12 cm in a water medium [43].

Thus, the main conclusions of this Chapter are as follows.

1. To evaluate quantitatively the synergistic interaction of two agents, we suggested the synergistic enhancement ratio defined as the ratio of the exposure time expected for an independent effect of both modalities to that observed from the experimental data obtained for the same effect after simultaneous application of interacting agents. The remarkable feature of this parameter is that it reaches the highest value for some optimal condition of treatment. This pattern was observed independently of the biological object and agents used.
2. The unusual triggered dose-effect relationship was demonstrated for haploid and diploid yeast cells exposed to UV light and α-particles: cell survival changed in oscillating way with one or several maxima and minima before it began to decrease monotonically. This effect was not observed in haploid and diploid yeast cells with mutation in RAD51 gene. This data can be considered as a direct indication to the participation of recovery processes in the manifestation of the phenomenon.
3. Some new experimental data for photoreactivation after combined action of ionizing radiation and visible light of bacterial and yeast cells have been presented: the dependence of photereactivation on the energy of low-LET radiation and the volume of irradiated suspension. These data are in line with early suggested hypothesis that the photoreactivable damage induced by ionizing radiation might be UV-like damage produced by Čerenkov light emitted by charged particles when their velocity exceeds the phase velocity of light in the medium.
4. Both the rate and extent of photoreactivation decreased in the following order of inactivating agents: UV light, 25 Mev X-ray, γ-ray of ^{60}Co and γ-ray of ^{137}Cs. The irreversible component of radiation damage increased with the same order of radiations whereas the probability of photorectivation per unit time was independent of the kind of radiation. It is concluded that the nature of photoreactivable damage produced by UV light and ionizing radiation may be similar to each other.
5. A simple mathematical model for determination of UV light dose concomitant to ionizing radiation of high energy is developed. It was shown that the value of UV dose for the same dose of ionizing radiation is increased with an increase both in the energy of ionizing radiation and volume of the irradiated cell suspension. Both these observations are support the idea that Čerenkov emission is responsible for the production of UV like damage and the phenomenon of photoreactivation observed after ionizing radiation exposure.
6. On the basis of the results obtained, two important issues are discussed: a possible synergistic interaction of the damage produced by ionizations and excitations and a probable participation of UV component of ionizing radiation in the mechanism of different low-dose responses: adaptive reaction, radiation hormesis, bystander effect.

REFERENCES

[1] Calkins, J; Ballard, B; Gillespie, M. Ultraviolet light-induced reactivation of alpha-irradiatedyeast cells. *Radiat. Res*., 1978; 73(3), 440-451.

[2] Calkins, J; Greenlaw, RH. Activated repair of skin: a damage-induced radiation repair system. *Radiology*, 1971; 100(2), 389-395.

[3] Carpentier, Y; Demange, L; Loirette, M et al. Chronology of combined chemotherapy (5FU) and radiotherapy. I. In vitro study. *Anticancer Res*., 1993; 13, 2177-2180.

[4] Chadwick, KH; Leenhouts, HP. *The Molecular Theory of Radiation Biology*. Berlin, Heidelberg, New York: Springer-Verlag, 1981.

[5] Chatzisymeon, E; Droumpali, A; Mantzavinos, D et al. Disinfection of water and wastewater by UV-A and UV-C irradiation: application of real-time PCR method. *Photochem. Photobiol. Sci*., 2011; 10(3), 389-395.

[6] Dewey, WC. *In vitro* systems: standardization of endpoints. *Int. J. Radiat. Oncol. Biol. Phys*., 1979; 5, 1165-1174.

[7] Dubrova, YE; Plumb, M; Brown, J et al. Radiation-induced germline instability at minisatellite loci. *Int. J. Radiat. Biol*., 1998; 74(6), 689-696.

[8] Elkind, MM; Sutton, H. Sites of action of lethal irradiation: overlap in sites for x-ray, ultraviolet, photoreactivation, and ultraviolet protection and reactivation in dividing yeast cells. *Radiat. Res*., 1959; 10(3), 296-312.

[9] Frankenberg, D. Repair of DNA double-strand breaks and its effect on RBE. *Adv. Space Res*., 1994; 14, 235-248.

[10] Frankenberg-Schwager, M; Frankenberg, D; Harbich, R. Repair of DNA double-strand breaks as a determinant of RBE of alpha-particles. *Br. J. Cancer*, 1984; 49, Suppl. VI, 169-173.

[11] Gerner EW; Leith JT. Interaction of hyperthermia with radiations of different linear energy transfer. *Int. J. Radiat. Biol*., 1977; 31, 283-288.

[12] Harm, W. Evidence for photoenzymatically repairable, lethal "non-dimer" photoproducts, formed in *E. coli* cells by 365-nm radiation or by sunlight greater than 360 nm. *Mutat. Res*., 1978; 51(3), 301-310.

[13] Haynes, RH. The interpretation of microbial inactivation and recovery phenomena. *Radiat. Res*., 1966; Suppl. 6, 1-29.

[14] Hijnen, WA; Beerendonk, EF; Medema, GJ. Inactivation credit of UV radiation for viruses, bacteria and protozoan (oo)cysts in water: a review. *Water Res*., 2006; 40(1), 3-22.

[15] Huang, L; Kim, PM; Nickoloff, JA et al. Targeted and nontargeted effects of low-dose ionizing radiation on delayed genomic instability in human cells. *Cancer Res*., 2007; 67(3), 1099-1104.

[16] Hug, O; Kellerer, AM. *Stochastik der Strahlenwirkung*. Berlin-Heidelberg-New York: Springer-Verlag, 1966.

[17] ICRU (International Commission on Radiation Units and Measurements). *Quantitative concepts and dosimetry in radiobiology*. Washington, DC: Report 30, 1979.

[18] Jagger, J. *Introduction to Research in Ultraviolet Photobiology*. NJ: Prentice Hall, Englewood Cliffs, 1967.

[19] Jelley, JV. *Čerenkov Radiation and Its Applications*. London: Pergamon Press, 1958.

[20] Joiner, MC. Induced radioresistance: An overview and historical perspective. *Int. J. Radiat. Biol.*, 1994; 65(1), 79-84.

[21] Kim, JK; Komarova, LN; Tkhabisimova, MD et al. Inhibition of recovery from potentially lethal damage by chemicals in Chinese hamster ells is realized through the production of irreversible damage. *Korean J. Environ. Biol.,* 2005; 23(4), 390-397.

[22] Krivova, AA; Petin, VG; Zhurakovskaya, GP. Ionizing radiation-induced formation of ultraviolet damages. *Nuclear Physics and Engineering,* 2012; 3(3), 239-242.

[23] Laskowski, W. Inaktivierungsversuche mit homozy-goten Hefestammen verschidenen Ploidiegrades. 1. Aufban homozygoter Stamme und Dosiseffektkurven fur ionisierende Strahlen, UV and organische Peroxyde. *Z. Naturforschung*, 1960; 15b, 495-506.

[24] Lea, DE. Action of Radiation on Living Cells. Cambridge University Press, 1946.

[25] Lisovsky, MA; Komarov, VP; Petin, VG. A mathematical description of synergistic effect of simultaneous cytotoxical action of chemical agents and hyperthermia. *Cytology,* 1994; 36, 1118-1122.

[26] Loshek, DD; Orr, JS; Solomonidis, E. Interaction of hyperthermia and radiation. *Br. J. Radiol.*, 1977; 50, 893-901.

[27] Luan, YC; Shieh, MC; Chen, ST et al. Re-examining the health effects of radiation and its protection. *Int. J. Low Radiat.*, 2006; 3(1), 27-44.

[28] Luckey, TD. *Radiation Hormesis*. Tokyo: Boca Raton Publisher, CRC Press, 1991.

[29] Luckey, TD. Radiation prevents much cancer. *Int. J. Low Radiat.*, 2007; 4(4), 336-344.

[30] Martignoni, KD; Smith, KC. The synergistic action of ultraviolet and X-radiation on mutants of *Escherichia coli* K-12. *Photochem. Photobiol.*, 1973; 18, 1-8.

[31] Maxwell, CA; Fleisch, MC; Costes, SV; Erickson, AC; Boissière, A; Gupta, R; Ravani, SA; Parvin, B; Barcellos-Hoff, MH. Targeted and nontargeted effects of ionizing radiation that impact genomic instability. *Cancer Res.*, 2008; 68(20), 8304-8311.

[32] Michael, BD; Harrop, HA; Held, KD. Photoreactivation of *Escherichia coli* after exposure to ionizing radiation: the role of U.V. damage by concomitant Čerenkov light. *Int. J. Radiat. Biol.*, 1981; 39(5), 577-583.

[33] Morgan, TL; Redpath, JL; Ward, JF. Further studies on Čerenkov induced photoreactivable damage in *E. coli*. *Radiat. Res.*, 1982; 89, 217-226.

[34] Morgan, TL; Redpath, JL; Ward, JF. Induction of lethal damage in *E. coli* by Čerenkov emission associated with high-energy X-rays: the effect of bromuracil substitution. *Int. J. Radiat. Biol.*, 1984; 45, 217-226.

[35] Morozov, II; Myasnik, MN. The relationship between the phenomenon of photoreactivation in *Escherichia coli* following ionizing radiation and the Čerenkov emission. *Radiat. Res.*, 1980; 82, 336-341.

[36] Moss, SH; Smith, KC. Čerenkov ultraviolet radiation (^{137}Cs γ- rays) and direct excitation (^{137}Cs γ -rays and 50 kVp X-rays) produce photoreactivable damage in *Escherichia coli*. *Int. J. Radiat. Biol.*, 1980; 38, 323-334.

[37] Mothersill, C; Seymour, G. Review radiation induced bystander effects: past history and future directions. *Radiat. Res.*, 2001; 155(6), 759-767.

[38] Munson, RJ. Is the major part of the photoreactivable lethal damage induced in Escherichia coli by X- or γ -rays due to Čerenkov ultraviolet radiatioin emitted by secondary electrons. *Int. J. Radiat. Biol.,* 1981; 40, 435-440.

[39] Munson, RJ; Neary, GJ; Bridges, BA; Preston, RJ. The sensitivity of *Escherichia coli* to ionizing particles of different LETs. *Int. J. Radiat. Biol.*, 1967; 13, 205-224.

[40] Myasnik, MN; Morozov, II. The phenomenon of photoreactivation in bacteria *E. coli* irradiated by ionizing radiation. *Int. J. Radiat. Biol.*, 1977; 31(1), 95-98.

[41] Myasnik, MN; Morozov, II; Petin, VG. The value of the photoreactivable component in *E. coli* B_{s-1} cells exposed to densely and sparsely ionizing radiations. *Int. J. Radiat. Biol.*, 1980; 37(1), 81-84.

[42] Pelevina, II; Saenko, AS; Gotlib, VN et al. *Survival of Irradiated Mammalian Cells and DNA Reparation*. Moscow: Energoatomizdat, 1985.

[43] Pennarun, AM; Kovoor, A; Heller, R. Čerenkov counting of ^{24}Na in suspension cultures – an economical use of scintillation apparatus for studing ion exchange. *Physiol. Plant*, 1977; 39, 323-327.

[44] Petin, VG. *Genetic Control of Cell Radiosensitivity Modification*. Moscow: Energoatomizdat, 1987.

[45] Petin, VG; Berdnikova, IP. Effect of elevated temperatures on the radiation sensitivity of yeast cells of different species. *Radiat. Environ. Biophys.*, 1979; 16, 49-61.

[46] Petin, VG; Berdnikova, IP. Responses of yeast cells to heat applied alone or combined with gamma-rays. *Int. J. Radiat. Biol.*, 1981; 39, 281-290.

[47] Petin, VG; Komarov, VP. Estimation of the contribution of ionization and excitation to the lethal effect of ionizing radiation. *Radiat. Environ. Biophys.*, 1982; 20, 79-87.

[48] Petin, VG; Komarov, VP. Photoreactivation of damage induced by ionizing radiation in yeast cells. *Radiat. Environ. Biophys.*, 1985; 24, 281-286.

[49] Petin, VG; Morozov, II; Kim, JK et al. Quantitative estimation of UV light dose concomitant to irradiation with ionizing radiation. *Radiation Physics and Chemistry*, 2011; 80(1), 33-37.

[50] Petin, VG; Zhurakovskaya, GP; Komarova, LN. *Radiobiological Basis of Synergistic Interaction in Biosphere*. Moscow: GEOS, 2012.

[51] Pohlit, W. Radiation sensitivity and repair capacity dependent on cell concentration? *Biophysik*, 1973; 10, 33-37.

[52] Pohlit, WE; Schafer, M. Recovery and repair in yeast cells after irradiation with densely ionizing particles. In: Welsh CW, editor. *Biological Effects of Neutron Irradiation*. Vienna: IAEA, 1974; 177–184.

[53] Raju, MR. *Heavy Particle Radiotherapy*. New-York: Academic Press, 1980.

[54] Redpath, JL; Zabilansky, E; Morgan, T et al. Čerenkov light and the production of photoreactivatable damage in X-irradiated *E. coli*. *Int. J. Radiat. Biol.*, 1981; 39(5), 569-575.

[55] Redpath, JL; Tortorello, ML. Photoreactivation of *E. coli* irradiated in the absence and presence of oxygen. *Int. J. Radiat. Biol.*, 1977; 32, 505-506.

[56] Redpath, JL; Zabilansky, E. Photoreactivation of ionizing-radiation-induced damage in *E. coli*. Influence of chemical and physical factors. *Int. J. Radiat. Biol.*, 1979; 35, 473-476.

[57] Reynolds, MC; Brannen, JP. Thermal enhancement of radiosterilization. In: *Radiation Preservation of Food*. Vienna: International Atomic Energy Agency, 1973; 165-176.

[58] Saeki, T; Machida, I; Nakai, S. Genetic control of diploid recovery after gamma-irradiation in the yeast *Saccharomyces cerevisiae*. *Mutat. Res.*, 1980; 73, 251-265.

[59] Scott, BR. Methodologies for predicting the expected combined stochastic radiobiological effects of different ionizing radiations and some applications. *Radiat Res.*, 1984; 98(1), 182-197.

[60] Skvortzov, VG; Myasnik, MN; Sokolov, VA et al. Action spectrum for photoreactivation of *Escherichia coli* B_{s-1} after gamma-irradiation. *Photochem. Photobiol.*, 1981; 33(2), 187-190.

[61] Steel, GG; Peckham, MJ. Exploitable mechanisms in combined radiotherapy-chemotherapy: the concept of additivity. *Int. J. Radiat. Oncol. Biol. Phys.*, 1979; 5(1), 85-91.

[62] Stewart, FA; Denekamp, J. Combined X-rays and heating: is there a therapeutic gain? In: Streffer C, editor. *Cancer Therapy by Hyperthermia and Radiation*. Baltimore-Munich: Urban & Schwarzenberg, 1978; 249-250.

[63] Trott, KR; Rosemann, M. Molecular mechanisms of radiation carcinogenesis and the linear, non-threshold dose response model of radiation risk estimation. *Radiat. Environ. Biophys.*, 2000; 39(2), 79-87.

[64] Trujillo, R; Dugan, VL. Synergistic inactivation of viruses by heat and ionizing radiation. *Biophys. J.*, 1972; 12, 92-113.

[65] Vinicombe, DA; Moss, SH; Davies, DJ. Photo-reactivation of gamma-radiation damage in *Escherichia coli* as evidence for the nature of the oxygen-enhancement effect. *Int. J. Radiat. Biol.*, 1978; 33(5), 483-492.

[66] Wang, TV; Smith, KC. Enzymatic photoreactivation of *Escherichia coli* after ionizing irradiation: chemical evidence for the production of pyrimidine dimmers. *Radiat. Res.*, 1978; 76, 540-548.

[67] Yarmonenko, SP; Konoplyannikov, AG; Vainson, AA. *Clinical Radiobiology*. Moscow: Medicine, 1992.

Chapter 4

CLUE FOR UNDERSTANDING THE MECHANISM INVOLVED IN SYNERGISTIC EFFECTS

ABSTRACT

Cell survival, recovery kinetics and inactivation forms after combined action of heat with ionizing or UV radiations were studied in diploid yeast cells of various genotype. It was shown that synergistic interaction of these agents can be observed only within a certain temperature range. Inside this range a specific temperature can be observed that maximizes the synergistic effect. The synergistic effect was expressed in great extent for cell capable of recovery from radiation damage. The magnitude and the rate of recovery were shown to be lowered with increasing of exposure temperature. It was shown that the probability of recovery per unit time was independent of various conditions of combined action of heat with ionizing or UV radiations while the irreversible component gradually increased with increase in thermal load. The rise of the irreversible component was accompanied by an increase of cell killing without postirradiation division. It is concluded on this basis that the synergistic interaction of heat with ionizing radiation and UV light is not related with the impairment of the recovery capacity itself and that it may be attributed to the enhanced yield of irreversible damage. Using data published by other authors, similar conclusion was made for mammalian cells: the inhibition of DNA repair and mammalian cell recovery was mainly realized through the production of irreversible damage. It is proposed that heat destabilization of the osmotic cell homeostasis, which is more profound after rapid heating of bacterial cells, may participate in heat induced cellular lethality.

INTRODUCTION

It is well established that elevated temperatures can enhance radiation sensitivity of a variety of biological system in a synergistic manner. The synergistic effect of heat and ionizing radiation has been demonstrated using cell culture of various origins and different model tumor systems [63]. It is assumed that thermal radiosensitization may be displayed by the inhibition of the repair of sublethal and potentially lethal damages on a cellular level [3; 4; 28; 54; 62]. On this basis, it has been claimed that combined exposure of heat and radiation may be very perspective for radiation therapy. The exact mechanisms of radiosensitization by heat have not yet been elucidated. At some conditions of hyperthermia treatments there was enhanced radiation killing due to elimination of the shoulder on the survival curve [4] and others high temperatures affected the final slope of the survival curve [36]. These modifications can be apparently due to inability of the cells to repair sublethal and potentially lethal damages resulting from the synergistic interaction between hyperthermia and ionizing radiation. The regularities of postradiation recovery have been established firstly for yeast cells [28; 49], the simplest model of eukaryotic cells. It would be worth to compare the effect of elevated temperature on radiation sensitivity and recovery of yeast cells.

Earlier the survival curves for diploid *S. cerevisiae* of wild-type (RAD/RAD) and radiosensitive mutant (*rad*52/*rad*52), defective in genetic recombination, have been measured after simultaneous action of ionizing radiation and high temperature (Petin and Berdnikova, 1981). The greatest value of the synergistic enhancement ratio was 2.7 for wild-type cells and was completely absent (additive interaction) for mutant cells. There are information concerning the synergistic interaction between UV light and hyperthermia [50; 55; 71]. It seems widely accepted now that the synergistic interaction of the UV light combined with other agent is displayed by decreased repair rates at molecular and cellular levels [5; 26]. These data indicate that the mechanism of synergistic interaction of heat with ionizing or UV radiations may involve the inhibition of cell capacity to recover damage resulted from the combined action of both modalities. However, the inhibition of the recovery after the combined action of hyperthermia with ionizing or UV radiations can be attributed to different reasons. It is not clear whether this is due to an actual impairment of the recovery process itself or to the increased portion of irreversible damage which reduces or prevents further recovery to occur. In the last case, the process of recovery may be either also damaged or stayed unchanged. Although the combination of hyperthermia with ionizing radiation or UV light is of considerable current interest, there have been no reports in the literature on a quantitative estimation of each of these reasons. It would be of interest to determine whether the synergistic interaction of high temperatures with ionizing radiation or UV light in cellular and molecular levels was related to the impairment of the recovery capacity itself or to the production of irreversible damages that cannot be repaired.

It was postulated [44] that higher cell sensitivity to rapid heating rates is the result of a more profound destabilization of osmotic homeostasis caused by rapid versus slow heating which leads to a larger increase in intracellular osmotic pressure in rapidly heated cells. This hypothesis is strongly supported by several reports illustrating the dependence of osmotic features in pro- and eukaryotic cells exposed to different temperatures in cell suspension [16; 22; 31; 39; 60]. Novitsky and Kushner [48] reported that growing bacterial cells required an increase in extracellular tonicity in order to maintain viability when incubation temperatures were increased while the reverse situation is true for the tonicity requirements of cell growing under reduced incubation temperature. Further support for the osmotic hypothesis was observation that protection of bacterial cells from thermal damages occurs when extracellular tonicity was increased as incubation temperature increased [43]. Herman et al. [19] reported an increase in cell volumes of rapidly heated CHO cells as opposed to slowly heated cells. These authors also proposed that the change in cell volumes observed in rapidly heated cells reflected damage to the plasma membrane with loss of ability to control the intracellular osmotic environment and, hence, cell volume.

It was postulated [19; 75] that heating rate induced cell membrane damage. There is compelling evidence that heat shock proteins are involved in the protection of cells from heat damage [1; 35]. Burns et al. [7] failed to detect the synthesis of heat shock proteins 70 in leukemia cells which were preheated over 3 h from 37 to 42 °C and then exposed to 42 °C for 1 h. Also Anderson et al. [2] reported that thermotolerance in Chinese hamster cells may occur under some conditions in the absence of elevated levels of heat shock proteins (HSP) 70. These authors as well as Tomasovich et al. [69] concluded that other factors within the cell can confer thermotolerance, in addition to the contribution made by HSP, especially during mild heat challenges such as occurs during slow heating rates. So, it was of interest to study the combined action of heat with chemical agent which is known to inhibit expression of heat shock proteins.

Thus, in this Chapter we are going to perform the following experiments and theoretical calculations.

1. To study survival and recovery of yeast cells after combined action of ionizing radiation and hyperthermia.
2. To analyze the impact of the recovery in yeast cells after simultaneous exposure to UV radiation and hyperthermia.
3. To calculate quantitatively parameters describing the process of DNA breaks repair in mammalian cells after thermoradiation action.
4. To evaluate behavior of mammalian cell recovery after combined influence of ionizing radiation and hyperthermia.
5. To elucidate the role of chloramphenicol on hyperthermic cytotoxicity and cell permeability under various heating rates.

4.1. Survival and Recovery of Yeast Cells after Combined Action of Ionizing Radiation and Hyperthermia

Here we compare the effect of hyperthermia on the radiosensitivity of yeast cells of different strains capable and incapable to recover radiation damages and investigate the liquid holding recovery of yeast cells exposed to ionizing radiation at high temperatures. Materials and methods were described above.

Figure 4.1 shows survival curves for diploid *S. cerevisiae*, strain XS800, of wild-type (RAD/RAD) (Figures 4.1A and 4.1C) and radiosensitive mutant XS806 (*rad*51/*rad*51), defective in genetic recombination (Figures 4.1B and 4.1D). Cells were exposed to simultaneous action of ionizing radiation and high temperature (Figures 4.1A and 4.1B) and heat alone (Figures 4.1C and 4.1D). Many investigators have inferred that the difference between the survival curves for wild-type and *rad* mutants can be attributed to the repair of potentially lethal damages in wild-type strains, and to the absence of such repair in the mutant. Comparing our data, one may note that *rad*51 mutant diploid cells are more sensitive both to ionizing radiation and heat treatment alone.

Figure 4.2 exhibits analogous survival curves for diploid yeasts *Endomyces magnusii* (RAD/RAD) (Figures 4.2A and 4.2D), *S. ellipsoideus* (RAD/RAD) (Figures 4.2B and 4.2E) and haploid *Z. bailii* yeast (RAD) (Figures 4.2C and 4.2E). Cells were exposed to simultaneous action of ionizing radiation and high temperature (Figures 4.2A, B, C) and heat alone (Figures 4.2D, E, F).

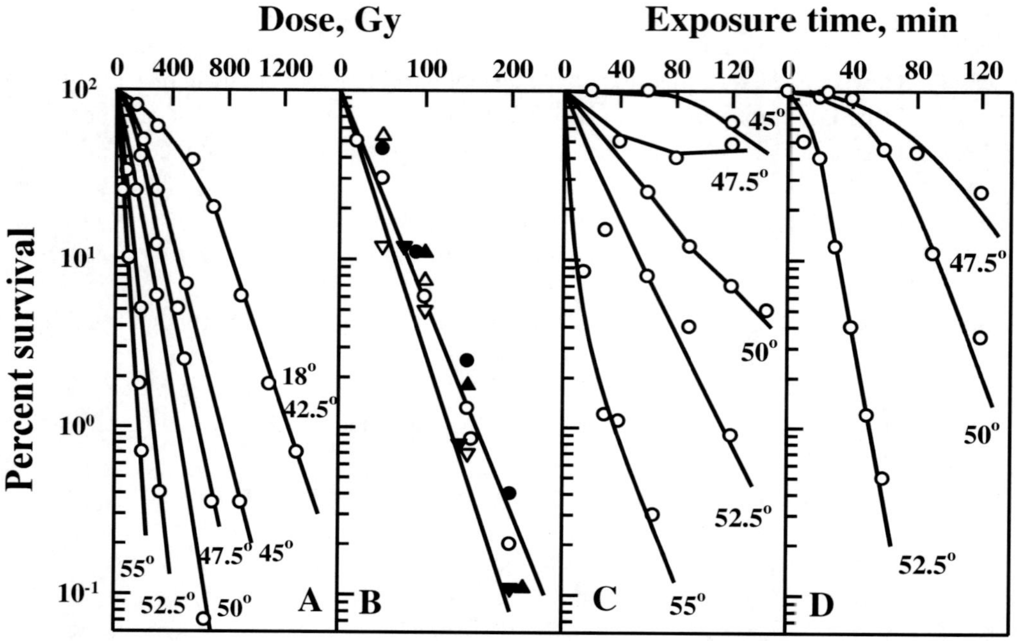

Figure 4.1. Survival curves for diploid *S. cerevisiae*, strain XS800, of wild-type (RAD/RAD) (A, C) and radiosensitive mutant XS806 (*rad*51/*rad*51), defective in genetic recombination (B, D). Cells were exposed to simultaneous action of ionizing radiation (γ-rays of ^{60}Co, 10 Gy/min) and high temperature (A, B) and heat alone (C, D).

As was described above, under independent action of two treatments the additive survival fraction can be calculated by the Eq. (3.1). Using this equation, we calculated the additive survival curves expected for these strains exposed by ionizing radiation at different temperatures. To estimate quantitatively the sensitization action of hyperthermia we have used the synergistic enhancement ratio k which was defined as the ratio of radiation doses in calculated (for additive effect) and experimental survival curves to produce the same effect (Eq. 3.5). The values of k calculated for the survival fraction of 0.1 are summarized for all investigated strains in Table 4.1. It is clear from these data that hyperthermia appeared to increase the radiation response.

The results summarized in Table 4.1 clearly indicate that thermal sensitization was observed in the temperature range of 40.0–52.5 °C only for wild-type diploid cells. Maximal value of the synergistic enhancement ratio (k) was 2.7. The synergistic interaction of heat and γ-rays for *rad*51 mutant cells was not established in a wide temperature range of 42.5–52.5 °C. The irradiation of this strain at 52.5 °C was more effective in comparison with irradiation at 30 °C however, the interaction between hyperthermia and ionizing radiation is described as additive, according which the net effect of both agents is equal to the sum of the logarithms of their separate action (Eq. 3.2). Thus, these results were interpreted in favour of the supposition that combined action of hyperthermia and ionizing radiation can damage recovery processes and/or produce more "severe" damages, which cell cannot repair. This supposition is supported by experiments with *Z. bailii* showing the smaller value of synergy

(Figure 4.2C, Table 4.1). It might be speculated that these results may be connected with the form of the survival curve, either with incapability of the haploid strain to recover lethal damage or with temperature sensitivity of the strain.

Some additional information about mechanisms of synergistic interaction of hyperthermia with ionizing radiation may be obtained using the Arrhenius representation of experimental data where cell inactivation rate is presented as a function of the inverse absolute temperature, with the inactivation rate represented as the reciprocal of the slope ($1/D_0$) of the exponential part of cell survival curves *versus* exposure time at a particular temperature. An Arrhenius plot for heat inactivation of mammalian cells shows an inflection point and the interaction of ionizing radiation and hyperthermia at temperatures above and below these points differs qualitatively. Murthy et al. [47] have carried out an Arrhenius analysis of heat inactivation of one strain of diploid yeast. It would be worth comparing their results with those obtained for yeast cells of various genotypes exposed to heat applied alone or combined with gamma-rays.

Here we recalculated the inactivation rate from the survival curves, most of which have been published by us elsewhere [50; 51; 56] and therefore not presented here. The Arrhenius plots of the effect of simultaneous action of ionizing radiation and hyperthermia for *S. cerevisiae* cells of various genotypes have been published [52]. For all investigated strains the break of the Arrhenius plot was near 50–52.5 °C. There was an additional break on the Arrhenius plots of thermosensitive strains at 42–43 °C (as for mammalian cells). Basing on the results published earlier [52], we calculated some parameters relevant to combined action of heat applied alone and combined with ionizing radiation. Table 4.2 summarizes the temperature ranges synergistically increasing the effect of ionizing radiation, the highest values of the synergy (k) and the activation energies, calculated for the action of hyperthermia alone and in combination with ionizing radiation, for yeast strains investigated.

Table 4.1. The dependence of the synergistic enhancement ratio (k) on the exposure temperature for various yeast strains exposed to ^{60}Co γ-rays (10 Gy/min)

Yeasts: species, strain, mutant repair loci	Exposure temperature, °C							
	30	40	42.5	45	47.5	50	52.5	55
S. cerevisiae, XS800 (RAD/RAD)	1.0	–	1.1	1.8	1.9	2.0	2.7	1.0
S. cerevisiae XS806 (*rad*51/*rad*51)	1.0	–	1.0	1.0	1.0	1.0	–	–
S, ellipsoideus, Megri 139-B(RAD/RAD)	1.0	1.0	–	1.2	1.8	1.9	2.3	1.0
Z. bailii K7 (RAD)	1.0	1.0	1.5	1.4	1.3	–	1.0	–
Endomyces magnusii K17 (RAD/RAD)	1.0	2.1	1.6	1.0	–	1.0	–	–

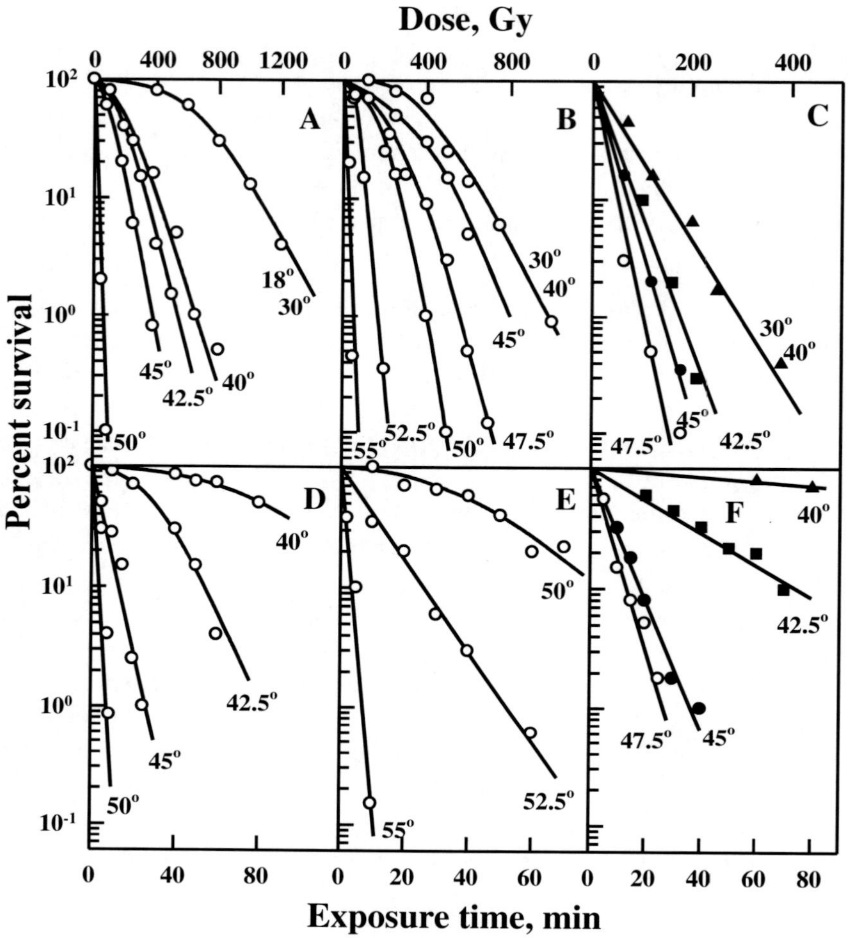

Figure 4.2. Survival curves for diploid yeasts *Endomyces magnusiis,* strain K17, (A, D), *S. ellipsoideus*, strain Megri 139-B, (B, E) and haploid yeast of *Z. bailii*, strain K7, (C, E). Cells were exposed to simultaneous action of ionizing radiation and high temperature (A, B, C) and heat alone (D, E, F).

Table 4.2. The synergistic enhancement ratio (k) and activation energies (kJ/mole) calculated from the Arrhenius plots for yeast cells of different strains treated with heat alone or combined with γ-rays

Yeasts	Temperature range, °C	k	Activation energy, kJ/mole	
			Heat alone	Heat plus γ-rays
S. cerevisiae, S288C (RAD)	50.0–52.5	1.2	360	100
S. cerevisiae, XS774-4d (*rad* 51)	50.0–52.5	1.1	360	80
S. cerevisiae, XS800 (RAD/RAD)	45.0–52.5	2.7	300	170
S. ellipsoideus (*vini*),Megri-139B (RAD/RAD)	45.0–52.5	2.3	330	190
Endomyces magnusii K17 (RAD/RAD)	40.0–45.0	2.1	220	140
Z. bailii K7 (RAD)	42.5–47.5	1.5	240	150

The data presented in Table 4.2 show that the synergistic enhancement ratio value lies between 2.1–2.7 for diploid strains capable of recovery and 1.2–1.5 for haploid strains. It is evident from Table 4.2 that the activation energies for thermal cell killing are similar for all strains and fall within the range 220– 350 kJ/mole, which may correspond to denaturation of some proteins and enzymes. In the last column of Table 4.2 activation energies for the combined action of radiation and hyperthermia are presented. They were also calculated for the temperature range, synergistically increasing the effect of ionizing radiation. These values prove to be lower and are comparable with those needed for different inactivation processes in DNA [42].

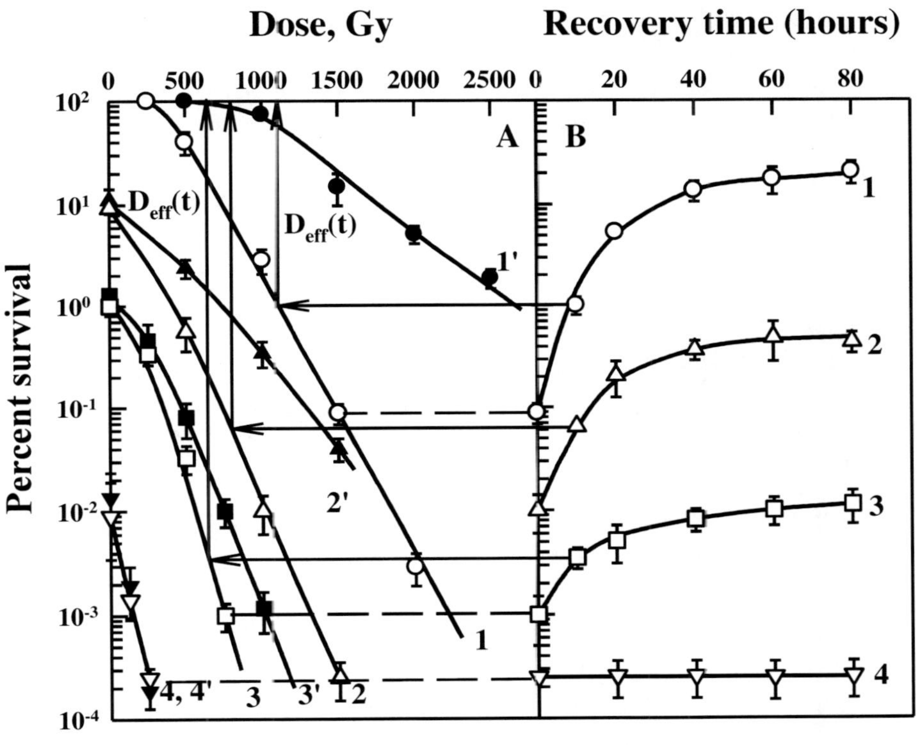

Figure 4.3. Survival curves (A) and LHR recovery kinetics (B) of diploid yeast cells *S. ellipsoideus* (*vini*), strain Megry 139-B. A: cells were exposed to graded doses of ionizing radiation alone (curves 1, 1′) or to a combined treatment with a prior heating at 60 °C for 3 (curves 2, 2′), 6 (curves 3, 3′) and 12 (curves 4, 4′) min. Cells were plated on nutrient agar immediately after irradiation (curves 1, 2, 3, 4) or after 80 h of LHR (curves 1′, 2′, 3′, 4′). B: cells were exposed to ionizing radiation alone (curve 1) or to a combined treatment of ionizing radiation with a prior heating at 60 °C for 3 (curves 2), 6 (curves 3) and 12 (curves 4) min. The dotted lines indicate survival points which were used to study recovery kinetic. The arrows indicate examples of the determination of the effective dose, $D_{eff}(t)$, after 10 h of recovery. Points are the means from at least three independent experiments. Error bars indicate standard errors of the mean. Curves were fitted to the data points by eye.

As was mentioned above, diploid cells are capable of recovery. The process of the LHR may serve as an indicator of this ability. It was shown in our previous publication [50] that if cells exposed to ionizing radiation at high temperature are kept under conditions promoting the LHR, the fraction of irreversibly damaged cells (Eq. 1.12) makes 0.29 at 30 and 40 °C; 0.39 at 45 °C; 0.59 at 50 °C and 1.00 at 55 °C. Diploid yeast *S. ellipsoideus* (*vini*) Megri-139B was used in these studies. These data show that the cell ability to recover radiation damage is inversely proportional to the temperature of irradiation. To resolve whether this reduced ability to recover was owing to the irreversible damage or to the influence of the combined action on the repair processes, the kinetics of recovery of Megri-139B strain irradiated with γ-rays at different temperatures were studied [50; 51; 53; 56; 57]. Here we present the averaged results of these papers.

Figure 4.3A shows the survival curves of yeast cells irradiated with γ-rays of ^{60}Co without heat or preheated at 60°C for 3, 6 and 12 min. An enhancing effect of hyperthermia on radiation-induced cell killing is observed. Preheating the cells for 3, 6 and 12 min markedly decreased their sensitivity to the subsequent exposure to ionizing radiation. The thermal enhancement ratio estimated by the ratio of slopes, increased with the duration of preheating: 1.2, 1.7 and 2.6 for heating during 3, 6 and 12 min.

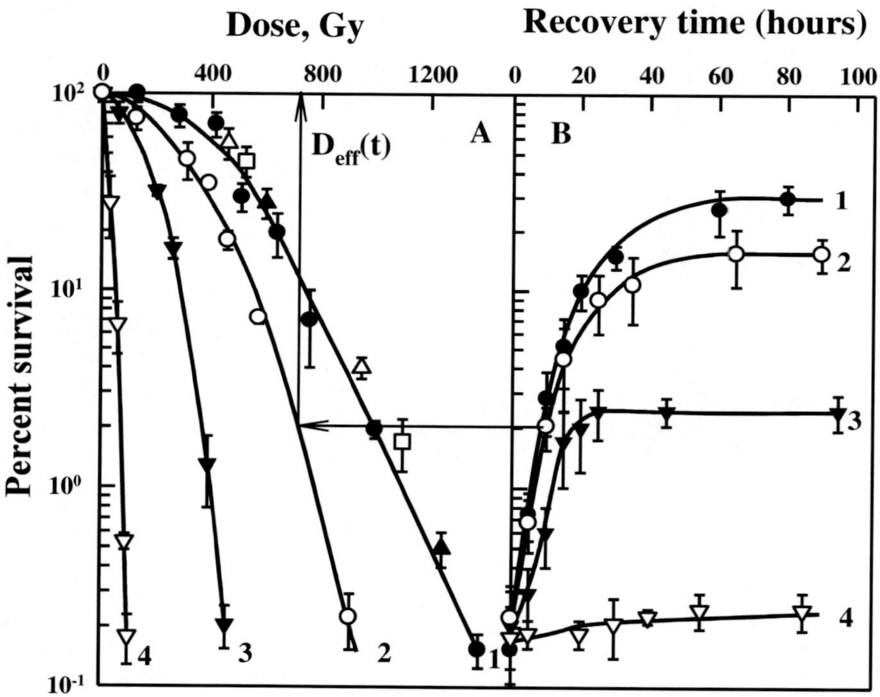

Figure 4.4. Survival curves (A) and LHR recovery kinetics (B) of diploid yeast cells *S. ellipsoideus* (*vini*) strain Megry 139-B. Cells were exposed to graded doses of ionizing radiation at various temperature: (curves 1: closed circles, 20 °C, open triangles, 30 °C, closed triangles, 35 °C, open squares, 40 °C), 45 °C (curves 2), 50 °C (curves 3), 55 °C (curves 4). The arrow indicates an example of the determination of the effective dose, $D_{eff}(t)$. Irradiated cells were plated immediately after irradiation (A) or after various recovery times (B, delayed plating). Points are the means from at least three independent experiments. Error bars indicate standard errors of the mean. Curves were fitted to the data points by eye.

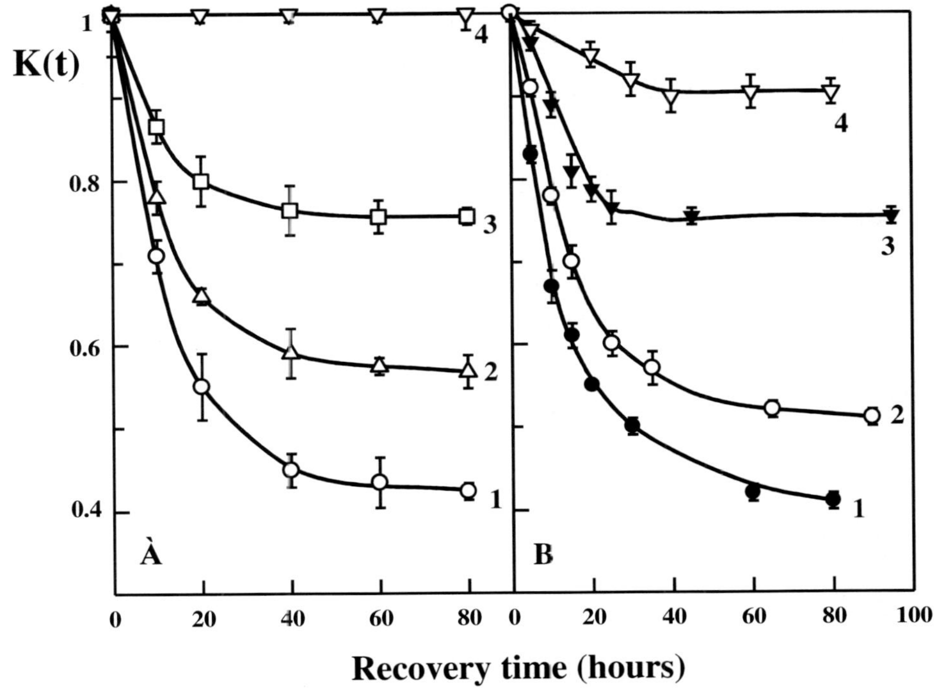

Figure 4.5. The decrease in the proportion of unrecovered radiation damage, $K(t)$, as a function of recovery time of diploid yeast cells *S. ellipsoideus* (*vini*), strain Megry 139-B, after various treatments with heat and radiation. A: sequential (heat + ionizing radiation) exposure. The prior heat treatment was 0 (curve 1), 3 (curve 2), 6 (curve 3), or 12 (curve 4) min at 60 °C. B: simultaneous treatment with heat and radiation. The radiation was delivered at 20 °C (curve 1), 45 °C (curve 2), 50 °C (curve 3), 55 °C (curve 4). Points are the means from at least three independent experiments. Error bars indicate standard errors of the mean. Curves were fitted to the data points by eye.

The patterns of recovery we observed are shown in Figure 4.3B. Preheating during 3 and 6 min slowed repair, and 12 min at 60 °C completely inhibited the recovery. The results of experiments with the same diploid yeast cells subjected to simultaneous treatment with heat and radiation are presented at Figure 4.4A. An increase in the exposure temperature in the range of 20–40 °C had no influence on survival (curve 1). A further increase in the temperature resulted in an increase in the cell radiosensitivity, as shown by both the increase in the slope and the decrease in the extrapolation number (curves 2–4). A comparison of the ability of yeast cells to undergo the LHR carried out at approximately equal levels of survival (i.e., at equal amounts of lethal lesions), is shown in Figure 4.4B. Again, the results indicate that the number of viable cells increased as a function of time, reaching a plateau after about 3 days. The present data suggest that both the rate and the extent of recovery decreased with increase in the temperature at which the irradiation occurred.

Using Eq. (1.11) and the results presented in Figures 4.3 and 4.4, we estimated the decrease in the relative amount of unrecovered radiation damage, $K(t)$, as a function of recovery time for various conditions of treatment with heat and radiation (Figure 4.5). It is evident the deceleration of the decrement in $K(t)$ with increases in both the duration of prior hyperthermia (Figure 4.5A) and the temperature at which the irradiation was delivered (Figure 4.5B).

It also appeared from these data that the limited quantities of $K(t)$, i.e., the values of irreversible damage, $K(plateau)$, increased with increasing of heat load. Figure 4.6A shows the increase in the irreversible component as a function of heat exposure (60 °C) for the successive action of heat and radiation while Figure 4.6B exhibits the increase in the irreversible component as a function of temperature at which irradiation occurred for simultaneous application of both modalities. It is obvious that the irreversible component increased progressively with thermal load.

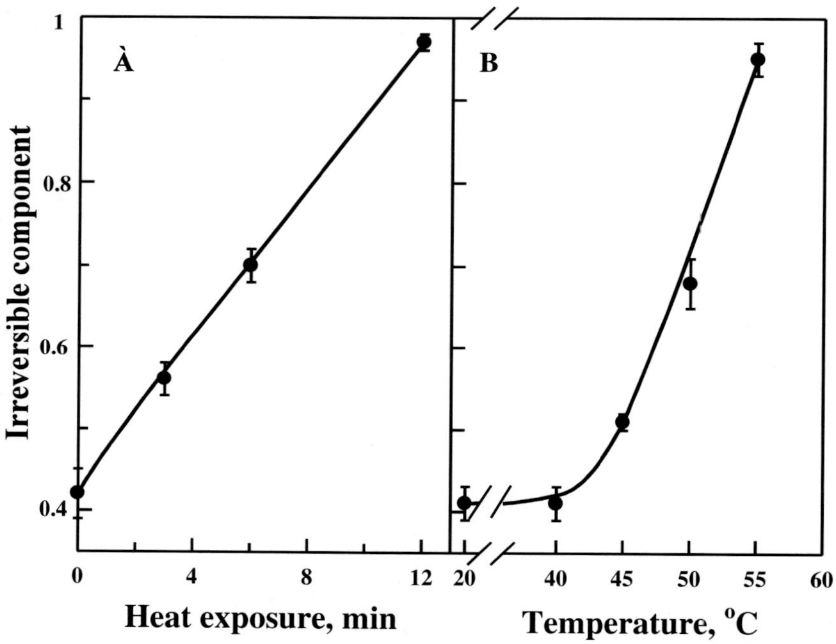

Figure 4.6. The increase of the relative proportion of irreversible radiation damage, $K(plateau)$, as a function of the effect of a prior heat exposure (60°C) on the effect of a sequential treatment with heat and radiation (A) and as a function of exposure temperature of a simultaneous treatment with heat and radiation (B) in diploid yeast cells *S. ellipsoideus* (*vini*), strain Megry 139-B. Points are the means from at least three independent experiments. Error bars indicate standard errors of the mean. Curves were fitted to the data points by eye.

The presented experimental data make it possible to calculate the relative part of the reparable damage, $A(t)$, defined by Eq. (1.14). The outcome is shown in Figure 4.7 for both the sequential (Figure 4.7A) and simultaneous (Figure 4.7B) treatment with heat and radiation. It is evident that this function decreases exponentially with the recovery time and does not depend on the duration of a prior heat treatment or the temperature at which irradiation occurs. Using Eq. (1.15) and the results shown in Figure 4.7, we obtained that the recovery constant $\beta = 0.07\ h^{-1}$ for sequential heat + radiation treatment and $\beta = 0.068\ h^{-1}$ for simultaneous application of both modalities. This means that about 6.9 % from the residual repairable damage are recovered every hour for both treatments. Thus, it can be concluded that the parameters characterizing the LHR depend differently on the conditions of thermoradiation action. While the irreversible component gradually increases both with the duration of a prior heat action and the irradiation temperature, the recovery constant is identical for all conditions investigated.

Studies *in vitro* have shown that the greatest amount of cell killing is obtained when radiation is delivered concurrently with hyperthermic treatment [62]. For simultaneous thermoradiation treatment, an independent interaction can be determined by Eq. (3.1) and the synergistic enhancement ratio by Eq. (3.5) The calculated in such a manner synergistic enhancement ratio is presented in Figure 4.8A as a function of the temperature at which the irradiation was delivered. One can see that the synergistic interaction of both modalities takes place only inside a certain temperature range and there is a specific temperature that maximizes the synergy. A quite different dependence is observed for irreversible component of radiation damage (Figure 4.8B) that increases continually with increase in the exposure temperature.

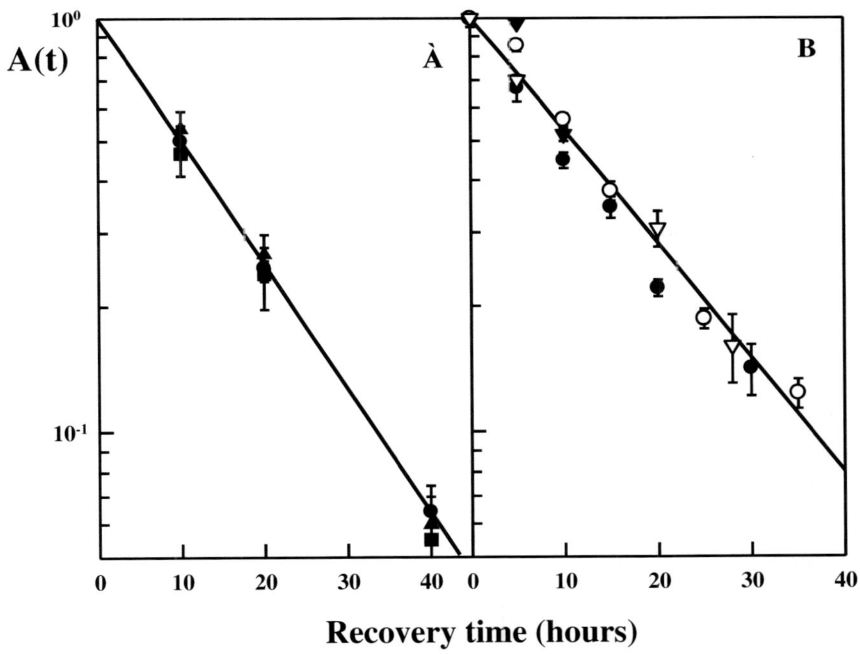

Figure 4.7. The decrease in the proportion of recovered radiation damage, $A(t)$, as a function of recovery time of diploid yeast cells *S. ellipsoideus* (*vini*), strain Megry 139-B, after various treatment with heat and radiation. A: sequential (heat + ionizing radiation) exposure. The prior heat treatment was 0 (circles), 3 (triangles), and 6 (squares) min at 60 °C. B: simultaneous treatment. Radiation was delivered at the following temperatures: 20 °C (closed circles), 45 °C (open circles), 50 °C (closed triangles), 55 °C (open triangles). Points are the means from at least three independent experiments. Error bars indicate standard error of the mean. Curves were fitted to the data points by the least-squares method.

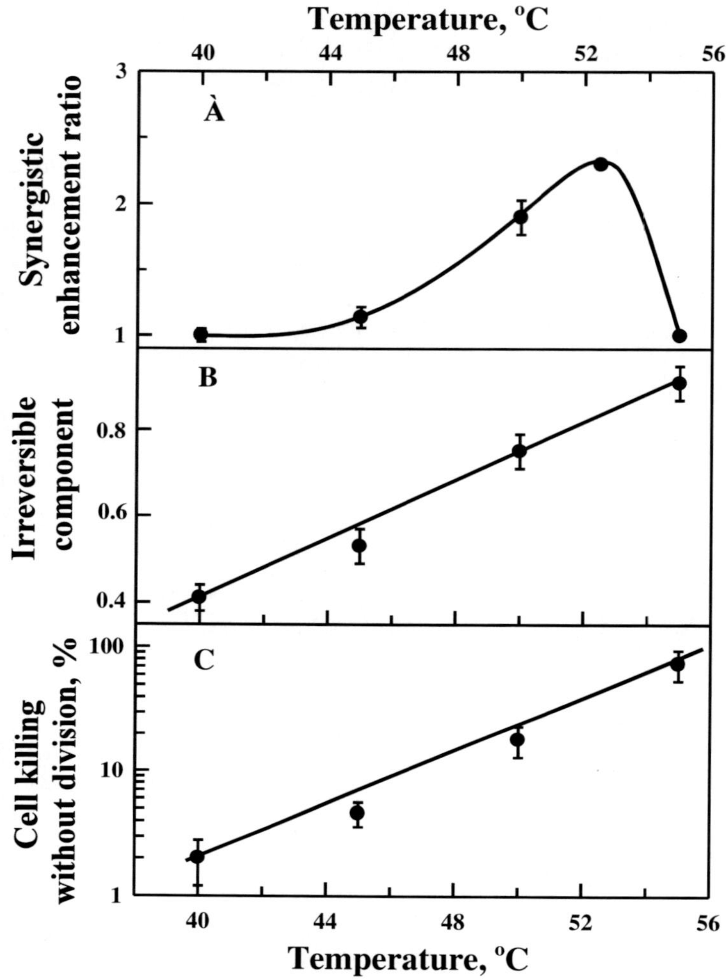

Figure 4.8. The dependencies of the synergistic enhancement ratio (A), irreversible component (B) and cell killing without division (C) on the exposure temperature for diploid yeast cells *S. ellipsoideus* (*vini*), strain Megry 139-B, irradiated with ionizing radiation at various temperatures.

The overwhelming majority of yeast organisms exposed to ionizing radiation is inactivated after at least one cell division cycle. Along with reproductive death, we analyzed microscopically interphase death, i.e., the yield of cell killing without any postirradiation division. In these experiments, equi-effective doses of thermoradiation action reducing cell survival to 10% were used. Figure 4.8C shows the yield of cell killing without division after simultaneous thermoradiation action in the dependence of temperature at which the irradiation occurred. It can be seen that this form of cell inactivation was rare at relatively low exposure temperatures at which no synergistic interaction of the modalities employed was obtained. For more higher temperatures, when the synergistic effect first increased, reached the highest value and then decreased with exposure temperature (Figure 4.8A), the fraction of irreversibly damaged cells (Figure 4.8B) as well as the part of cells killing without division (Figure 4.8C) were constantly increased both reaching 90% at 55 °C. In other words, within the temperature range, synergistically enhancing the effect of ionizing radiation, the cell death was gradually transformed from the reproductive to the interphase death (cell killing without any postradiation division).

Experiments described in this section were carried out to determine the effect of hyperthermia alone or combined with γ-rays on the survival and recovery of yeast cells of different species. The main results of experiments presented may be formulated as follows. (i) Simultaneous action of ionizing radiation and high temperatures appeared to increase the radiation response expected from the additive sum of the effects produced by each agent by a factor of approximately 3.0 for diploid yeast cells capable of the diploid-specific recovery. (ii) Synergistic effect of thermoradiation action was observed only within a certain temperature range inside which an optimal temperature exists that maximizes the synergy. (iii) The magnitude and the rate of the LHR is lowered with an increase in both the duration of heat treatment that was followed by radiation and the exposure temperature used for simultaneous treatment with heat and radiation. On this basis, it can be concluded that possible mechanism for thermal sensitization of yeast cells may involve the reduced capacity of cells to recover damage resulted from the combined action of both modalities.

One of our main tasks was to determine whether the synergistic interaction of heat and ionizing radiation may be causally related to the impairment of the recovery capacity itself or to the production of irreversible damage, which cannot be repaired. It was shown that the irreversible component increased substantially as a function of the duration of heat treatment (60 °C) for a sequential treatment with heat and radiation and as a function of the exposure temperature after simultaneous treatment with heat and radiation. For the latter case, the increase in the irreversible component was accompanied by an increase in cell killing without division. To elucidate whether the retardation of the yeast cell recovery occurred only due to the increase in the irreversible component or was related to the impairment of the LHR, the recovery constant describing the probability of recovery per unit time was estimated. This parameter appeared to be unchanged (about 0.07 h^{-1}) for either the duration of a prior heat or the exposure temperature. Hence, the reduced rate and the volume of recovery takes place not because the process of the LHR is damaged but because of the increased yield of irreversible damages.

4.2. The Impact of Simultaneous Exposure to UV Radiation and Hyperthermia to the Recovery in Yeast Cells

Whereas considerable knowledge about repair processes in UV-irradiated cells has been gained during past decades, the situation is far less clear for the combined action of UV light with other inactivating factors. In the previous section it was shown that the probability of recovery was independent of conditions of the combined action of ionizing radiation with heat while the irreversible component gradually increased with an increase in thermal load. The rise of the irreversible component was accompanied by an increase of cell killing without postirradiation division. It is concluded on this basis that the synergistic interaction of ionizing radiation and hyperthermia in yeast cells is not related with the impairment of the recovery capacity *per se* and may be attributed to the enhanced yield of irreversible damages. Without doubt it would be of interest to check whether the same conclusions may be revealed after the simultaneous action of UV radiation and heat.

Cell suspension (1.5 ml, 10^6 cells/ml) in an open quartz vessel was irradiated with a germicidal lamp that emitted predominantly UV light of a wavelength of 254 nm (far-UV radiation) at fluence rate of 1.5 W/m^2 at different temperatures. The temperatures were thermostatically controlled for their variations to be within 0.2 °C. The source of UV light was a low-pressure mercury vapour lamp (a General Electric germicidal lamp, G15T8) with an integral filter that provided more than 95% of the emitted radiation at 254 nm. The UV light source was calibrated by potassium ferrioxalate actinometry and the fluence rate was checked with a germicidal meter (General Electric Co., Fairfield, CT).

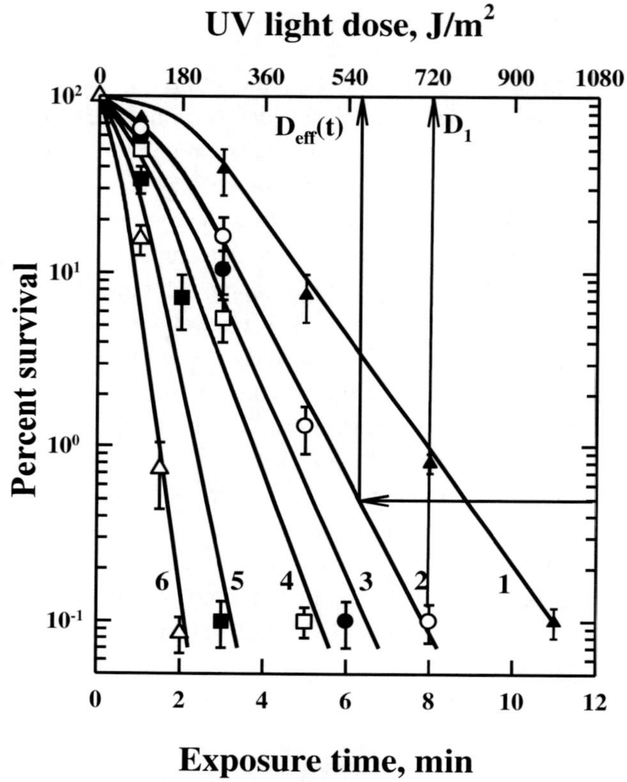

Figure 4.9. Survival curves of diploid yeast cells *S. cerevisiae* strain XS800. Cells were irradiated with UV light radiation alone (at room temperature, curve 1) and at the following temperatures: 53, 54, 55, 56 and 57 °C (curves 2, 3, 4, 5 and 6, respectively). Arrows indicate the examples of the initial, D_1, and effective, $D_{eff}(t)$, doses determination. Irradiated cells were plated immediately after irradiation at red light. Points are the means from at least three independent experiments. Error bars indicate standard error of the mean.

The results of experiments when wild-type diploid yeast cells (*S. cerevisiae*, strain XS800) were subjected to simultaneous treatments with the UV light and heat are presented in Figure 4.9. Curve 1 at this figure represents the survival curve obtained after cell exposure with UV light applied alone (at room temperature). The increase in the exposure temperature up to 50 °C has no or small influence on the survival. It is apparent that the further increase in the exposure temperatures (53–56 °C, curves 2–5) resulted in an increase in the cell UV sensitivity as shown by both the increase in the slope and the decrease in the extrapolation number. Finally, the survival curve became almost exponential at 57 °C (curve 6). As shown in previous publications [28; 50; 67], survival after a combined treatment with the UV light and hyperthermia may be enhanced beyond that expected on the basis of simple additivity of the effects. However, for the highest temperature investigated here, the cell sensitivity was almost completely determined by its thermal sensitivity and the contribution of UV light damage in cell inactivation was negligible. In fact, for temperatures of 20–50 °C cell death is determined only by the UV light whereas at temperature 57 °C the cell death is mainly caused by heat. This means that the synergistic interaction of UV light and hyperthermia was obtained only within a certain temperature range.

The survival curves of cells exposed to high temperatures for various periods of time are shown in Figure 4.10. At temperatures above 53 °C, cell sensitivity to hyperthermia progressively increased. All the survival curves were of exponential forms. These curves will be used here to calculate the extent of synergistic interaction. The increase in the slope and the decrease in the extrapolation number with the exposure temperature increase (Figure 4.9) reflect the reduced ability of the cell to recover from potentially lethal and sublethal radiation damage, respectively and thus can be considered as an evidence of the recovery processes participation in the heat sensitization.

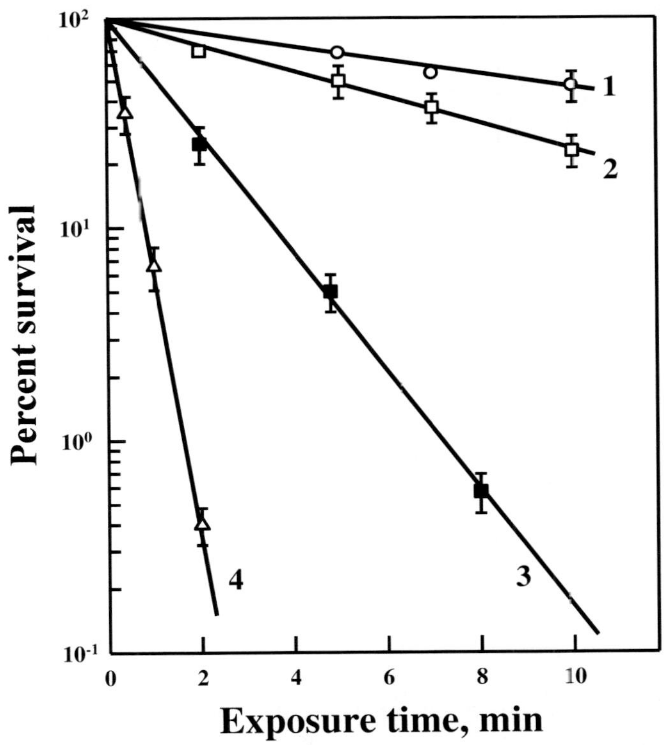

Figure 4.10. Survival curves of *S. cerevisiae*, strain XS800, after heat treatment alone: 53 °C (curve 1), 55 °C (curve 2), 56 °C (curve 3), 57 °C (curve 4). Points are the means from at least three independent experiments. Error bars indicate standard error of the mean. Curves were fitted to the data points by the least-squares method.

This suggestion is supported by the data indicating the luck of the synergistic interaction of heat and UV light in the UV sensitive (*rad*18/*rad*18) yeast mutant [50]. That is why it is of interest to study the effect of hyperthermia applied during the UV light irradiation on the kinetics of the LHR of yeast cells. A comparison of the ability of yeast cells to undergo LHR was carried out at equal levels of survival (0.1%), i.e., at equal amounts of lethal lesions. The results are shown in Figure 4.11, where the survival of diploid yeast cells irradiated at equieffective doses of the UV light radiation delivered at different temperatures is depicted in the dependence of the recovery time. These equieffective doses were 990, 720, 540, 460, 280 and 180 J/m^2 for UV light applied alone and combined with temperatures 53, 54, 55, 56 and 57 °C, respectively.

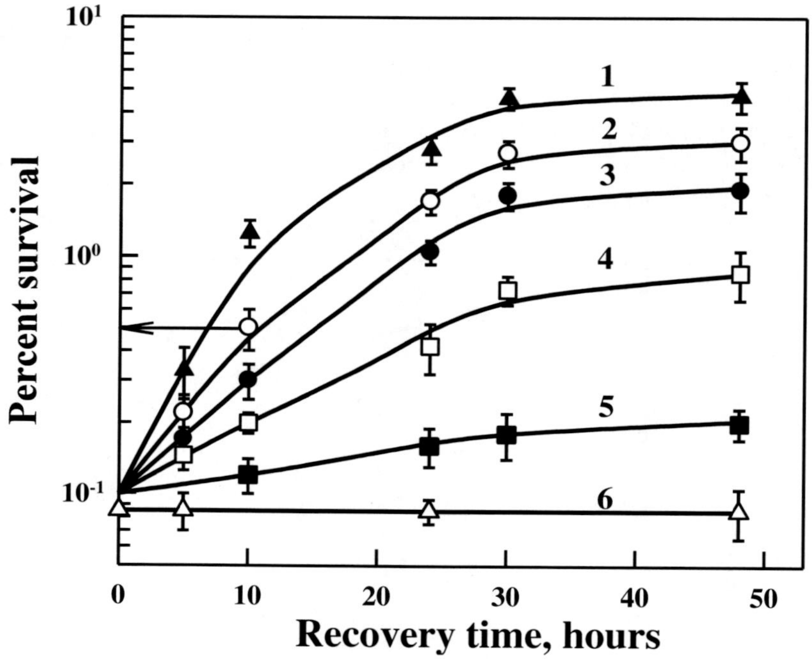

Figure 4.11. The LHR recovery kinetics of diploid yeast cells *S. cerevisiae* strain XS800. Cells were irradiated in equieffective doses of UV light radiation alone (at room temperature, curve 1) and at the following temperatures: 53, 54, 55, 56 and 57 °C (curves 2, 3, 4, 5 and 6, respectively). Arrow indicates the example of $S(t)$ determination after 10 h LHR of cells UV illuminated at 53 °C. This value of $S(t)$ is used in Figure 4.9 to show the example of the effective dose determination. Irradiated cells were plated after various recovery time (delayed plating). Points are the means from at least three independent experiments. Error bars indicate standard error of the mean.

One can see that when samples were plated at different times after the treatment, the number of viable cells increased as a function of time, reaching a plateau after about 2–3 days. It is apparent that both the slope and the plateau of the recovery curves depend on the exposure temperature. The present data suggest that both the rate and the extent of recovery are decreased with increase in the temperature at which the UV illumination occurred.

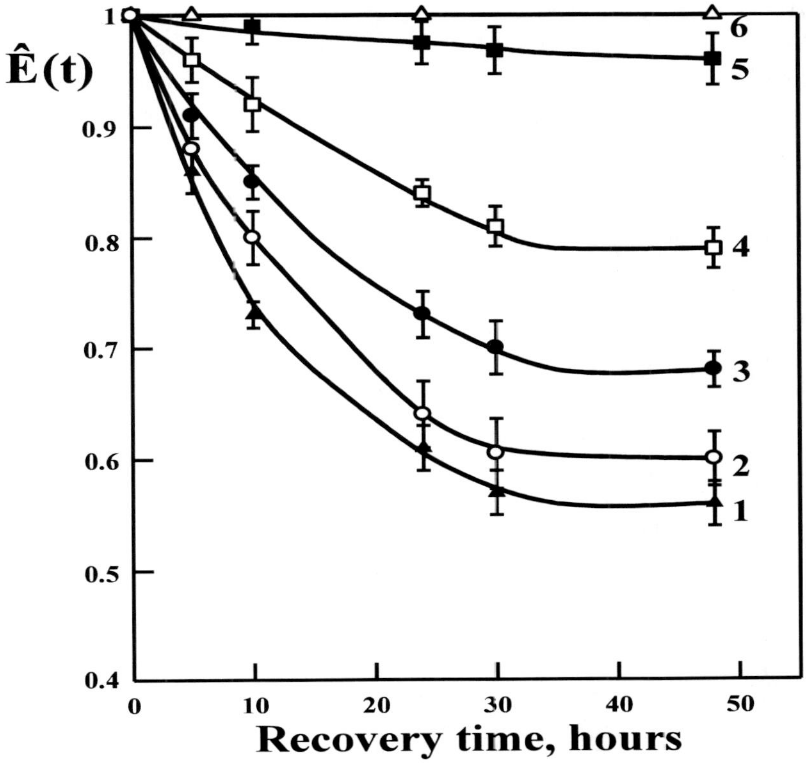

Figure 4.12. The decrease in the proportion of unrecovered UV damage, $K(t)$, as a function of recovery time for diploid yeast cells *S. cerevisiae*, strain XS800. Cells were irradiated in equieffective doses of UV light radiation alone (at room temperature, curve 1) and at the following temperatures: 53, 54, 55, 56 and 57 °C (curves 2, 3, 4, 5 and 6, respectively). Points are the means from at least three independent experiments. Error bars indicate standard error of the mean.

Using the results presented in Figures 4.9 and 4.11, we estimated (Eq. 1.11) the change in the relative part of unrecovered radiation damage, $K(t)$, as a function of recovery time for various conditions of treatment with UV light and heat (Figure 4.12). These data demonstrate that the values of $K(t)$ decreased uniformly, reaching a plateau after about 2–3 days of recovery. In the figure, one can see the deceleration of the decrease in this parameter with increase in the temperature at which the UV illumination was delivered. It also appeared from these data that the limited quantities of $K(t)$ (Eq. 1.12), i.e., the values of irreversible damage, $K(plateau)$, increased with the exposure temperature. Figure 4.13A shows the increase in the irreversible component as a function of the exposure temperature.

The experimental data presented make it possible to calculate the relative part of the reparable damage, $A(t)$ (Eq. 1.14), in the dependence of the recovery time. The outcome is shown in Figure 4.14. It is evident that this function decreases exponentially with the recovery time and does not depend on the temperature at which the UV illumination occurs. Using Eq. (1.15) and the results shown in Figure 4.14, we obtain that the recovery constant β = 0.07 h^{-1} independently of the exposure conditions used here. It means that 7.0 % of the residual reparable damage is recovered every hour. Thus it can be concluded that the parameters describing the LHR process depend differently on the conditions of the treatment with UV light and heat. Although the irreversible component increases gradually with the exposure temperature, the recovery constant is identical for all the conditions investigated.

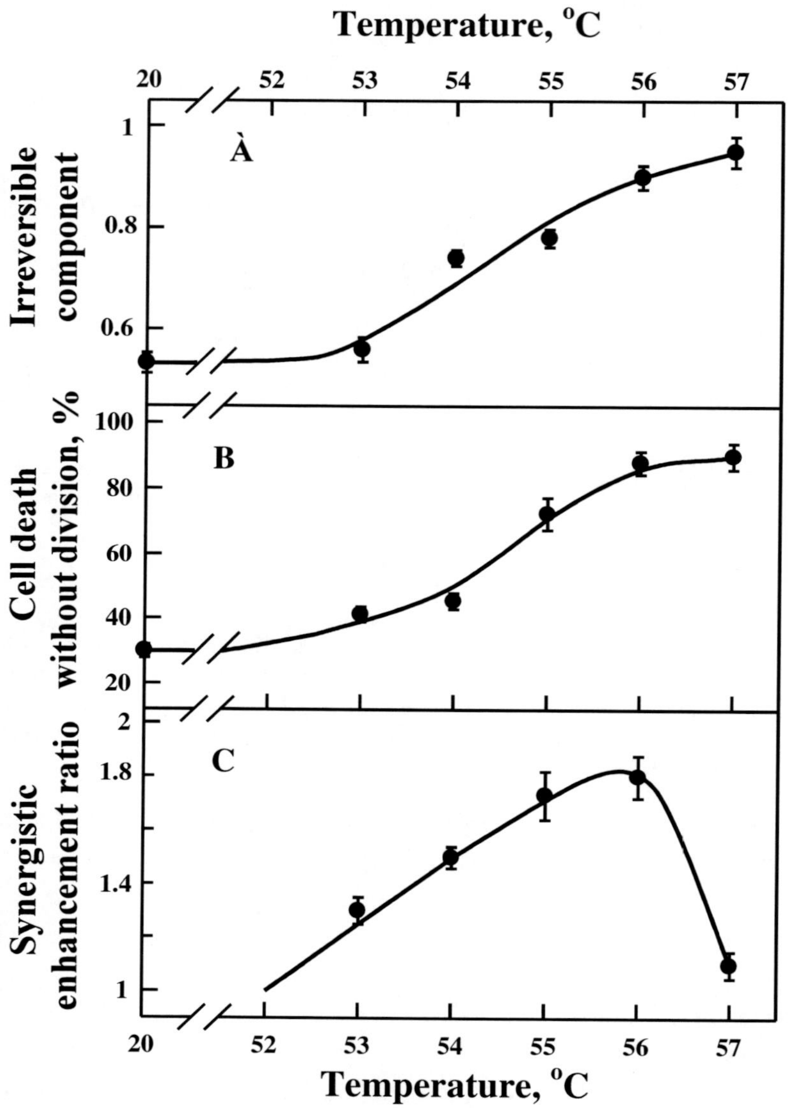

Figure 4.13. The dependencies on the exposure temperature the irreversible component (A), cell death without division (B) and the synergistic enhancement ratio (C) for diploid yeast cells *S. cerevisiae* strain XS800. Cells were irradiated with UV light radiation (254 nm) at various temperatures. Points are the means from at least three independent experiments. Error bars indicate standard error of the mean.

Along with the reproductive death, we analysed microscopically interphase death, i.e., the yield of cell death without any postirradiation division. In these experiments, equieffective doses of treatment with heat and UV light, which reduced the cell survival to 10% (i.e., to equal amounts of lethal lesions), were used. These equieffective doses were 440, 290, 260, 230, 160 and 70 J/m^2 for UV light applied alone (at room temperature) and combined with temperatures 53, 54, 55, 56 and 57 °C, respectively. Figure 4.13B shows the yield of cell death without division after simultaneous treatment with heat and UV light as a function of the temperature at which the illumination took place. It can be seen that this form of cell inactivation was about 30% at low exposure temperatures, at which no synergistic interaction of the modalities was observed. For higher temperatures, the fraction of cells that died without postirradiation division was constantly increased (Figure 4.13B), reaching 90% at 57 °C when all inactivated cells were died without any postirradiation division.

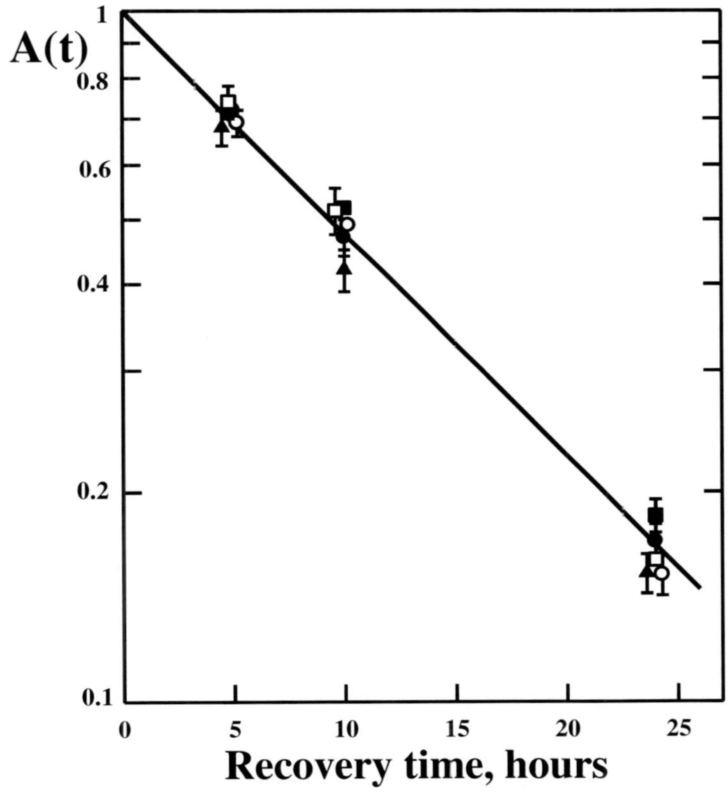

Figure 4.14. The decrease in the proportion of recovered radiation damage, $A(t)$, as a function of recovery time of diploid yeast cells *S. cerevisiae*, strain XS800. Cells were irradiated in equieffective doses with UV light radiation alone (at room temperature, 20 °C, dark triangles) and at the following temperatures: 53 °C (open circles), 54 °C (closed circles), 55 °C (open squares), 56 °C (closed squares) and 57 °C (open triangles). Points are the means from at least three indepeendent experiments. Error bars indicate standard error of the mean. Curve was fitted to the data points by the least-squares method.

We calculated the synergistic enhancement ratio at 1% survival using the results obtained here (Figures 4.9 and 4.10) and some results of our previous publications [50; 55; 66]. The synergistic enhancement ratio calculated in this manner is presented in Figure 4.13C as a function of the temperature at which the UV irradiation was delivered. One can see that the synergistic interaction of the both modalities occurs only within a certain temperature range and there is a specific temperature that maximizes the synergy. Any deviation of this temperature from the optimal value results in a decrease of synergy. Within the temperature range, synergistically enhancing the effect of the UV radiation, both the irreversible component and cell death without any postirradiation division were increased substantially with the exposure temperature.

4.3. Repair of DNA Breaks in Mammalian Cells after Thermoradiation Action

The combination of ionizing radiation with different chemical and physical agents including hyperthermia is attaining considerable interest for the treatment of cancer [17; 20; 62]. Understanding of the underlying mechanism will hopefully lead to an optimization of the synergistic interaction of these modalities and to the increase in the efficient procedures used for cell inactivation. As was shown above, the inhibition of the recovery from potentially lethal damage in yeast cells exposed to hyperthermia and ionizing radiation or hyperthermia and UV light was proceeded through the enhanced yield of the irreversible damage whereas the recovery capacity itself was not damaged or impaired. Repair of DNA, as reflected in rejoining of strand breaks, may be of relevance to the cell recovery from sublethal and potentially lethal damage [14; 17]. Hyperthermic inhibition of the rate and extent of repair of radiation-induced DNA damage is observed with all the assays currently used for DNA damage assessment. It seems to be generally accepted that DNA repair inhibition is one of the most important mechanisms of the synergistic interaction of ionizing radiation with hyperthermia and chemical compounds [64]. The inhibition of DNA repair is expressed both as a reduced repair rate and a lesser extent of reparation [17; 21]. It is obvious that these observations may be caused by the following reasons: a damage of the recovery process itself, an increase in the portion of irreversible damage, both of these reasons. Although there are considerable interests in combining of ionizing radiation with hyperthermia, there were no reports in the literature on a separate estimation of hyperthermia influence on the process of DNA repair itself and the production of irreversible damage. Quantitative approach describing the recovery process as a decrease in the effective dose is well-known and was described above. It allowed us to determine whether the process of recovery was destroyed or damaged itself or that the decrease in the rate and extent of recovery was entirely due to an increase in the fraction of irreversible damage with thermal load increase. Here this approach will be applied to experimental results published by others [21] to estimate quantitatively the irreversible component of radiation damage and the probability of recovery per unit time. The authors themselves did not estimate these parameters. Material and methods of experiments have been described in this paper. It is worth of noting that a part of our results has been published previously [27] and here we present the averaged results from these publications.

Figure 4.15 shows the dependence of the number of breaks of Ehrlich ascites cells on the X-ray dose (A) and the effect of hyperthermia at 44 °C on the repair of DNA strand breaks (B). In some cases the data was obtained by a linear extrapolation of the curves published in the original paper. It is evident from Figure 4.15 that both the extent and the rate of repair were decreased by an exposure to hyperthermia prior to ionizing radiation. These results indicate that the number of DNA breaks decreased as a function of the repair time, reaching a plateau after about 90–120 minutes.

Using the results published in the original paper [21], a number of curves, similar to those depicted in Figure 4.15, have been calculated for other hyperthermic conditions at 42, 43 and 45 °C. A whole set of this data was used to determine whether this inhibition may causally be related to an impairment of the repair capacity itself or to the production of irreversible damage, which cannot be repaired. We calculated (Eq. 1.11) the dependence of the relative fraction of the unrepaired radiation damage, $K(t)$, on the duration of the repair time of Ehrlich ascites cells repairing after various treatments with heat and radiation.

The results are shown in Figure 4.16. From these data it is evident that unheated cells exposed to ionizing radiation subjected to post-irradiation DNA repair showed an appreciable decrease in $K(t)$ whereas this effect became gradually worse as the exposure temperature and its duration increased. It appears that the inhibition of DNA repair depends on the thermal load. On this basis the limited values of $K(t)$, i.e., in the values of the irreversible component of radiation damage, were estimated (Eq. 1.12). They are listed in Table 4.3.

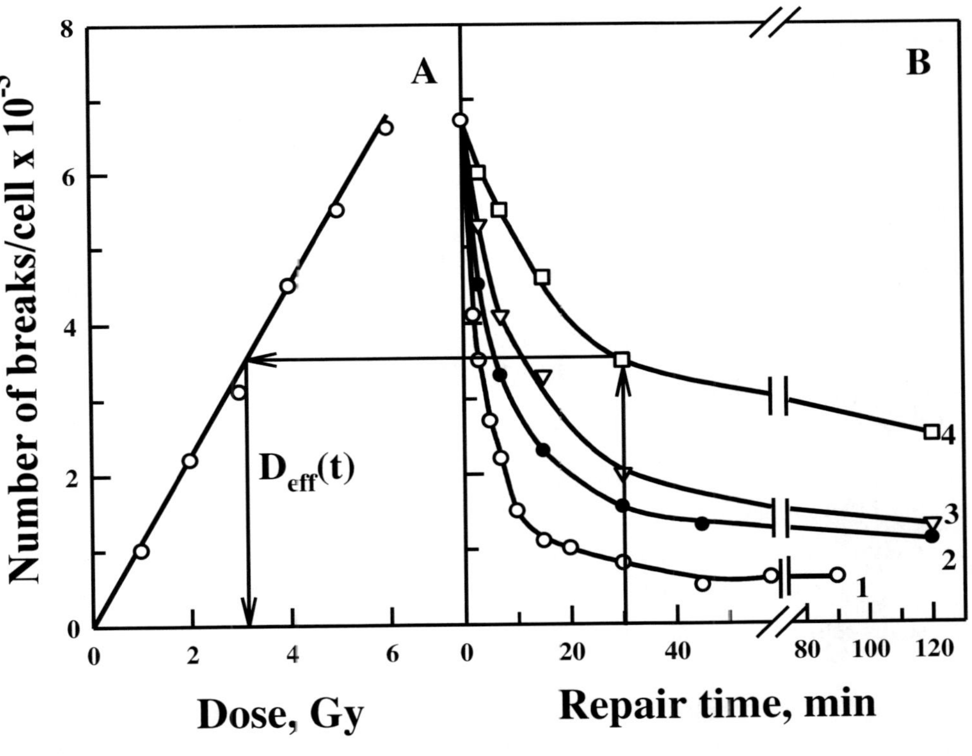

Figure 4.15. The dependence of the number of breaks of Ehrlich ascites cells of the X-ray dose (A) and the effect of hyperthermia at 44 °C on the DNA strand breaks repair kinetics of Ehrlich ascites cells irradiated at 0 °C with 6 Gy of X-rays (B): curve 1 – unheated control cells irradiated at 0 °C with 6 Gy of X-rays; curves 2, 3 and 4 – pretreatment at 44 °C during 15, 30 and 60 min, respectively, and then irradiated at 0 °C with 6 Gy of X-rays. The arrow indicates the example of the determination of the effective dose, D_{eff}, after 30 min of repair. Curves were fitted to the data points by eye.

A progressive increase in the irreversible component was found within the temperature range investigated when longer heat treatments were applied. For instance, the values of the irreversible component are equal to 0.07, 0.17, 0.22, 0.40 and 1.00 for the cells repairing from radiation damage without a preheating and preheated at 44 °C during 15, 30, 60 and 120 min, respectively. It means that 93% of the radiation damage can be repaired by unheated cells while this portion is decreased to 83, 78 and 60% after a preheating during 15, 30 and 60 min. No repair was observed for the cells preheated at 44 °C during 120 min. Qualitatively similar results were obtained for other temperatures (Table 4.3). The obvious increase in the irreversible component with the thermal load should lead to a decrease in the repair rate because of the decrease in the number of cells capable of DNA damage repair.

Table 4.3. Radiobiological parameters of DNA strand breaks repair of Ehrlich ascites cells after the successive action of heat and X-ray (6 Gy)

Treatment	Irreversible component K	Repair constant β, min^{-1}		Percent of slowly repairing DNA damage
		Fast	Slow	
X-ray	0.07	0.18	0.06	25
42 °C 60 min + X-ray	0.11	0.14	0.07	40
42 °C 150 min + X-ray	0.17	–	0.06	80
43 °C 30 min + X-ray	0,11	0.18	0.07	40
43 °C 60 min + X-ray	0.14	0.15	0.06	73
43 °C 90 min + X-ray	0.40	–	0.03	100
43 °C 120 min + X-ray	0.66	–	0.02	100
44 °C 15 min + X-ray	0.17	0.14	0.05	55
44 °C 30 min + X-ray	0.22	–	0.05	80
44 °C 60 min + X-ray	0.40	–	0.05	100
44 °C 120 min + X-ray	1.00	–	–	–
45 °C 7 min + X-ray	0.11	0.17	0.05	35
45 °C 15 min + X-ray	0.17	–	0.05	70
45 °C 30 min + X-ray	0.42	–	0.02	100

To clarify whether or not this decrease could completely explain the observable deceleration of the repair rate, we estimated the probability of repair per unit time (β) for various conditions of cell treatment. The experimental data makes it possible to calculate the part of recovered radiation damage $A(t)$ (Eq. 1.14) in the dependence of the repair time of Ehrlich ascites cells repairing after irradiation with and without preheating. Figure 4.17 shows the decrease in the function of $A(t)$ as a function of repair time for Ehrlich ascites cells repairing after various treatments with heat and radiation. It is obvious that the DNA damage repair process is clearly divided into fast and slow components. It is of interest to estimate separately the influence of heat treatment to the probability of repair for both of these components. By taking Eq. (1.15) and the results shown in Figure 4.17, we calculated the repair constant for all the conditions of heat and radiation treatment which had been tested.

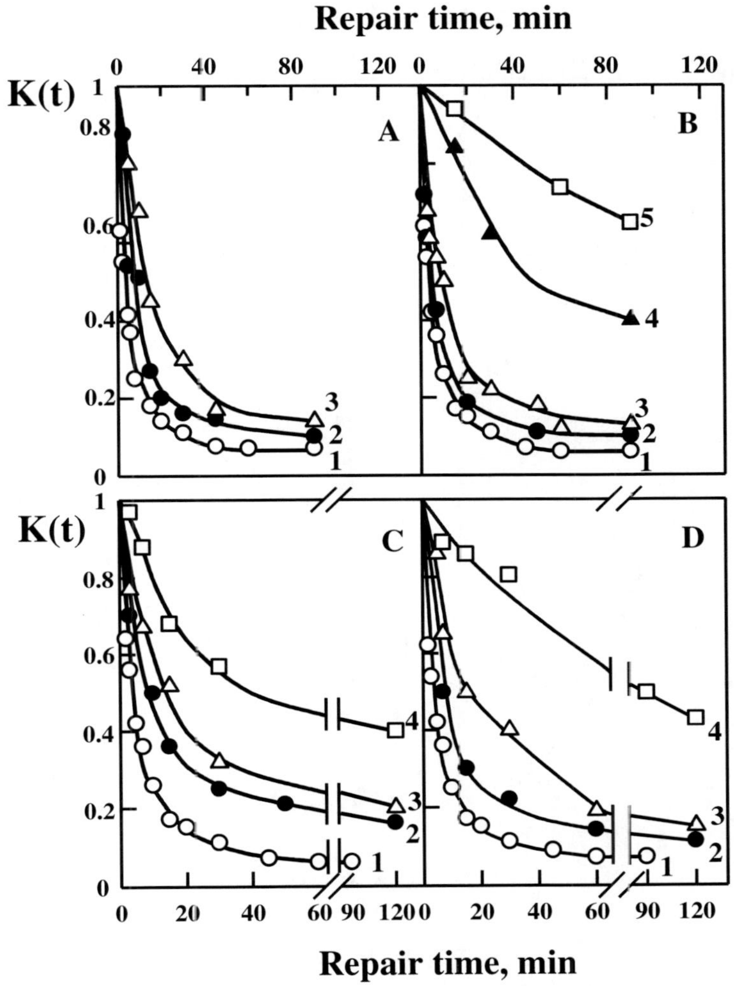

Figure 4.16. The decrease in the relative fraction of unrepaired DNA radiation damage, $K(t)$, as a function of repair time for Ehrlich ascites cells repairing after various treatments with heat and radiation. Curves were fitted to the data points by eye.
A, unheated control cells irradiated at 0 °C with 6 Gy of X-rays (curve 1); pretreatment at 42 °C during 60 (curve 2) and 150 min (curve 3) and then irradiated at 0 °C with 6 Gy of X-rays.
B, unheated control cells irradiated at 0 °C with 6 Gy of X-rays (curve 1); pretreatment at 43 °C during 30, 60, 90 and 120 min and then irradiated at 0 °C with 6 Gy of X-rays (curves 2, 3, 4 and 5, respectively).
C, unheated control cells irradiated at 0 °C with 6 Gy of X-rays (curve 1); pretreatment at 44 °C during 15, 30 and 60 min and then irradiated at 0 °C with 6 Gy of X-rays (curves 2, 3 and 4, respectively).
D, unheated control cells irradiated at 0 °C with 6 Gy of X-rays (curve 1); pretreatment at 45 °C during 7, 15 and 30 min and then irradiated at 0 °C with 6 Gy of X-rays (curves 2, 3 and 4, respectively).

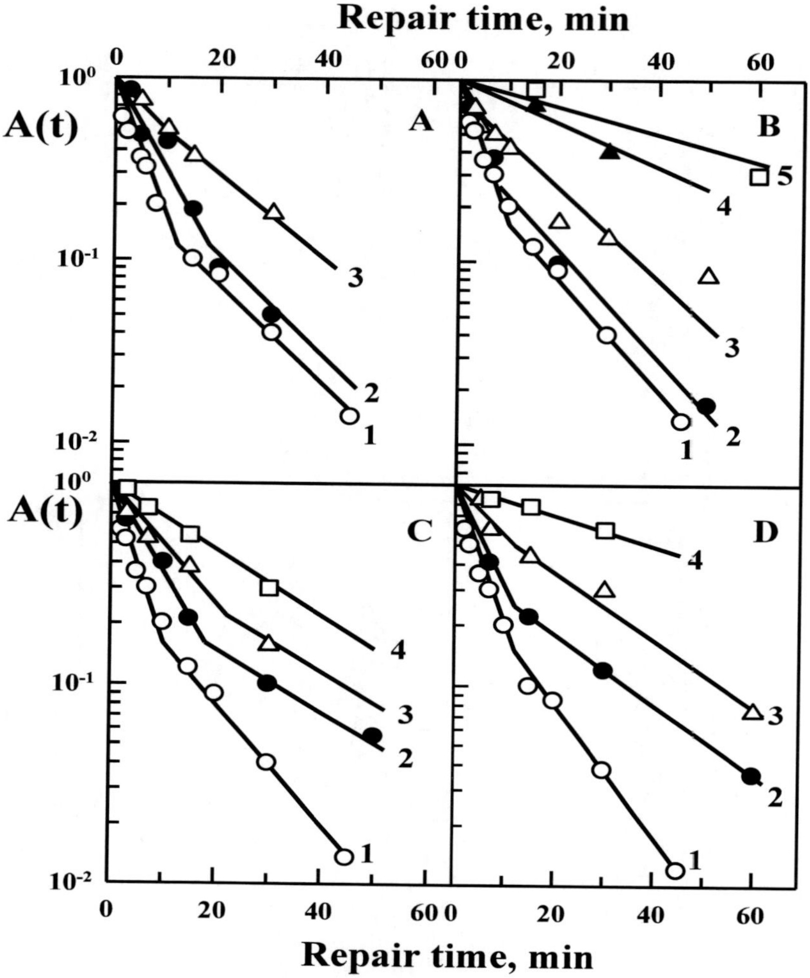

Figure 4.17. The decrease in the proportion of recovered DNA radiation damage, $A(t)$, as a function of repair time of Ehrlich ascites cells repairing after various treatments with heat and radiation. Curves were fitted to the data points by the least-squares method.
A, unheated control cells irradiated at 0 °C with 6 Gy of X-rays (curve 1); pretreatment at 42 °C during 60 (curve 2) and 150 min (curve 3) and then irradiated at 0 °C with 6 Gy of X-rays.
B, unheated control cells irradiated at 0 °C with 6 Gy of X-rays (curve 1); pretreatment at 43 °C during 30, 60, 90 and 120 min and then irradiated at 0 °C with 6 Gy of X-rays (curves 2, 3, 4 and 5, respectively).
C, unheated control cells irradiated at 0 °C with 6 Gy of X-rays (curve 1); pretreatment at 44 °C during 15, 30 and 60 min and then irradiated at 0 °C with 6 Gy of X-rays (curves 2, 3 and 4, respectively).
D, unheated control cells irradiated at 0 °C with 6 Gy of X-rays (curve 1); pretreatment at 45 °C during 7, 15 and 30 min and then irradiated at 0 °C with 6 Gy of X-rays (curves 2, 3 and 4, respectively).

A full set of the parameters describing the DNA repair of Ehrlich ascites cells is collected in Table 4.3. It is worth noting that the repair constants of the fast components were estimated quantitatively for only a relatively low duration of heat treatment (60 min at 42 °C, 30 and 60 min at 43 °C, 15 min at 44 °C and 7 min at 45 °C).

Further increase in the duration of the applied temperature resulted in a remarkable increase (up to 70-100 %) in the portion of cells showing a slow repair. This was a reason why the probability of the fast recovery couldn't be estimated with a high accuracy for some of the conditions. The averaging of all the calculated values of the fast repair constant resulted in the value β = 0.16 ± 0.02 min^{-1}. It means that 16% of the residual damage which can be repaired by the fast component is being repaired every minute. It can be concluded on this basis that the fast repair of DNA breaks does not depend on the condition of thermoradiation action on Ehrlich ascites cells. It can be inferred on this basis that the fast repair does not influence the deceleration of the observed DNA break repair after the successive heat and ionizing radiation treatment on Ehrlich ascites cells.

The averaging of most of the calculated values of the slow repair constant (see Table 4.3) resulted in the value of β = 0.06 ± 0.01 min^{-1}. It means that 6% of the residual damage which can be repaired by the slow component is being repaired every minute. One could also conclude on this basis that the slow repair of DNA breaks would not provide an appreciable contribution to the deceleration of the observed DNA break repair for most cases of the successive heat and ionizing radiation treatment on Ehrlich ascites cells. However, this parameter became considerably less (β= 0.02–0.03 min^{-1}, Table 4.1) after the cells were preheated at 43 °C during 90 and 120 min as well as at 45 °C during 30 min when the whole cell population was repaired due to this component of repair.

4.4. Recovery of Mammalian Cells after Combined Influence of Ionizing Radiation and Hyperthermia

In this section, the dose-effect survival curves and the recovery curves obtained and published by other authors [34; 36; 58; 72] will be used to calculate the irreversible component and the probability of recovery per unit time for various mammalian cell lines and conditions of thermoradiation action. It should be noted that the authors themselves did not carried out such calculations. Using the array of experimental data published for Chinese hamster cells [36] for sequential treatment with ionizing radiation and heat, we calculated the dependence of the relative fraction of unrecovered radiation damage, $K(t)$ (Eq. 1.11), as a function of recovery time for various conditions of thermoradiation damage. The results are shown in Figure 4.18A. One can see that cells subjected to postirradiation recovery showed an appreciable decrease in the fraction of irreversible damage, the irreversible component of radiation damage being dependent on the condition of exposure. The data presented here make it possible to calculate the proportion of recovered damage, $A(t)$ (Eq. 1.14), charactering the change of repairable damage in the dependence of repair time. The results are shown in Figure 4.18B.

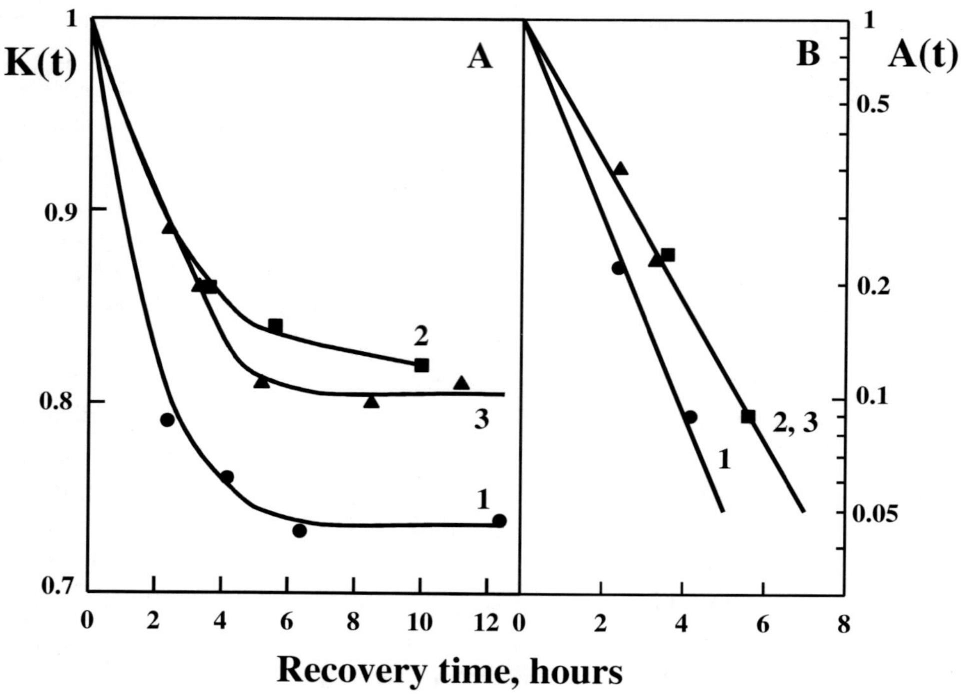

Figure 4.18. The proportion of unrecovered radiation damage, $K(t)$, (A) and recovered radiation damage, $A(t)$, (B) as a function of recovery time at 37 °C for Chinese hamster HA-1 cells after X-ray irradiation alone (curves 1) and the subsequent combined action at first X-rays of 16 Gy and then hyperthermia at 43 °C for 30 and 45 min (curves 2 and 3, respectively).

Taking into account the results shown in Figure 4.18, we calculated the irreversible component of radiation damage K (Eq. 1.12) and the probability of recovery per time unit (recovery constant β) (Eq. 1.15) for all conditions of treatment tested. The results are presented in Table 4.4. The data reveal some non-trivial conclusions. The inhibition of repair of potentially lethal radiation damage of Chinese hamster (line HA-1) cells by postirradiation hyperthermia is realized by the both opportunity: the increase in the value of irreversible component and the decrease in the probability of recovery. One can see (Table 4.4) that the irreversible component progressively increased with thermal load: from 0.74 for X-ray alone to 1.00 after the sequence X-rays 16 Gy + 43 °C heat during 60 min. The last value indicates that after such treatment the recovery is completely inhibited due to the lack of repairable damages. It is also apparent that simultaneously with a significant increase in the proportion of irreversibly damaged cells a decrease in the recovery constant from 0.59 h^{-1} after X-radiation alone to 0.43 h^{-1} after the successive exposure to X-ray and hyperthermia during 30 and 45 minutes takes place. The mean value from three values presented in the Table 4.4 is (0.48 ± 0.07) h^{-1}, i.e., deviation from the mean value is around 15%, which can be in the range of possible statistical errors. These data suggest that the recovery inhibition could be mainly occurred through the increase in the irreversibly damaged cells.

Table 4.4. Recovery parameters of Chinese hamster mammalian cells, line HA-1, after the sequential combined action of X-rays and hyperthermia

Treatment	Irreversible component, K	Recovery constant, β, h^{-1}		Percent of slowly recovered cells
		Fast	Slow	
X-ray alone (16 Gy)	0.74	0.59	–	–
X-rays, 16 Gy, + 43 °C heat during 30 min	0.82	0.43	–	–
X-rays, 16 Gy, + 43 °C heat during 45 min	0.81	0.43	–	–
X-rays, 16 Gy, + 43 °C heat during 60 min	1.00	0.00	–	–
43 °C heat during 15 min + X-rays, 16 Gy	0.66	0.49	0.10	26
43 °C heat during 30 min + X-rays, 16 Gy	0.70	0.49	0.10	39
43 °C heat during 45 min + X-rays, 16 Gy	0.66	0.49	0.08	42

More complex pattern was obtained for the reverse sequence when hyperthermic treatment was applied at first and then the cells were irradiated with X-rays (Figure 4.19, Table 4.4). The main feature of this sequence consisted in the appearance of the slow component of radiation recovery, which was absent when the hyperthermia was applied after irradiation. Recovery constant were 0.49 and about 0.1 h^{-1} for fast and slow recovery, respectively. The mean value from six values of fast recovery constant presented in this Table is (0.49 ± 0.03) h^{-1}, i.e., deviation from the mean value is around 6% that can be in the range of possible statistical errors. These data suggest that the recovery inhibition could be realized mainly through the increase in the irreversibly damaged cells. It can be seen from the data presented that the fraction of slowly recovered cells increases from 26 to 42% with the increase in the duration heat treatment from 15 to 45 min. Another feature of this sequence is that the irreversible component of radiation injury was 0.67 ± 0.02, i.e., the proportion of irreversibly damaged cells decreased in comparison with that for the reverse exposure (K = 0.81–1.00) and even for X-irradiation alone (K = 0.74). This may occurred due to the operation of two systems of radiation recovery: fast and slow repair with specific constants of recovery. It is possible that, due to this fact, the cell radiosensitivity modification was more effective when hyperthermia used after irradiation than the reverse order of these agents.

Similar calculations have been carried out for Chinese hamster lung fibroblasts (line V79-S171) according to data published by others [58]. The authors investigated potentially lethal radiation damage repair and its inhibition by hyperthermia in normal hamster cells, mouse cells, and transformed mouse cells. Cells were recovered from potentially lethal damage after only X-ray irradiation (250 kV, 3 Gy/min) and after the sequential action of hyperthermia at 42.5 °C for 30 min, and irradiation at a dose of 8 Gy. Cells were recovered from potentially lethal radiation damage after X-ray irradiation (250 kV, 8 Gy, 3 Gy/min) alone and after the sequential action of hyperthermia at 42.5 °C for 30 min plus X-ray alone (8 Gy) and the reverse order. Our calculations (Figure 4.20) of both $K(t)$ (Eq. 1.11) and $A(t)$ (Eq. 1.14) based on these data showed the irreversible component increased from 0.57 after irradiation alone to 0.80 for the combined action of X-ray + heat. For the reverse order of these modalities K = 0.92 (Table 4.5) while the recover constant β = 0.019 min^{-1} could be calculated only for the first combination and was independent on the treatment conditions. This means that about 2% of residual recoverable damage is recovered every minute. Hence, these data suggest that the inhibition of Chinese hamster lung fibroblasts recovery is realized through an increased proportion of irreversibly damaged cell rather than through the impairment of recovery processes themselves.

Hyperthermia in combination with ionizing radiation is widespread in radiation therapy for increasing the sensitivity of radioresistant tumors. It is of important to evaluate the quantitative parameters of human cell recovery. To carry out this task, we used data published by Raaphorst et al. [59] for cultured cells of human glioma (line U-87MG). There again the authors themselves did not make such kind of calculation.

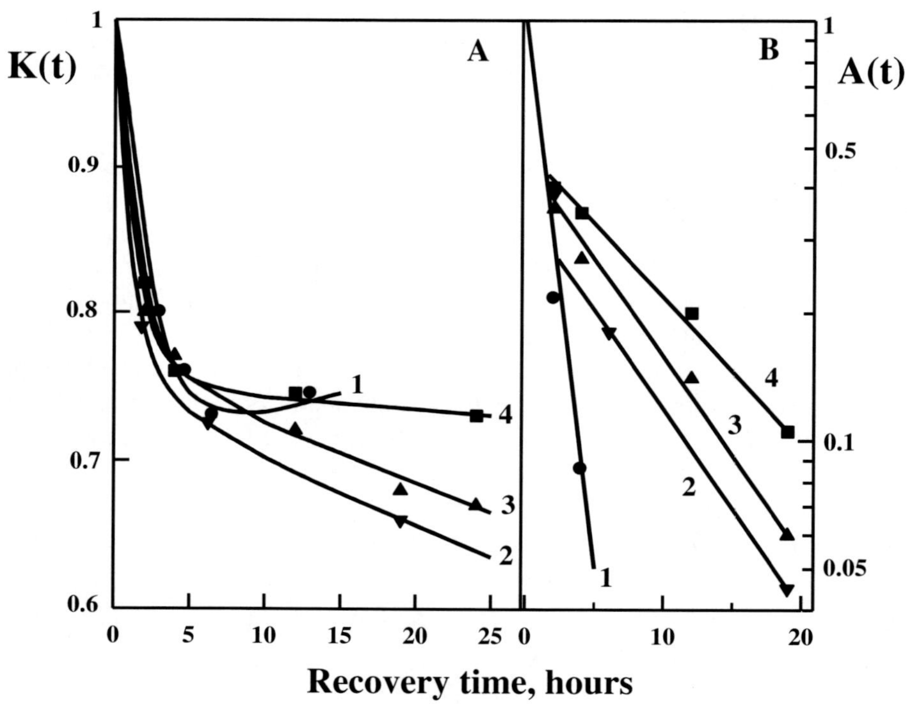

Figure 4.19. The proportion of unrecovered radiation damage, $K(t)$, (A) and recovered radiation damage, $A(t)$, (B) as a function of recovery time at 37 °C for Chinese hamster HA-1 cells after X-ray irradiation alone (curves 1) and after the subsequent combined action at first hyperthermia at 43 °C for 15, 30, and 45 min (curves 2, 3 and 4, respectively) and then X-rays of 16 Gy.

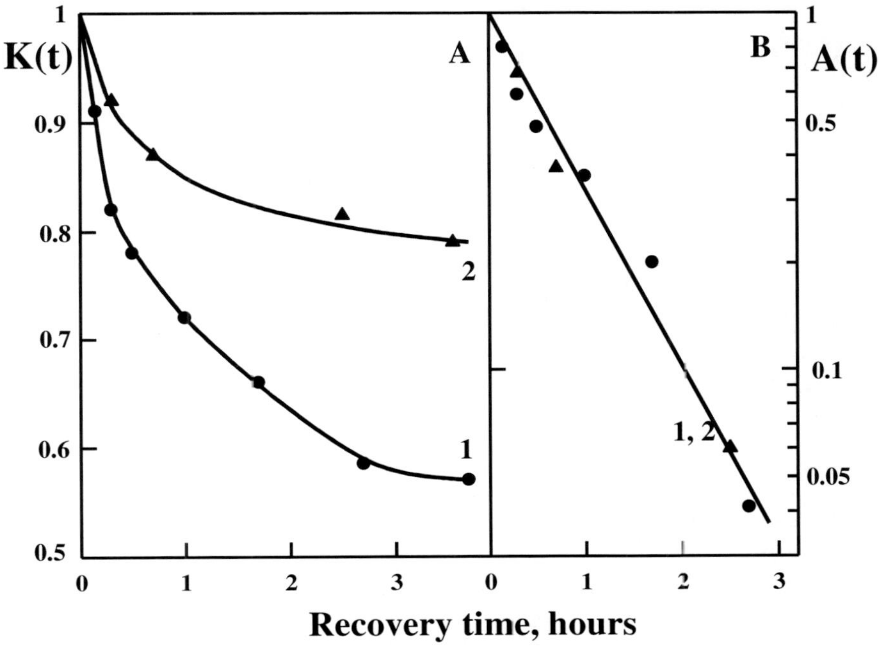

Figure 4.20. The proportion of unrecovered radiation damage, $K(t)$, (A) and recovered radiation damage, $A(t)$, (B) as a function of recovery time at 37 °C for Chinese hamster lung fibroblasts (line V79-S171) after X-ray irradiation alone (8 Gy, curves 1) and after the subsequent exposure to heat (42.5 °C, during 30 min) and then to X-irradiation at dose 8 Gy.

Table 4.5. Recovery parameters of Chinese hamster lung fibroblasts (line V79-S171) after X-ray irradiation alone (8 Gy) and after the sequential exposure to heat (42.5 °C during 30 min) and then to X-irradiation at dose 8 Gy as well as the reverse order

Treatment	Irreversible component K	Recovery constant β, min^{-1}
X-rays, 8 Gy	0.57	0.019
42.5 °C during 30 min + X-rays, 8 Gy	0.80	0.019
X-rays, 8 Gy + 42.5 °C during 30 min	0.92	–

Cells were recovered from potentially lethal damage after irradiation with ^{60}Co γ-rays (1.3 Gy/min) at doses 5 and 9 Gy as well as after the combined action of γ-ray at dose of 5 Gy and the subsequent exposure to heat at 45 °C during 10, 30 and 60 min. The outcome of our calculation is depicted in Figure 4.21.

These data allowed us to evaluate the irreversible component K (Eq. 1.12) and the recovery constant β (Eq. 1.15) after various treatment conditions. The results are listed in Table 4.6.

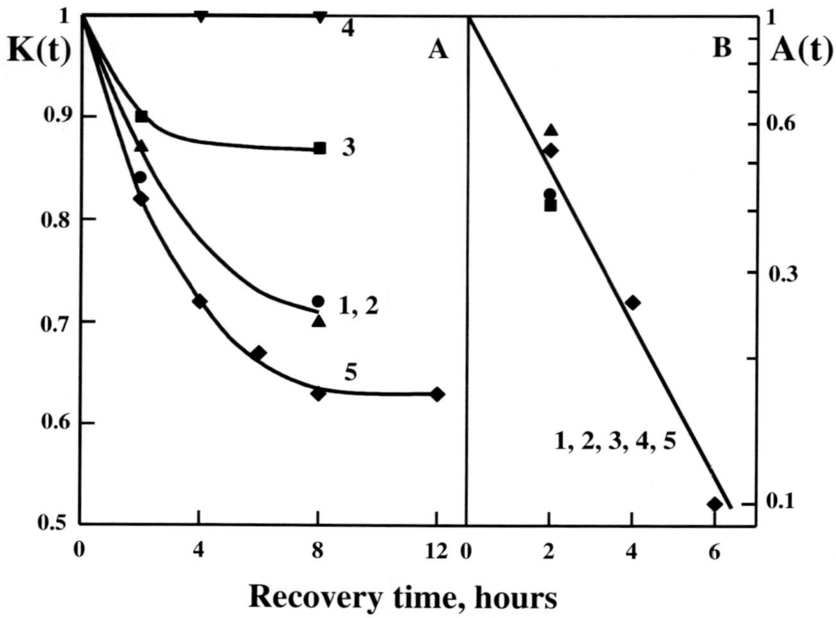

Figure 4.21. The proportion of unrecovered radiation damage, $K(t)$, (A) and recovered radiation damage, $A(t)$, (B) as a function of recovery time at 37 °C for cultured cells of human glioma (line U-87MG) after γ-ray irradiation alone at doses of 5 (circles, curve 1) and 9 Gy (curve 5) and after the combined action of γ-ray at dose of 5 Gy and the subsequent exposure to heat at 45 °C during 10, 30 and 60 min (curves 2, 3 and 4, respectively).

Table 4.6. Recovery parameters of cultured cells of human glioma (line U-87MG) after γ-ray irradiation alone at doses of 5 and 9 Gy and after the combined action of γ-rays at dose of 5 Gy and the subsequent exposure to heat at 45 °C during 10, 30 and 60 min

Treatment	Irreversible component K	Recovery constant β, min^{-1}
γ-ray, 5 Gy	0.72 (0.67 ± 0.04)	0.006
γ-ray, 9 Gy	0.63 (0.67 ± 0.04)	0.006
γ-ray, 5 Gy + 45 °C during 10 min	0.70	0.006
γ-ray, 5 Gy + 45 °C during 30 min	0.83	0.006
γ-ray, 5 Gy + 45 °C during 60 min	1.00	–

We first discuss the recovery parameters only after exposure to ionizing radiation. It is seen that the recovery constant β = 0.006 min^{-1} for both doses of ionizing radiation. It also shows that the irreversible component of radiation damage K is equal to 0.72 and 0.63 for cells exposed to doses of 5 and 9 Gy. As a rule, it is independent or even increase slightly with increase in dose of ionizing radiation [65]. One can believe that the two different values for the irreversible component are a simple reflect of possible variability in this parameter. Then the average value of the irreversible component K = (0.67 ± 0,04) and this parameter is increased to 0.70, 0.83 and 1.00 for the combined effect of ionizing radiation (5 Gy) and hyperthermia at 45 °C during 10, 30 and 60 min. This implies that the increase in duration of hyperthermia results in significantly increase in the proportion of irreversible radiation damage from which cells are incapable to recover. At the very long temperature exposure cell recovery completely suppressed. However, the recovery constant characterizing the probability of recovery per unit time is 0.006 min^{-1} both after X-ray irradiation alone and the preliminary use of hyperthermia and subsequent irradiation, i.e., 0.6% of the remaining repaired damage is recovered every minute. On the whole, these results indicate that the recovery inhibition after the combined application of ionizing radiation and hyperthermia on cultured human glioma cells (line U-87MG) could be proceeded through the increase in the irreversible damage, i.e., through he formation of a "severe" damage, from which the cells are not capable to recover, while the recovery processes themselves remained undamaged.

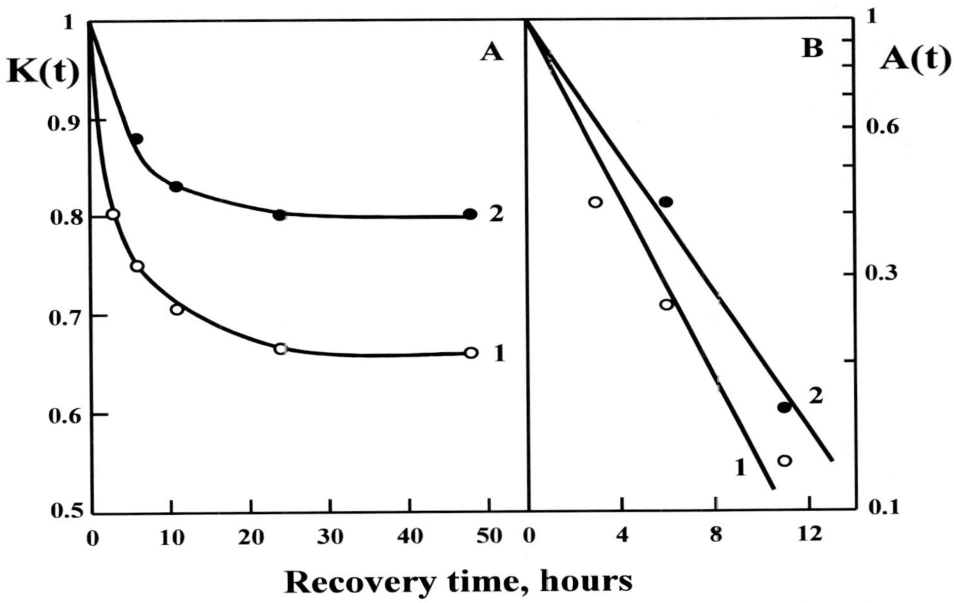

Figure 4.22. The proportion of unrecovered radiation damage, $K(t)$, (A) and recovered radiation damage, $A(t)$, (B) as a function of time after irradiation of EMT-6 tumor cells growing *in vivo* as a solid tumors after irradiation with ionizing radiation applied alone at dose of 20 Gy (curve 1) and the subsequent exposure to heat at 43 °C during 60 min (curve 2).

It is of interest to determine whether these patterns persist for cells from solid tumors growing *in vivo*. Close results were published by others [34] for mouse mammary sarcoma (line EMT-6). The tumor itself appeared to be radioresistant. The combination of hyperthermia and ionizing radiation, as author supposed, should result in greater cytotoxicity than can be accounted for by the addition of the cytotoxic effects of either agent employed alone. Cells were growing *in vivo* as solid tumors that were irradiated *in situ*, and a number of surviving cells was assayed by their ability to form clones in tissue culture *in vitro*. They were given either 20 Gy of X-irradiation alone or 20 Gy plus 43 °C for 1 hour immediately after X-irradiation. Tumors were then explanted for assessment of colony formation *in vitro* at various times after treatment and the amount of potentially lethal damage measured. The authors believed that hyperthermia was not only complimentary to radiation in its cell cycle specific killing response, but also acts to interfere with two different radiation damage recovery mechanisms (recovery from sublethal and potentially lethal radiation damage). They concluded that the interference of hyperthermia in these radiation repair systems may be a prime reason for the observed synergistic interaction of the two modalities. Because of large scattering of the initial experimental results we made some approximation of survival response of EMT-6 cells published. Then the proportion of unrecovered and recovered radiation damage was calculated in the dependence of time after treatment. The outcome is presented in Figure 4.22.

The data in this figure have allowed us to calculate the irreversible component of radiation damage K (Eq. 1.12) and the recovery constant β (Eq. 1.15) (Table 4.7). Note that the authors of the original papers did not make such kind of computation. It can be seen from these data that the irreversible component of radiation damage K, i.e., the proportion of irreversibly damaged cells, increases from 0.66 after X-ray application alone to 0.80 after combined thermoradiation action. Parallel, insignificant decrease of the recovery constant is observed, the mean value of $\beta = (0.19 \pm 0.1)\ h^{-1}$. It is quite probable that this decrease only reflects the possible variability of the parameter. This means that 18% from the residual repair damage is recovered every hour. This calculation show that the synergistic interaction of ionizing radiation and high temperature in solid tumors is mostly realized through the increase in the irreversible damage while the recovery processes may stay undamaged.

Table 4.7. Recovery parameters of EMT-6 tumor cells growing *in vivo* as a solid tumors after irradiation with ionizing radiation applied alone at dose of 20 Gy (curve 1) and the subsequent exposure to heat at 43 °C during 1 hour

Treatment	Biological object	Irreversible component K	Recovery constant β, h^{-1}
X-rays alone, 20 Gy	Mouse mammary sarcoma (EMT6)	0.66 ± 0.02	0.20 ± 0.02
X-rays 20 Gy + 43 °C for 1 hour	Mouse mammary sarcoma (EMT6)	0.81 ± 0.03	0.18 ± 0.02

As a whole, the data of this section demonstrate that the recovery inhibition of mammalian cells of various origins can be mainly accomplished due to formation of irreversible damage from which cells are incapable to recover while the recovery processes themselves stayed almost undamaged.

4.5. The Role of Chloramphenicol on Hyperthermic Cytotoxicity and Cell Permeability under Various Heating Rates

It would be expected that hypotonic culture medium should increase the differentiated cytotoxic effects of various heating rates and, on the contrary, hypertonic medium should decrease these effects. Thus, our aim here was to investigate the interaction between tonicity of the culture media, hyperthermic cell killing and cell permeability changes. As was mentioned above, some authors concluded that other factors within the cell can confer thermotolerance, in addition to the contribution made by heat shock protein, especially during mild heat challenges such as occurs during slow heating rates. So, it was of interest to study the combined action of heat with chemical agent which is known to inhibit expression of heat shock proteins. In addition, the results of the influence of protein synthesis inhibitor (chloramphenicol) on the development of thermotolerance at different heating rates were studied.

The repair-proficient bacterial strain *E. coli* B/r, the repair-deficient mutant *E. coli* $B_{s\text{-}1}$ and haploid yeast strain of *Z. bailii* were used in the present investigation. Before heating, the strains were grown in YEP broth (yeast extract – 1 %, glucose – 1 %, peptone – 1 %, NaCl – 1 %) at 37 and 30 °C during 18 and 48 h of incubation for bacterial and yeast cells, respectively. Cells were harvested 2x by centrifugation and washed in saline (0.85 % NaCl, pH 7.0). Final suspensions prepared in this manner contained approximately 10^8 cells per ml for bacterial and 10^6 cells per ml for yeast cells. Cellular permeability was estimated by measurement of optical activity (λ = 260 nm) of the supernatant of cell samples. Prior to heat treatment, cells from plateau phase were diluted in 0.01 M phosphate buffer, pH 7.0, 0.01 M phosphate buffer containing 0.9 or 6 % of NaCl, or YEP broth with or without chloramphenicol (1O μg/ml). This concentration of chloramphenicol is known to inhibit protein synthesis [23]. Transfer of cell samples to solution with chloramphenicol was made at least 30 min before thermal treatment [41]. The cells were then cooled at 0 °C for 20 min prior to heating in order to increase heating rate and inhibit cell metabolism. This procedure was checked in special experiments and does not affect cell killing in subsequently heated cells. After this treatment, the cell samples were heated at different heating rates (mean heating rates were 0.015, 0.25 and 1.5 °C per s) from 0 to 52 °C, over time period of 1 h, 3.5 min, and 35 s. After having reached the peak temperature (52 °C), the cell survival was close to 100% independently of cell species and heating rate. Then the cell samples were heated at 52 °C for different time periods. The following methods of heat treatment were used: (i) 5 ml plastic tubes with the cooled at 0 °C cell suspension placed into a water circulator with water cooled at 0 °C and then the temperature of water circulator and cell suspension was increased up to 52 °C in 60 min (mean heating rate was about of 0.015 °C/s); (ii) 5 ml plastic tubes with cell suspension cooled at 0 °C placed into a water circulator previously heated to 52 °C and then the cell samples were heated from 0 to 52 °C over 3–5 min (mean heating rate was about of 0.25 °C/s); (iii) 0.05 ml of cell suspension at 0 °C was put into the tubes containing 5 ml suspension liquid preheated to 52 °C; in this case, the cells were heated from 0 to 52 °C at 35 s (mean heating rate was about of 1.5 °C/s). After having reached the peak temperature (52 °C), the cell samples were heated at 52 °C for different time periods. Here we presented the averaged results published in our earlier works [45; 46].

Figure 4.23 exhibits the effect of different rates of heating on the *E. coli* B/r cell heat survival curves. In all figures, the error bars show the standard deviation of the mean from different experiments. It is clear that the bacterial cells were much more sensitive to heat after heating at the highest heating rate (1.5 °C/s) in isotonic phosphate buffer.

To test the effect of heat shock proteins on the different cell responses to various heating rates, chloramphenicol (10 μg/ml) was added before and was present during hyperthermic treatment of cells in YEP broth and isotonic phosphate buffer. As shown in Figure 4.24, slowly heated cells in nutrient broth reveal a higher level of thermotolerance than cells slowly heated in phosphate buffer.

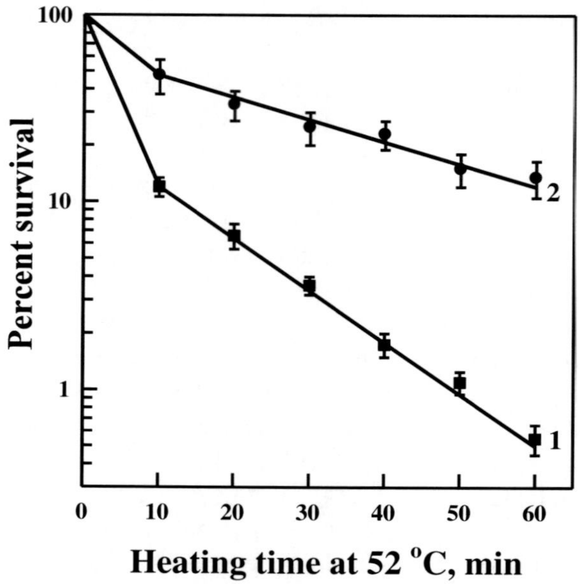

Figure 4.23. The effect of different heating rates (closed squares: 1.5 °C per s; closed circles: 0.015 °C per s) on survival curve of *E. coli* B/r cells heated in isotonic 0.01 M phosphate buffer (pH 7.0 and 0.9 % NaCl) at 52 °C.

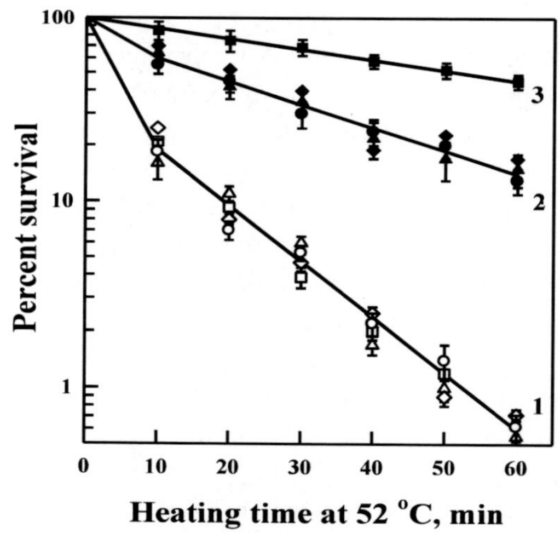

Figure 4.24. The influence of the heating time at 52 °C on thermal cytotoxicity of *E. coli* B/r cells heated at different heating rates (open signs: 1.5 °C per s; closed signs: 0.015 °C per s) in isotonic 0.01 M phosphate buffer (with 0.9 % NaCl) with (open and closed circles) and without (open and closed triangles) chloramphenicol (10 µg/ml) and in YEP broth without (open and closed squares) and with (open and closed rhombs) chloramphenicol.

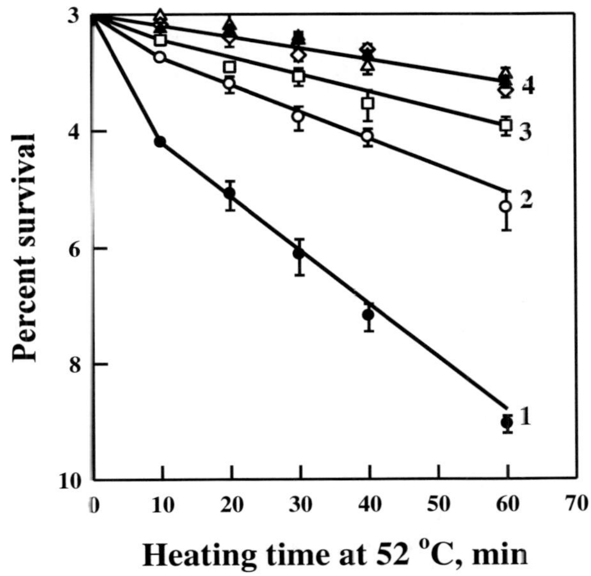

Figure 4.25. The influence of the heating time at 52 °C on thermal cytotoxicity of *E. coli* B/r cells heated at different heating rates (closed circles and triangles: 1.5 °C per s; open circles and triangles: 0.25 °C per s; open squares and rhombs: 0.015 °C per s) in the heat media with various tonicity (closed and open circles, open squares: hypotonic 0.01 M phosphate buffer, pH 7.0; closed and open triangles, open rhombs: hypertonic 0.01 M phosphate buffer, pH 7.0, containing 6% sodium chloride).

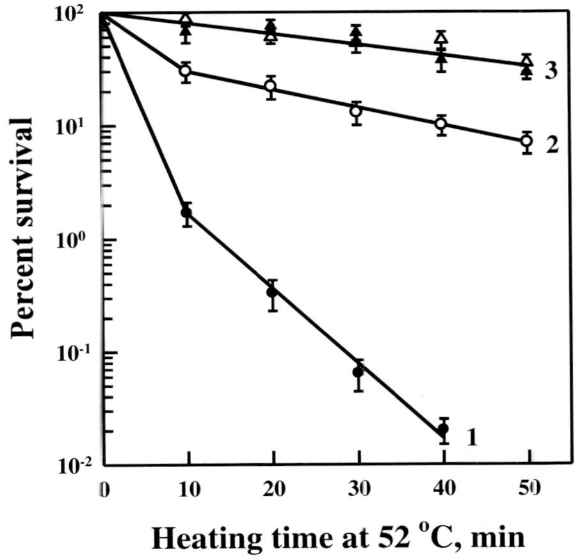

Figure 4.26. The influence of the heating time at 52 °C on thermal cytotoxicity of *E. coli* B_{s-1} cells heated at different heating rates (closed circles and triangles: 1.5 °C per s; open circlers and triangles: 0.015 °C per s) in the heat media with various tonicity (closed and open circles: hypotonic 0.01 M phosphate buffer, pH 7.0; closed and open triangles: hypertonic 0.01 M phosphate buffer, pH 7.0, containing 6% sodium chloride).

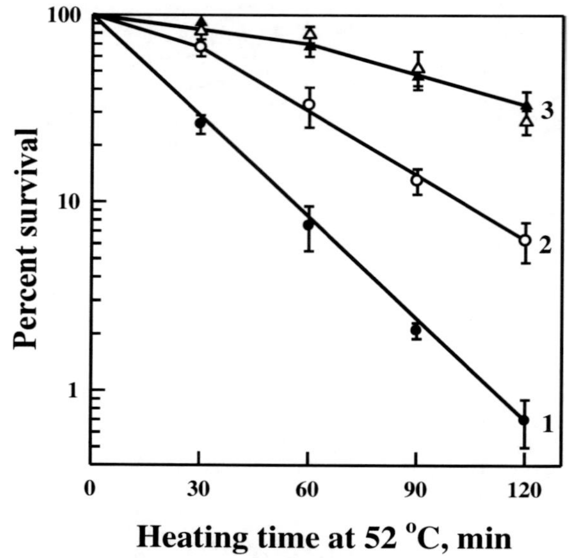

Figure 4.27. The influence of the heating time at 52 °C on thermal cytotoxicity of *Z. bailii* haploid yeast cells heated at different heating rates (closed circles and triangles: 1.5 °C per s; open circlers and triangles: 0.015 °C per s) in the heat media with various tonicity (closed and open circles: hypotonic 0.01 M phosphate buffer, pH 7.0; closed and open triangles: hypertonic 0.01 M phosphate buffer, pH 7.0, containing 10 % sodium chloride).

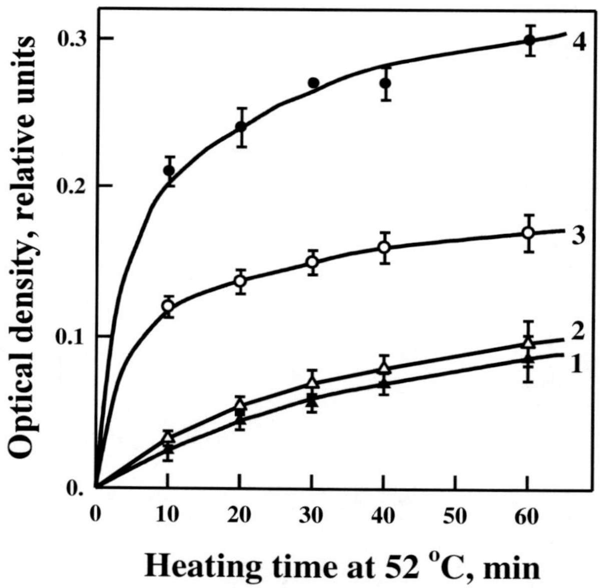

Figure 4.28. The effect of heating rates (open circles and triangles: 0.015 °C per s; closed circles and triangles: 1.5 °C per s) and tonicity of heat media (open and closed circles: hypotonic 0.01 M phosphate buffer, pH 7.0; closed and open triangles: hypertonic 0.01 M phosphate buffer, pH 7.0, containing 6% sodium chloride) on optical activity ($\lambda = 260$ nm) of the supernatant of cell samples of *E. coli* B/r cells.

This additional thermotolerance was inhibited by chloramphenicol. However, even in this case, the majority of the induced thermotolerance was expressed in the presence of chloramphenicol in the YEP broth media. In addition, it can be seen that the compound had no effect on cell thermoresponse after heating in phosphate buffer. Figure 4.25 illustrates the dependence of *E. coli* B/r survival on the tonicity of heated cell suspensions in cultures exposed to different heating rates. These data shows that the difference in cytotoxicity at various heating rates was more profound in hypotonic media and disappeared in hypertonic media.

The absence of heating rate effect in hypertonic conditions seems to be due to thermoprotective actions of hypertonicity which were more expressed in cells exposed to the highest heating rate. Comparable results also were observed for the repair-deficient mutant *E. coli* B_{s-1} (Figure 4.26) and haploid yeast *Z. bailii* (Figure 4.27).

The effect of heating rates and tonicity of medium on cell permeability was estimated by measurement of optical activity (λ = 260 mm) of the supernatant of cell samples. Cell lysis *per se* was not observed in these experiments since total cell count before and after heating was identical. Hence, the optical density changes of the supernatant was not related to cell lysis but only to loss of intracellular solution. As shown in Figure 4.28, cell membrane permeability was increased most profoundly when hyperthermic treatment of *E. coli* B/r cells were performed at the highest heating rate in hypotonic solution. The cells heated at hypertonic conditions exhibited no such response. Membrane permeability of cells in hypertonic media was little altered and was essentially the same when heating occurred slowly or rapidly and was far less affected even than in cells heated slowly in hypotonic media.

The present results demonstrates the influence of heating rate on thermal cytotoxicity in bacterial and yeast cell systems and substantiates the data of other reports in mammalian cells [19; 75]. Taking our results and the data of these other authors as a whole, we can conclude that the effect of heating rate is expressed in both pro- and eukaryotic cells.

CONCLUSION

Thus, in our experiments with diploid yeast cells exposed to hyperthermia and ionizing radiation it was shown that both the extent and the rate of the liquid holding recovery were considerably decreased with an increase in the duration of the heat treatment (60 °C) followed by radiation and with the exposure temperature after a simultaneous treatment with heat and radiation. The probability of recovery was independent of the conditions of the thermoradiation action while the irreversible component gradually increased as a function of the duration of the heat treatment after a sequential thermoradiation action and as a function of the exposure temperature after a simultaneous treatment with heat and radiation. The increase in the irreversible component was accompanied by an increase in the cell death without any postirradiation division. It can be concluded on this basis that the synergistic interaction of hyperthermia with ionizing radiations in yeast cells is not related to the impairment of the recovery capacity itself and that it may be attributed to an increased yield of the irreversible damage. This means that some general mechanism of radiosensitization or synergistic effects may underlie the interaction of heat with ionizing radiation.

Similar results have been obtained for *S. cerevisiae* yeast cells simultaneously exposed to heat and UV light radiation. We analyzed whether the synergistic effect may be causally related to the impairment of the recovery capacity itself or to the production of irreversible damages, which cannot be repaired. The extent and the rate of recovery were shown to be gradually decreased with an increase in the exposure temperature. A quantitative approach, describing the recovery process as a decrease in the effective dose, was applied to estimate the irreversible component and the recovery constant after the simultaneous action of heat and UV light. It was shown that the irreversible component increased greatly as a function of the exposure temperature. Moreover, the increase in the irreversible component was accompanied by an increase in cell death without division. To elucidate whether the reduction in the yeast cell recovery was only due to the increase in the irreversible component or was also related to the impairment of the LHR, the recovery constant describing the probability of recovery per unit time was estimated. This parameter appeared to be unchanged (0.071 h^{-1}) with the exposure temperature. It means that the same portion (about 7.0 %) of reparable radiation damage is eliminated each hour independently of the conditions of treatment with heat and UV light radiation. Hence, the reduced rate of recovery takes place not because the process of the LHR is damaged itself but only because of the enhanced yield of irreversible damage.

It would be of interest to speculate on the basis of this study about the mechanism of synergistic interaction of heat with ionizing radiation or UV light. The bell-shaped curve for synergistic interaction of heat and UV light observed here and in our previous publications [50, 55] has been also obtained for other agents, including UV light and methotrexate [6], UV light and cadmium chloride [18], X-rays and chemical mutagen (1,2-dibromoethane) [8; 33], for interaction of some chemical agents [38]. It follows that the synergistic interaction may take place for factors after which the cells are incapable of recovery at all. These data clearly indicate that the mechanism of synergistic interaction unlikely related with the damage of recovery processes. This implies also the need to develop another mechanism of synergistic interaction. Then such an interaction can be considered as an alternative explanation for the production of irreversible damage.

Because the synergistic effects are not related to direct damage to the recovery process itself, it is not excluded that synergism would be expected to result from some additional lethal lesions arising from the interaction of sublesions induced by both modalities. These sublesions should be considered ineffective after each agent is applied alone. This hypothesis has been put forward and applied in the interpretation and mathematical modeling of biological effects produced by radiation and heat applied separately and in combination with each other by many authors [9; 28; 32; 33; 47; 54; 55; 68]. Based on the results of this Chapter, the hypothesis of sublesion interaction could be extended to include the possibility that this additional lethal damage could not be repaired or was less repairable. It was concluded that the reduced rate and volume of recovery is not related with damage of repair and recovery processes after combined treatments, but rather they can be expected and predicted as a consequence of irreversible damage production.

It would be of interest to put forth some hypothesis regarding the yield of irreversible component of damage. In our LHR experiments, after heat and radiation treatment the yeast cells were hold for up to 100 h, in the absence of nutrients. We observed that the recovery rate and the extent of recovery decrease with increasing heating temperature or duration of heating. It is not excluded that the increase in the irreversible component may be related to the available energy resources remaining in the cells and that the extent of recovery may be proportional to those energy reserves. It is not the case for ionizing radiation alone where the irreversible component does not increase with radiation dose [30; 65] excluding tremendous radiation doses resulting in interphase cell killing. However, it is possible that the energy reserves after treatment with both heat and radiations may be depleted in proportion to the severity of the treatment and the energy required for recovery may increase in proportion to the severity of the damage.

The formulated hypothesis of synergistic interaction may be attractive for interpreting a number of the characteristics of the synergistic interaction of heat and radiation seen in yeast and other cells. For example, the lack of the synergy in radiosensitive yeast mutants defective in repair ability [52] may be a result of the reduced number of sublesions that could interact, because sublesions for wild-type cells can be lethal for radiosensitive mutants. A similar explanation can be given for reduced synergy seen in haploid yeast cells [51], which are known to be incapable of repairing of DNA double-strand breaks because two homologous duplexes are required [37]. The increase in the synergy at relatively low temperatures before the maximum level of synergy has been achieved can be explained by the relatively small amount of both lethal damages and sublesions produced by heat. Conversely, the decrease in synergy at relatively high temperatures after the maximum level of synergy has been achieved can be explained by the relatively small amount of lethal damages and sublesions produced by radiation. Thus it can be expected that the greatest synergy would be obtained when the equal number of sublesions are produced by heat and by radiation. Since the magnitude of the synergistic effect would depend directly on the amount of sublesions available, sublesion repair would contribute to a decrease in the effect. Hence, the explanation that simultaneous exposure is more effective than sequential exposures becomes evident.

We studied quantitative parameters describing the repair of DNA breaks in mammalian cells after successive treatment with heat and ionizing radiation. Original experimental results have been published by others [21]. The aim of our study was to determine whether the inhibition of DNA repair of radiation-induced strand breaks by hyperthermia may causally be related to the impairment of the repair capacity itself or to the production of irreversible damage, which cannot be repaired. The findings revealed that (i) the irreversible component of radiation damage was gradually enhanced as the duration of heat exposure increased; (ii) the probability of the fast repair, was independent of whether the repair process happened after ionizing radiation applied alone or combined with a preheating sensitizing the radiation effect; (iii) the portion of slow repairing damage increased with an increase in the thermal load; (iv) the probability of the slow DNA breaks repair became considerably less after an exposure to a high thermal load when the whole cell population was repaired due to only this component of repair. It would seem probable that the first inference may be explained by the conversion of the radiation induced repairable damage by hyperthermia so that the enzymes could not deal with the lesions or the hyperthermia could interfere with the synthesis of the requisite enzymes. The second inference would imply that the same portion of the fast repairable radiation damage is eliminated for a unit time independently of the preheating conditions investigated. This result strongly suggests that the combined actions of preheating and ionizing radiation do not impair or damage the fast repair process itself. As regards to the third inference, several authors [13] have considered the fast and slow phases of repair to represent the restoration of single- and double-strand breaks in DNA, respectively. Others have proposed that different repair rates may represent a repair of the lesions in high accessible and low accessible regions of DNA or the involvement of different enzymes (polymerase α or β) [14]. The last inference means that the combination of ionizing radiation and a high thermal load when the contribution of the fast repair is negligible may result in some impairment of the slow repair process. Thus, the observed inhibition of DNA repair by hyperthermia may be realized through two processes. The basic process took place at all the conditions of the thermoradiation action tested, which appears to be a reduction in the number of cells capable to repair owing to an increase in the irreversible damage. The second process was due to some impairment of the slow DNA repair. It was shown by a decrease of the slow repair constant and realized only after a high thermal load, when the cell inactivation is mostly determined by a heat rather than by the synergistic interaction of heat and radiation.

Detailed analysis of molecular mechanisms involved is beyond the main aim of this study. However, it is noted that many mechanisms have been discussed by different authors [3; 10; 12; 14; 17; 21; 54; 62]. Only as examples, the following possibilities can be mentioned. Lesions produced by one type of agent could affect the lesions already present in the DNA in such a way that a repair becomes less feasible and more difficult. Evidence has been accumulated that it might be due to a protein aggregation when the damaged DNA becomes less accessible to the DNA repair mechanism and this (rather than repair enzyme inactivation *per se*) is responsible for the observed thermal inhibition of the DNA repair. Because of the well-known fact that ionizing radiation inhibits a replicative DNA synthesis [14], it can be admitted that a decrease in a quantity of these enzymes wouldn't interfere with the probability of recovery but result in a greater portion of irreversible damage. Similar situation may be realized with an insufficient energy metabolism.

We have used a number of other results published in literature [34; 36; 58; 59] in order to estimate the parameters of a cultured mammalian cell recovery and to reveal whether the interaction of ionizing radiation and hyperthermia on cell survival was related to the impairment of the recovery capacity itself or to the production of irreversible damage that cannot be recovered at all. It was shown that the portion of the cells incapable of a recovery after a consecutive thermoradiation action was increased with a thermal load increase under an insignificant change of the repair constant.

The molecular mode of the repair inhibition is still unclear. There is now abundant evidence pointing that hyperthermic inhibition of the rate and/or extent of repair of radiation-induced DNA damage which was observed with all the assays currently used for DNA damage assessment may reflect the production of irreversible damage rather than the impairment of recovery mechanisms themselves after the combined exposures. In fact, evidence has been accumulated that hyperthermic radiosensitization might be due to the nuclear protein aggregation that damaged DNA becomes less accessible for the DNA repair machinery and this (rather than repair enzyme inactivation *per se*) is responsible for the observed thermal inhibition of DNA repair [24; 25; 61; 76]. Thermal repair inhibition is generally described as reduced repair rates leading to an increased number of residual strand breaks [73]. It was previously shown by Mills and Meyn [40] that for a combined treatment with X-irradiation and heat, the decrease in cell survival correlated with the increase in the number of non-rejoined strand breaks. The combined treatment appeared to affect the repair rates but also lead to the formation of additional strand breaks [12]. Residual additional strand breaks then can result in additional chromosome aberrations [11; 74] finally producing the enhanced cell killing. Considering all these data as a whole, we can conclude that they don't contradict the results obtained in this study. However, detailed analysis of the molecular mechanism of synergistic interaction and production of irreversible damage is beyond the main aim of this paper. Moreover, it can be envisaged that they might be different for various biological objects, test effects and acting agents.

We tested the possibility that the effect of heating rate on cytotoxicity might be caused by differential induction of heat shock proteins. This assumption was presumed by other authors [1; 2; 35]. To test the influence of HSP on the effects of heating rate, chloramphenicol was added before and was present during hyperthermic treatment of *E. coli* B/r cells to YEP broth and isotonic phosphate buffer. Chloramphenicol is known to be an inhibitor of protein synthesis [15] and bacterial cell thermotolerance [23; 77]. In our experiments, however, the presence of chloramphenicol did not affect the results when cells were heated in isotonic phosphate buffer and only partly prevented the development of thermotolerance in cells heated in nutrient medium. Cellular thermotolerance phenomena have mainly been investigated in nutrient media [2; 41; 70; 77]. Our results in bacterial cells demonstrated a minor additional degree of thermotolerance is induced during slow heating in nutrient medium compared with that observed in phosphate buffer and that this additional thermotolerance was prevented by chloramphenicol. These results suggest that no thermotolerance related to heat shock proteins levels is present in our experiments, although heat shock proteins levels were not directly measured. These results also suggests that other reasons, besides the induction of heat shock proteins, may be responsible for differences in thermosensitivity of cells heated at various rates in non-nutrient media. Thus, it is apparent from the data presented that even in nutrient media the greater portion of thermotolerance was induced by slow heating even when chloramphenicol was present and, presumably, heat shock proteins induction was prevented. As a resume, this conclusion is supported by the following findings: (i) effects of heating rate on thermal cytotoxic action that more expressed in hypotonic conditions; (ii) when similar experiments were carried out using hypertonic conditions no significant difference in cellular responses to heat treatment related to various heating rates were observed; (iii) suppression of the heating rate effects in hypertonic media was found to be due to the protection from thermal injury induced at the highest heating rate, but not to an alteration in the thermoresistance of slowly heated cells; (iv) the magnitude of cell killing and cell permeability were altered in the same direction and were directly proportional to heating rate; and (v) the major portion of the thermotolerance induced by slower heating rates is not due to the altered production of heat shock proteins in bacterial cells heated in nutrient media and thermotolerance appears to be unrelated to the induction of these proteins when cells are heated in non nutrient media.

Based on various experimental data, other authors [2; 69] also concluded that further factors within the cell can confer differences in thermosensitivity of rapidly and slowly heated cells in addition to the contribution made by heat shock proteins. It seems logical to suppose the significant role of tonicity in the effects observed may be explained by a more intensive increase in intracellular osmotic pressure after high heating rate in comparison with that which occurs after a low heating rate. The enlarged cell volume and increased cholesterol synthesis observed by Herman et al. [19] may have resulted in high cell membrane rigidity during rapid heating as well as in increased cell permeability and cell killing [19; 75] and may represent inevitable cellular responses to thermogenic increase in intracellular osmotic pressure induced by high heating rates.

In addition, the mathematical approach used here may be helpful to search thermoradiation regimens or other inactivating agents combined with ionizing radiation, selectively acting on the probability of recovery and the yield of irreversible radiation damage. The recognition that specific inhibitors of recovery may exist, such as an inhibitor of recovery process itself and that resulting in the increased yield of irreversible damage, provides both a possibility to analyze the mechanism of agents combined with ionizing radiation from this point of view and an expectation that useful regimens in cancer research may be devised to make use of these inhibitors.

Hence, the following new fundamental data have been obtained here.

1. The recovery constant characterizing the probability of yeast cell recovery per unit time, after combined action of heat with ionizing radiation or UV light, was independent of various conditions of the thermoradiation action. Hence, the synergistic interaction of ionizing or UV radiations with heat is not related with the impairment of the recovery capacity itself.
2. The irreversible component of radiation damage gradually increased as a function of thermal load after combined action of heat with ionizing or UV radiations on yeast cells. The rise of the irreversible component was accompanied by an increase of cell killing without postirradiation division. This means that mechanism of synergistic interaction may be attributed to the enhanced yield of irreversible damages.
3. The quantitative estimation of mammalian cell recovery parameters strongly suggests that the recovery processes themselves are not damaged after combined action of various chemical agents and ionizing radiation. It was concluded that the synergistic effect of combined action of various chemical sensitizers is mainly due to the enhanced yields of the irreversible damages, while the probability of recovery per unit time was almost independent on the exposure conditions.
4. Synergistic interaction of ionizing radiation with hyperthermia on cultured mammalian cells of various origins is not related with the damage of repair systems themselves but can be explained by the production of irreversible damages which cells are incapable of recovery.
5. The reduced rate and volume of recovery is not related with the damage of repair and recovery processes after combined treatments, but rather they can be expected and predicted as a consequence of irreversible damage production. Therefore some general mechanism of radiosensitization or synergistic effects may underlie the interaction of heat with ionizing radiation.
6. The role of chloramphenicol on hyperthermic cytotoxicity and cell permeability under various heating rates was elucidated. Our results with bacterial cells suggest that cell thermotolerance is not always related to heat shock protein levels and other reasons may be responsible for differences in thermosensitivity of cells heated at various rates in non-nutrient media.

REFERENCES

[1] Alexandrov, VY; Kislyuk, IM. Cell responses to the heat shock: physiological aspect. *Cytology*, 1994; 36, 5-59.

[2] Anderson, RL; Herman, TS; van Kersen, BA et al. Thermotolerance and heat shock protein induction by slow rates of heating. *Int. J. Radiat. Oncol. Biol. Phys.*, 1988; 15, 717-725.

[3] Ben-Hur, E. Mechanisms of the synergistic interaction between hyperthermia and radiation in cultured mammalian cells. *J. Radiat. Res.,* 1976; 17, 92-98.

[4] Ben-Hur, E; Elkind, MM; Bronk, BV. Thermally enhanced radio-response of cultured Chinese hamster cells – Inhibition of repair of sublethal damage and enhancement of lethal damage. *Radiat. Res.*, 1974; 58, 38-51.

[5] Bodel, WJ; Cleaver, JE; Roti, JL. Inhibition by hyperthermia of repair synthesis and chromatin reassembly of ultraviolet-induced damage to DNA. *Radiat. Res.*, 1984; 100, 87-95.

[6] Borchers, AH; Kennedy, KA; Straw, JA. Inhibition of DNA excision repair by methotrexate in Chinese hamster ovary cells following exposure to ultraviolet irradiation or ethylmethanesulfonate. *Cancer Res.*, 1990; 50, 1786-1789.

[7] Burns, CP; Lambert, BJ; Haugstad, BN et al. Influence of rate of heating on thermosensitivity of L1210 leukemia: membrane lipids and Mr 70,000 heat shock protein. *Cancer Res.*, 1986; 46, 1882-1887.

[8] Cebulska-Wasilewska, A; Leenhouts, HP; Chadwick, KH. Synergism between EMS and X-rays for the induction of somatic mutations in Tradescantia. *Int. J. Radiat. Biol.*, 1981; 40, 163-173.

[9] Chadwick, KH; Leenhouts, HP. *The Molecular Theory of Radiation Biology*. Berlin, Heidelberg, New York: Springer-Verlag, 1981.

[10] Dahm-Daphi, J; Brammer, I; Dikomey, E. Heat effects on the repair of DNA double-strand breaks in CHO cells. *Int. J. Radiat. Biol.*, 1997; 72(2), 171-179.

[11] Dewey, WC; Sapareto, SA; Betten, DA. Hyperthermic radiosensitization of synchromous Chinese hamster cells: Relationship between lethality and chromosomal aberrations. *Radiat. Res.*, 1978; 76, 48-59.

[12] Dikomey, E; Jung, H. Correlation between thermal radiosensitization and slowly rejoined DNA strand breaks in CHO cells. *Int. J. Radiat. Biol.*, 1995; 8(3), 227-233.

[13] Frankenberg, D; Brede, HJ; Schrewe, UJ et al. Induction of DNA double-strand breaks in mammalian cells and yeast. *Adv. Space Res.*, 2000; 25(10), 2085-2094.

[14] Friedberg, EC; Walker, GC; Siede, W; Wood, RD; Schultz, RA; Ellenberger, T. *DNA Repair and Mutagenesis, 2nd edition*. Washington DC: ASM Press, 2006.

[15] Gillies, NE; Alper, T. Reduction in the lethal effects of radiations on *E. coli* by treatment with chloramphenicol. *Nature*, 1959; 183, 237-238.

[16] Grishchenko, VI; Mejhidov, SH; Moiseev, VA et al. Effect of temperature and concentration of different substances on erythrocyte microviscosity cytosole. *Biophysica*, 1995; 40, 106-109.

[17] Hall, EJ. *Radiobiology for the Radiologist*. Lippincott: Williams and Wilkins, 2000.

[18] Hartwig, A; Beyersmann, D. Comutagenicity and inhibition of DNA repair by metal ions in mammalian cells. *Biol. Trace Element Res.*, 1989; 21, 359-365.

[19] Herman, TS; Gerner, EW; Magun, BE et al. Rate of heating as a determinant of hyperthermia cytotoxicity. *Cancer Research*, 1981; 41, 3519-3523.

[20] Jain, VK; Pohlit, W. *Byocybernetic of Cancer. Optimizing Cancer Treatment with Ionizing Radiation*. Banglore: INSDOC, 1986.

[21] Jorritsma, JB, Konings, AW. Inhibition of repair of radiation-induced strand breaks by hyperthermia, and its relationship to cell survival after hyperthermia alone. *Int. J. Radiat. Biol.*, 1983; 43(5), 505-516.

[22] Jozwiak, Z; Palecz, D. Influence of heat treatment on osmofic fragility of carp erythrocytes. *Int. J. Radiat. Biol.*, 1988; 54, 299-303.

[23] Kagramanova, VK. Inhibition of protein synthesis of archaebacteria. *Progress in Modern Biology*, 1995; 115, 563-572.

[24] Kampinga, HH. Cell biological effects of hyperthermia alone or combined with radiation or drugs: a short introduction to newcomers in the field. *Int. J. Hyperthermia*, 2006; 22, 191-196.

[25] Kampinga, HH; Dynlacht, JR; Dikomey, E. Mechanism of radiosensitization by hyperthermia (> or = 43 degrees C) as derived from studies with DNA repair defective mutant cell lines. *Int. J. Hyperthermia*, 2004; 20, 131-139.

[26] Kiefer, J. The effect of caffeine on survival of UV-irradiated diploid yeast strains of different sensitivities. *Mutat. Res.*, 1975; 30, 317-326.

[27] Kim, JK; Komarova, LN; Kim, SH et al. Inhibition of DNA repair from radiation damages by hyperthermia is mainly realized through the production of irreversible damage. *Ind. J. Radiat. Res.,* 2007; 4(3-4), 188-195.

[28] Kim, JK; Petin, VG; Zhurakovskaya, GP. Exposure rate as a determinant of synergistic interaction of heat combined with ionizing or ultraviolet radiations in cell killing. *J. Radiat. Res.*, 2001; 42, 361-365.

[29] Korogodin, VI. Some regularities of the post-irradiation changes in resting yeast cells. *Biofizika,* 1958; 3, 703-710.

[30] Korogodin, VI. The study of post-irradiation recovery of yeast: the premolecular period. *Mutat. Res.*, 1993; 289, 17-26.

[31] Kushner, DJ. *Microbial Life in Extreme Environments*. Edition. London, New York, San Francisco: Academic Press, 1978.

[32] Leenhouts, HP; Chadwick, KH. An analysis of synergistic sensitization. *Brit. J. Cancer*, 1978; 37, Suppl. III, 198-201.

[33] Leenhouts, HP; Sijsma, MJ; Cebulska-Wasilewska, A et al. The combined effect of DBE and X-rays on the induction of somatic mutations in Tradescantia. *Int. J. Radiat. Biol.*, 1986; 49, 109-119.

[34] Leith, JT; Miller, RC; Gerner, EW et al. Hyperthermic potentiation. Biological aspects and applications to radiation therapy. *Cancer*, 1977; 39(2 Suppl), 766-779.

[35] Li, GC. Elevated levels of 70,000 Dalton heat shock protein in transiently thermotolerant Chinese hamster fibroblasts and their stable heat resistant variants. *Int. J. Oncol. Biol. Phys.*, 1985; 11, 165-177.

[36] Li, GC; Evans, RC; Hahn, GM. Modification and inhibition of potentially lethal X-ray damage by hyperthermia. *Radiat. Res.*, 1976; 67, 491-501.

[37] Luchnik, AN; Glaser, VM; Shestakov, SV. Repair of DNA double-strand breaks requires two homologous DNA duplexes. *Mol. Biol. Rep.*, 1977; 3, 437-442.

[38] Maehara, H; Iwami, Y; Mayanagi, H et al. Synergistic inhibition by combination of fluoride and xylitol on glycolysis by mutants streptococci and its biochemical mechanism. *Caries Res.*, 2005; 39, 521-528.

[39] Mejhidov, SH; Moiseev, VA; Influence of temperature on erythrocyte hydration. *Biophysica*, 1991; 36, 294-297.

[40] Mills, MD; Meyn, RE. Effects of hyperthermia on repair of radiation-induced DNA strand breaks. *Radiat. Res.*, 1981; 87, 314-328.

[41] Mizzen, LA; Welch, WJ. Characterization of the thermotolerant cell. I. Effects on protein synthesis activity and the regulation of heat-shock protein 70 expression. *J. Cell Biol.*, 1988; 106, 1105-1116.

[42] Mondovi, B; Agro, FA; Rotilio, G et al. The biochemical mechanism of selective heat sensitivity of cancer cells II. Studies on nucleic acids and protein synthesis. *Eur. J. Cancer*, 1969; 5, 137.

[43] Morozov, II. Cellular injury by heating and osmotic homeostasis. *Medical Radiology*, 1992; 37, 39-42.

[44] Morozov, II; Dergacheva, IP; Ansimova, NS. The effect of an osmotic shock on the viability, optic density and permeability of heated Escherichia coli cells. *Microbiology*, 1986; 55, 278-281.

[45] Morozov, II; Petin, VG. Modification of cell damage, caused by variable rate heating, by means of changes in osmotic pressure of the medium or using chloramphenicol. *Tsitologiia*, 1998; 40(2-3), 178-184.

[46] Morozov, II; Petin, VG; Dubovick, BV. Influence of tonicity and chloramphenicol on hyperthermic cytotoxicity and cell permeability under various heating rates. *Int. J. Hyperthermia*, 1997; 13, 49-57.

[47] Murthy, MS; Deorukhakar, VV; Rao, BS. Hyperthermic inactivation of diploid yeast and the interaction of damage caused by hyperthermia and ionizing radiation. *Int. J. Radiat. Biol.*, 1979; 35, 333-341.

[48] Novitsky, TJ; Kushner, DJ. *Planococcus halophilus* sp. nov., a facultatively halophilic coccus. *Int. J. Systemic Bacteriology*, 1976; 26, 53-57.

[49] Patrick, MH; Haynes, RH. Dark recovery phenomena in yeast. II. Conditions that modify the recovery process. *Radiat. Res.*, 1964; 23, 564-579.

[50] Petin, VG. *Genetic Control of Cell Radiosensitivity Modification*. Moscow: Energoatomizdat, 1987.

[51] Petin, VG; Berdnikova, IP. Effect of elevated temperatures on the radiation sensitivity of yeast cells of different species. *Radiat. Environ. Biophys.*, 1979; 16, 49-61.

[52] Petin, VG; Berdnikova, IP. Responses of yeast cells to heat applied alone or combined with gamma-rays. *Int. J. Radiat. Biol.*, 1981; 39, 281-290.

[53] Petin, VG; Kim, JK. Survival and recovery of yeast cells after combined treatment with ionizing radiation and heat. *Radiat. Res.*, 2004; 161, 56-63.

[54] Petin, VG; Kim, JK; Zhurakovskaya GP et al. Some general regularities of synergistic interaction of hyperthermia with various physical and chemical inactivating agents. *Int. J. Hyperthermia*, 2002; 18, 40-49.

[55] Petin, VG; Kim, JK; Zhurakovskaya; GP et al. Mathematical description of synergistic interaction of UV light and hyperthermia for yeast cells. *J. Photochem. Photobiol. B: Biology*, 2000; 55, 74-79.

[56] Petin, VG; Komarov, VP. *Quantitative Description of Radiosensitivity Modification*. Moscow: Energoatomizdat, 1989.

[57] Petin, VG; Zhurakovskaya, GP; Komarova, LN. *Radiobiological Basis of Synergistic Interaction in Biosphere*. Moscow: GEOS, 2012.

[58] Raaphorst, GP; Azzam, EI; Feeley, M. Potentially lethal radiation damage repair and its inhibition by hyperthermia in normal hamster cells, mouse cells, and transformed mouse cells. *Radiat. Res.*, 1988; 113, 171-182.

[59] Raaphorst, GP; Feeley, MM; Danjoux, CE et al. Hyperthermia enhancement of radiation response and inhibition of recovery from radiation damage in human glioma cells. *Int. J. Hyperthermia*, 1991; 7(4), 629-641.

[60] Richieri, GV; Mel, HC. Temperature effects on osmotic fragility, and the erythrocyte membrane. *Biochimica et Biophysica Acta*, 1985; 813, 41-50.

[61] Sakkers, RJ; Filon, AR; Brunsting, JF et al. Selective inhibition of repair of active genes by hyperthermia is due to inhibition of global and transcription coupled repair pathways. *Carcinogenesis,* 1995; 16, 743-748.

[62] Streffer, C; Vaupel, P; Hahn, G. *Biological Basis of Oncologic Thermotherapy*. Berlin, Heidelberg, New York, London, Paris, Tokyo, Hong Kong: Springer, 1990.

[63] Sugahara, T. *Modification of Radiosensitivity in Cancer Treatment*. Edition. Tokyo: Academic Press, Inc., 1984.

[64] Takahashi, S; Takeda, E; Kubota, Y et al. Inhibition of radiation-induced DNA double –strand breaks by nickel and arsenite. *Radiat. Res.*, 2000; 154(6), 686-691.

[65] Timofeeff-Ressovsky, NV; Ivanov, VI; Korogodin, VI. *Die Anwendung des Trefferprinzip in der Strahlengbiologie*. Vienna: VEB Gustav Fischer Verlag, 1972.

[66] Tkhabisimova, MD; Komarova, LN; Petin, VG. Dark recovery of diploid yeast cells after simultaneous exposure to UV-irradiation and hyperthermia. *Cytology*, 2002; 44, 555-559.

[67] Tkhabisimova, MD; Komarova, LN; Petin, VG. Quantitative analysis of parameters describing recovery of mammalian cells exposed to thermoradiation action. *Molecular Medicine*, 2007; 2, 29-32.

[68] Tobias, CA. The repair-misrepair model in radiobiology: comparison to other models. *Radiat. Res.,* 1985; Suppl. 8, 77-95.

[69] Tomasovic, SP; Steck PA; Heitzman, D. Heat-stress proteins and thermal resistance in rat mammary tumor cells. *Radiat. Res.*, 1983; 95, 399-413.

[70] Tsuchido, T; Vanbogelen, RA; Neidhardt, FC. Heat shock response in *Escherichia coli* influences cell division. *Proc. Nat. Acad. Sci. USA*, 1986; 83, 6959-6963.

[71] Tyrrell, RM. Synergistic lethal action of ultraviolet radiations and mild heat in *Escherichia coli*. *Photochem. Photobiol.*, 1976; 24, 345-351.

[72] Wang, Y; Iliakis, G. Effects of 5'-iododeoxyuridine on the repair of radiation induced potentially lethal damage interphase chromatin breaks and DNA double strand breaks in Chinese hamster ovary cells. *Int. J. Radiat. Oncol. Biol. Phys.*, 1992; 23, 353-360.

[73] Warters, RL. Persistence of radiation-induced double-strand breaks in the DNA of heated CHO cells. *Int. J. Radiat. Biol.*, 1993; 64, 669-676.

[74] Weissenborn, U; Obe, G. Modification of X-ray induced chromosome aberration frequency by pre- and postirradiation hyperthermia of human peripheral lymphocytes. *Int. J. Radiat. Biol.*, 1991; 59, 973-984.

[75] Wu, A; Lin, PS. Peak temperature influences on heating rate effect in hyperthermic cytotoxicity. *Int. J. Radiat. Oncol. Biol. Phys.,* 1985; 11, 983-986.

[76] Wynstra, JH; Wright, WD; Roti, JL. Repair of radiation-induced DNA damage in thermotolerant and nonthermotolerant HeLa cells. *Radiat. Res.*, 1990; 85, 85-89.

[77] Yamamori, T; Yura, T. Genetic control of heat-shock protein synthesis and its bearing on growth and thermal resistance in *Escherichia coli* K-12. *Proc. Nat. Acad. Sci. USA*, 1982; 79, 860-864.

Chapter 5

THE MATHEMATICAL APPROACH FOR SYNERGISTIC EFFECT: PROGNOSIS AND OPTIMIZATION

ABSTRACT

Various cellular systems have been analyzed for synergistic interactions of the simultaneous action of hyperthermia with physical agents including ionizing radiation, ultraviolet light, and ultrasound, and with some chemical agents. Several universal rules of synergistic interaction were revealed. (1) For any constant rate of exposure, the synergy can be observed only within a certain temperature range. (2) The temperature range synergistically increasing the effects of radiations is shifted to lower temperature for thermosensitive objects. (3) Inside this range, there is a specific temperature that maximizes the synergistic effect. (4) A decrease in the exposure rate results in a decrease of this specific temperature to achieve the greatest synergy and *vice versa*. (5) For a constant temperature at which the irradiation occurs, synergy can be observed within a certain dose rate range. (6) Inside this range an optimal intensity of physical agent may be indicated, which maximizes the synergy. The conceptual idea and mathematical model is proposed to explain these experimental data. The model is based on the supposition that synergism takes place due to the additional effective damage arisen from the interaction of non-lethal sublesions induced by both agents. This process is assumed to proceed until the sublesions of the less frequent type is used up. The model predicts the greatest value of the synergistic effect and the condition under which it can be achieved. The synergistic effect appeared to decline with any deviation from the optimal value of the ratio of the effective damages produced by each agent applied separately. The model predicts the irreversible component of radiation damage after combined action and the dependence of synergy on the intensity of agents applied.

INTRODUCTION

It is a well known fact that interactions can take place between hyperthermia and other numerous physical and chemical agents found in modern life. Some of these combinations lead to a synergistic interaction when the final biological effects are greater than the addition of the effects from separate exposures to the individual inactivating agents. Heat is known to be extensively utilised to enhance cell inactivation induced by ionizing radiation, ultraviolet light, ultrasound and chemical agents [10; 14; 16; 41; 42]. From many publications it is easily recognized that the degree of synergistic effect appears to be influenced by many factors such as cell line, thermotolerance, recovery and repair, phase of cell cycle, temperature, exposure time, dose, intensity of the physical factors, concentration of the chemical agents, and treatment sequence. The greatest synergistic interaction is observed when hyperthermia and another agent are applied simultaneously; i.e., the amount of synergism decreases as the time interval that separates the two modalities increases.

Despite the potential importance of synergistic interaction, information regarding the synergistic interaction of hyperthermia with various physical and chemical inactivating agents is largely missing. To reveal some new general rules describing the response of various unicellular organisms to the combined action of hyperthermia with another inactivating agent, synergistic interaction of the simultaneous action of hyperthermia with ionizing radiation, ultraviolet light, ultrasound and some chemical agents should be analyzed in a wide range of acting temperature and the intensity of agent applied in combination.

The effect of ionizing radiation combined with heat on bacterial, plant and animal viruses has been extensively investigated by many authors. The phenomenon of synergistic viral inactivation was observed when viruses were exposed to the simultaneous application of heat and ionizing radiation [61]. In this work, the mathematical model based on the laws of chemical kinetics was used to provide a description of observed dose rate and temperature dependences. The authors believe that the successful application of the model to diverse types of viral systems inactivated by heat and ionizing radiation emphasizes the generality of the kinetic model in describing inactivation phenomena and may well serve as an extension of target theory in describing the radiobiological effects of heat and ionizing radiation. Similar version of kinetic analysis was realized for spore inactivation in a composite heat and gamma radiation environment [13]. Mathematical optimizing thermal and radiation effect for bacterial inactivation was described [56]. On this basis, a temperature- and dose rate-dependent model for the kinetics of cellular response to ionizing radiation was published [4].

A cybernetic model has been developed for radiation reaction in living cells [23]. Two types of radiation damage are considered: lethal and potential lethal damage. The latter is also lethal if the cells are forced to divide immediately after irradiation. But this damage can be repaired in the cells if a suitable interval elapses between irradiation and cell division. The radiation reaction constants, characterizing the probability of transitions between different states of cells, have been determined for ionizing radiation of various qualities. This model was adopted [20] to optimize cancer treatment with combined action of ionizing radiation and 2-deoxy-D-glucose, an inhibitor of glucose transport and glycolysis, which has been shown to differentially inhibit repair of radiation damage in normal and cancer cells, leading to an enhancement of the radiation damage in transformed cells and tumors.

The survival of asynchronous CHO cells after thermoradiation action [5] was approximately described [37] by a model comprising three components of cell killing: the unperturbed radiation component, the unperturbed hyperthermia component and the interaction component. The temperature sensitivities of the hyperthermia component and of the interaction component were investigated by using different hyperthermic temperatures, revealing that the interaction component is significantly less sensitive to a change in temperature than the unperturbed hyperthermia component. The authors drew a conclusion that the interaction component of the hyperthermia might arise from radiation and hyperthermia effects on a single target rather than separate effects on two complementary systems essential to cellular reproductive integrity.

Using the molecular theory of radiation action a mechanism was suggested which predicts that synergistic interaction can arise between chemical mutagens and ionizing radiation [9]. The synergism was expected to emerge from the interaction of single-strand lesions induced in the DNA by the chemical mutagen and radiation. The mathematical derivation resulted in three conditions which must be satisfied for logical analysis. The model was applied to explain qualitatively the synergism of various chemical agents and ionizing radiation for some mutations, chromosomal aberration and survival. The theoretical analysis was developed further to the induction of cancer. It was concluded that the synergistic interaction between various agents would be expected to lead to an increase of cancer at a low radiation doses [9].

Most of the agents used in cancer therapy are known to damage DNA, and DNA repair is one of the most important factors in determining cellular sensitivity in a particular agent [38]. Dikomey [11] described a mathematical approach for the suppression of cell recovery from sublethal damage by hyperthermia. In comparison to 37 °C, the rate of single-strand break repair was increased by hyperthermia at 42 °C, but decreased at 45 °C. The model does not pretend to describe synergy, but only demonstrates the influence of hyperthermia on cell recovery and differential cytotoxic effects of hyperthermia below and above 43 °C alone or combined with X-radiation.

Theoretical model for the estimating dose-effect relationship after combined exposure to cytotoxicants has been analyzed [1; 57]. One of the benefits of this model is that it can predict the ultimate effect in the dependence of sequence application of agents. The model does not pretend to describe the synergistic effects. The best fit of the model to experimental data was shown versus the model of the synergistic effects of sequential exposure to different types of radiation [63], based on the theory of dual radiation action [26].

Summarizing this short analysis of the well-known publications related with synergistic interaction, we can conclude that a variety of mathematical models has been suggested that reflect different aspects of the overall cell reaction to the combined application of various agents. They based on a number of assumptions and postulates and in some cases describe experimental data. In the most of the mentioned models, to predict the effects of combined exposure it is necessary to identify a set of unknown parameters, i.e., one need to perform a large number of preliminary experimental studies. Moreover, none of these models can predict the value of the greatest synergistic effect, criteria under which it can be achieved, the volume of cell recovery after combined exposures as well as the role of sequence application of agents and the influence of dose rate to synergistic interaction at the same time. Therefore it is of importance to develop a simple mathematical approach to synergistic interaction which would be able to predict several universal patterns of synergistic effect display that do not depend on biological object and test being studied as well as on the agents applied.

The exposure rate constitutes a significant parameter in radiation research. While it is generally accepted that lowering the dose rate leads to a decrease in the cell-killing efficiency per unit dose [15; 17], the matter is less clear for the synergistic interaction of agents combined together [4; 36]. At the same time, any risk assessment must consider the question of whether combined effects will influence the health outcome. Of all possible situations of combined action, the long-term exposure of living objects to low levels of the agents widely presented in nature is especially important [3; 7]. However, an assessment of the potential significance of synergistic interactions between adverse environmental factors acting together at the level of intensity and concentration found in the biosphere is still an intriguing and unresolved problem. Real experiments with low intensities found in environmental and occupational settings are prone to large uncertainties. A feasible approach to this problem is to analyze the dependence of the efficiency of the synergistic interaction on the intensity of the agents used. Hypothetically, some possibilities could be realized. The case when synergism is decreasing with the intensities of applied deleterious agents is unimportant. The same is true for the situation when a decrease in the intensity of one factor should be accompanied by an increase in the intensity of another to retain their synergistic interaction at the same level. The only possibility would be of great importance: if the lesser intensity of one of the agents is applied, then the smaller intensity of another agent should be employed to display the highest, or some definite, level of synergy. In such a case, it would be expected that environmental agents, even at low intensity, may, in principle, interact with each other in a synergistic manner, and thereby enhance their harmful action. In this section, we present conclusive evidence confirming the last opportunity.

Therefore, the main aims of this Chapter may be formulated as follows.

1. To reveal some common features of synergistic effects display, which would not depend on interactive agents, biological object and test.
2. To analyze sublesions retaining in the distant descendants of cells surviving after irradiation.
3. To suggest a unified biophysical conception for synergistic interaction mechanism.
4. To develop a mathematical model of synergy for simultaneous action of agents.
5. To prove the dependence of synergy on the intensity of applied agents.
6. To explore a possibility of the model to predict the irreversible component of radiation damage after combined action of agents.

5.1. Some Common Features of Synergistic Effects Display: Order among the Chaos of the Effects Observed

Much experimental evidence and theoretical considerations demonstrated the potential importance of synergistic interaction of radiations combined with other environmental agents including heat. The combined thermoradiation action is known to result in a synergistic interaction when the final biological effects are greater than the addition of the effects from separate exposures to the individual inactivating agents.

In view of the emerging importance of the combined actions of ionizing radiation with other environmental factors, the potential significance of the synergy, among other radioecological problems, has to be carefully assessed. Knowledge of common regularities of synergistic effects derived from experimental results obtained at molecular and cellular levels as well as with laboratory animals should enhance our understanding of the risk of radiation exposure and their interaction with other agents and possible strategies to reduce them. Moreover, the elaboration of the quantitative assessment of combined effects and common rules of synergistic effects display is urgently needed for many practical and theoretical reasons. First of all, to bridge the gap between differing conceptual approaches in the mechanisms of synergistic interaction. Secondly, these results will be useful to develop the unified concept and mathematical approach to predict and optimize the synergy. Thirdly, they will promote the extrapolation to low exposures. And finally, practical improvement of combined treatments in radiotherapy and sterilization methods can be expected. Our purpose here is to review and discuss some new general rules of synergistic effect display. The response of various biological objects to the simultaneous combined action of hyperthermia with ionizing radiation, ultraviolet light, ultrasound and some chemical agents will be analyzed using the experimental data obtained by authors. To check the universality of the rules obtained, the results published by others will be involved. Figures 3.1 and 3.2 provide examples of the basic experimental data used for synergistic interaction estimation. To calculate the synergistic effect we used the synergistic enhancement ratio (Eq. 3.5).

Figure 5.1 show some experimental examples obtained with viruses (Figure 5.1A) and bacterial spores (Figure 5.1B). Data on the inactivation of viruses and dry bacterial spores exposed to ionizing radiation at different temperature were published by others [56; 61]. In these papers, the authors did not calculate the synergistic effect so we used their data to obtain results presented in Figure 5.1. One can see that the synergistic interaction between hyperthermia and ionizing radiation is realized only within a certain temperature range independently of the object analyzed. For temperatures below this temperature range, the synergistic effect is not observed and cell killing was mainly determined by the damages induced by ionizing radiation. For temperatures above this temperature range, the synergistic effect was also not observed but cell killing was chiefly caused by hyperthermia.

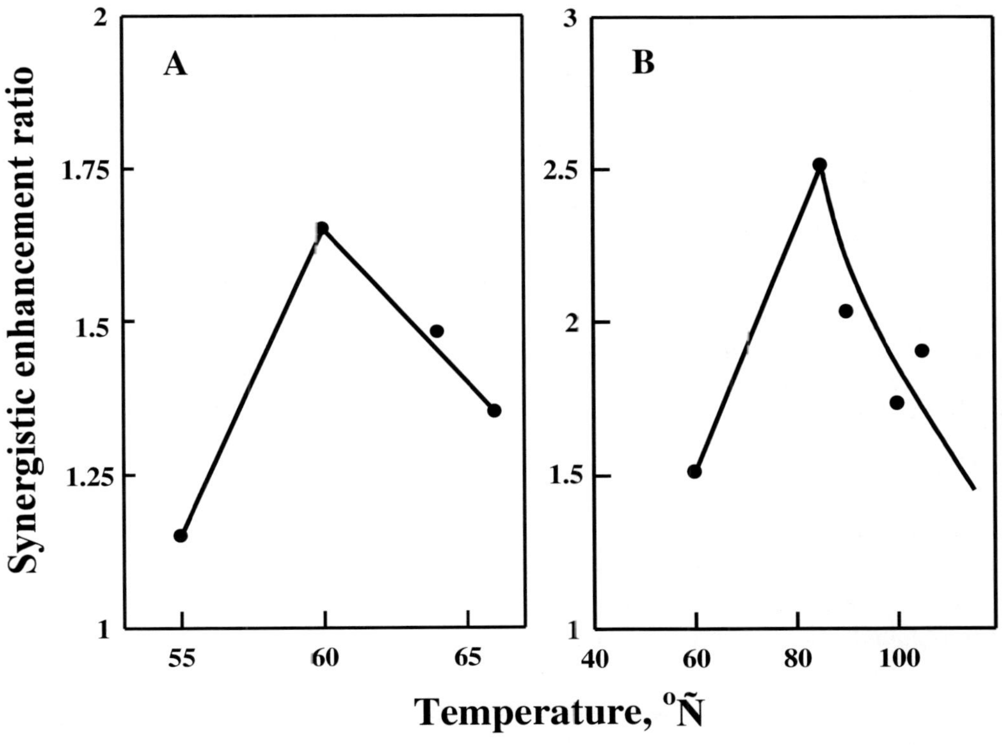

Figure 5.1. The dependence of the synergistic enhancement ratio upon the exposure temperature for bacteriophage (A) and bacterial spores (B). Cells were simultaneously treated with high temperatures and ionizing radiation.

Noteworthy is the fact that just such a dependence of the synergistic effect on temperature under which the exposure occurred was obtained for diploid yeast cells, the simplest representative of eukaryotic cells, upon the simultaneous combination of hyperthermia with ^{60}Co γ-rays (Figure 5.2A), 254 nm UV light (Figure 5.2B) and 20 kHz ultrasound (Figure 5.2C).

The dependence of the synergistic enhancement ratio upon the exposure temperature for Chinese hamster cells simultaneously exposed to heat and chemicals is depicted in Figure 5.3. This figure includes our calculations of data published by other authors for cis-diamminedichloroplatinum [62] and tri(1-aziridinyl)phosphine sulphide [21]. It is obvious that the synergy is observed for both chemicals within a certain temperature range, inside of which there is an optimal temperature resulting in the greatest synergistic interaction.

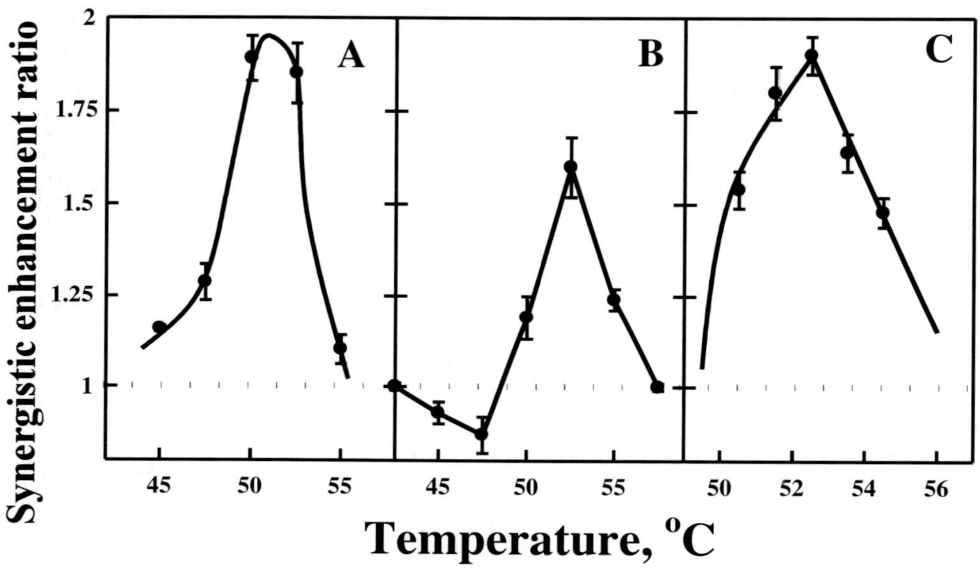

Figure 5.2. The dependence of the synergistic enhancement ratio upon the exposure temperature for diploid *S. cerevisiae* yeast cells (strain XS800) simultaneously exposed to heat and ^{60}Co γ-rays, 10 Gy/min (A), UV light, 254 nm, fluence rate 0.25 W/m^2 (B) and 20 kHz ultrasound, intensity 0.053 W/cm^2 (C). Error bars show inter-experimental standard errors.

Another example of similar synergy pattern is shown in Figure 5.4. Here again, the dependence of the synergistic enhancement ratio upon the exposure temperature is depicted for relatively thermosensitive (*Endomyces magnusii*) and thermoresistance (*S. ellipsoideus*) diploid yeast strains. Thermoresistance of haploid strain *Z. bsilii* was intermediate between them. One can see that the temperature range strengthening the effect of ionizing radiation has been shifted toward the lower temperatures for temperature-sensitive cell lines. It is evident also that the effectiveness of synergistic interaction was smaller for haploid cells incapable of LHR recovery since repair of DNA double strand breaks requires two homologous DNA duplexes. It is in agreement with the fact that the mechanism of synergy many authors connect with reduction of cell ability to repair radiation damage.

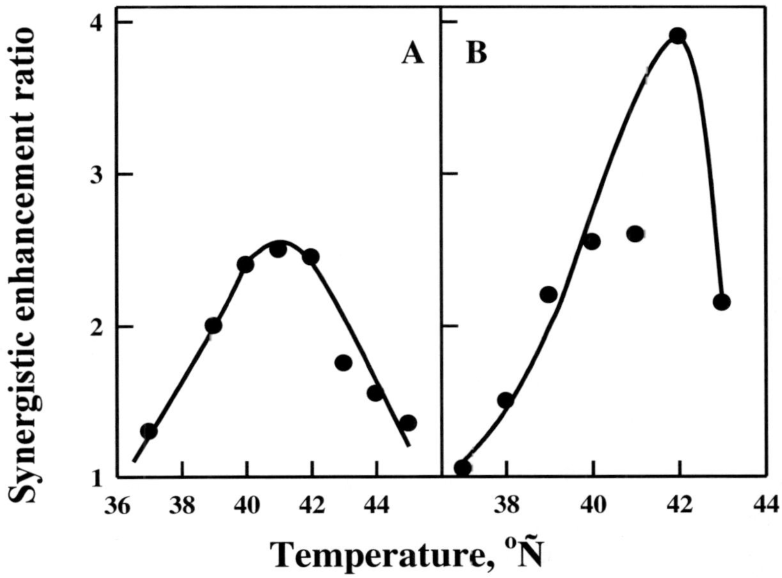

Figure 5.3. The dependence of the synergistic enhancement ratio upon the exposure temperature for inactivation of Chinese hamster cells simultaneously exposed to heat and chemicals: cis-diamminedichloroplatinum (A), tri(1-aziridinyl)-phosphine sulphide (B).

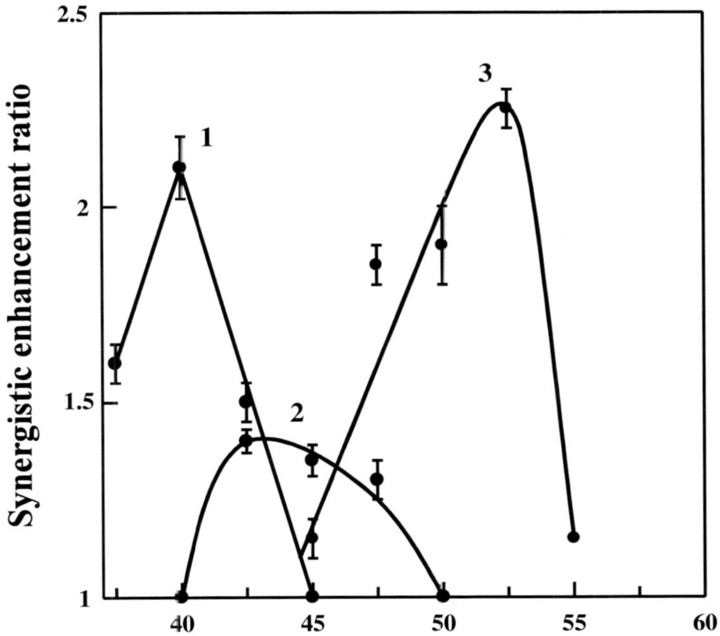

Figure 5.4. The dependence of the synergistic enhancement ratio upon the exposure temperature for inactivation of diploid yeast cells *Endomyces magnusii* (curve 1), haploid yeast cells *Z. bailii* (curve 2) and *S. ellipsoideus* (curve 3). Cells were simultaneously exposed to heat and γ-radiation (^{60}Co, 10 Gy/min).

We have studied the radiation enhancement by hyperthermia in combination with α-particles of ^{239}Pu. It is generally assumed that a cell can repair the damage caused by low-LET radiation more efficiently than that caused by densely ionizing particles. Then, if the thermal potentiation of low-LET radiation is connected with recovery processes, one would expect a reduction of thermal potentiation of high-LET radiation effect. We have studied the combined simultaneous action of hyperthermia and α-particles on survival of the fission yeast *Endomyces magnusii.* It was found that interaction between hyperthermia and high-LET radiation was slightly synergistic. In fact, the maximal value of the synergistic enhancement ratio for high-LET radiation was only 1.33 (Table 5.1). It would appear that hyperthermia when used with high-LET radiation has a little added advantage in cell inactivation. Thus, these results were interpreted in favor of the supposition that combined action of hyperthermia and ionizing radiation can damage recovery processes and/or produce more "severe" damages, which cell cannot repair.

Table 5.1. The dependence of the synergistic enhancement ratio by hyperthermia upon the temperature of irradiation for *Endomyces magnusii* yeast cells irradiated with α-particles of ^{239}Pu (25 Gy/min)

Yeasts	Temperature of irradiation, °C							
	30	40	45	74.5	50	52.5	55	57.5
Endomyces magnusii	1.00	1.00	1.14	1.20	1.33	1.29	1.21	1.00

It is curious that in all these examples the synergistic enhancement ratio at first increases, then reaches a maximum, which is followed by a decrease independently of biological object tested and agent applied with heat. This implies that a synergistic interaction of hyperthermia with ionizing radiation, UV light, ultrasound and some chemical compounds is observed only within a certain temperature range. Hence, one can conclude that for a given intensity of physical factors or concentration of chemical agents there should be a specific temperature that maximizes the synergistic interaction. Any deviation of temperature from the optimal value results in a decrease of synergism. The temperature range strengthening the effect of ionizing radiation has been varied with different cell systems and shifted toward lower temperatures for temperature-sensitive cell lines. Our experiments and estimations showed that this range was about 55–70 °C for viruses, 95–105 °C for the most thermoresistant bacterial spores, 50–60 °C for bacterial cells, 45–55 °C for relatively thermoresistant *S. ellipsoideus* diploid yeast cells and 37.5–45 °C for thermosensitive diploid *Endomyces magnusii* yeast cells.

Some of the data presented above are difficult to interpret from the widespread point of view explaining the synergy as a consequence of cell repair inhibition. Actually, this suggestion can easily explain the increase in the synergistic effect with an increase in acting temperature. But this hypothesis meets difficulty to explain the decrease in the synergistic effect with further increase in acting temperature after reaching the greatest synergy. Moreover, the cell recovery is unknown for ultrasound and chemical compounds. These data indicates the need to elaborate a new presentation for mechanism of synergy.

On the basis of the results presented the following conclusion should also be valid: for any constant ambient temperature (or concentration of chemical agent) there should be an optimal intensity of ionizing or UV light radiation resulting in the greatest synergy. Some selected experimental results confirming this prediction are presented in Figure 5.5. The data on the simultaneous thermoradiation inactivation of *S. cerevisiae* yeast cells exposed to electron beam from a 25 MeV pulsed linear accelerator [53] were used to estimate the dependence of the synergistic enhancement ratio on the dose rate for 51 °C exposure temperature. The results are depicted in Figure 5.5A. One can see that for a constant temperature, at which the irradiation occurs, synergy can be obtained within a certain dose rate range. Inside this range an optimal dose rate of ionizing radiation may be indicated, which maximizes the synergy.

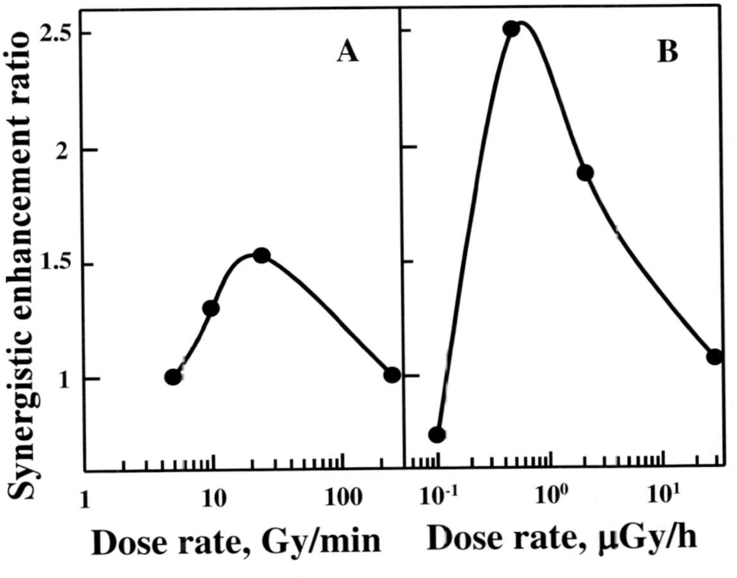

Figure 5.5. The dependencies of the synergistic enhancement ratio upon the dose rate of ionizing radiation: A, inactivation of diploid *S. cerevisiae* yeast cells exposed to ionizing radiation at 51 °C. B, the yield of mutation in *Arabidopsis thaliana* seeds after combined action of lead nitrate and chronic irradiation.

Figure 5.5B demonstrates the relationship between the synergistic enhancement ratio and the dose rate of ionizing radiation for a combined effect of lead nitrate and chronic irradiation of *Arabidopsis thaliana* seeds. This relationship was calculated using the experimental results obtained by others [12]. The seeds were selected in wild populations growing for five years in places with different levels of radioactive pollution inside the 30 km zone around the Chernobyl nuclear power station and then treated with lead nitrate (3.39 g/l).The frequency of mutant embryos and the proportion of lethal embryos were estimated. For this case, the synergistic enhancement ratio was defined as the ratio between the increment of the mutant or lethal embryos after the combined action and the sum of these increments after separate action f each agent. As it can be seen from Figure 5.5B, the synergistic effect has a pronounced maximum at a certain dose rate of ionizing radiation. Thus, the data presented in Figure 5.5 indicate that at a constant temperature there is an optimal intensity of ionizing radiation combined with heat (Figure 5.5A) or with lead nitrate concentration (Figure 5.5B) that provide the greatest synergistic interaction.

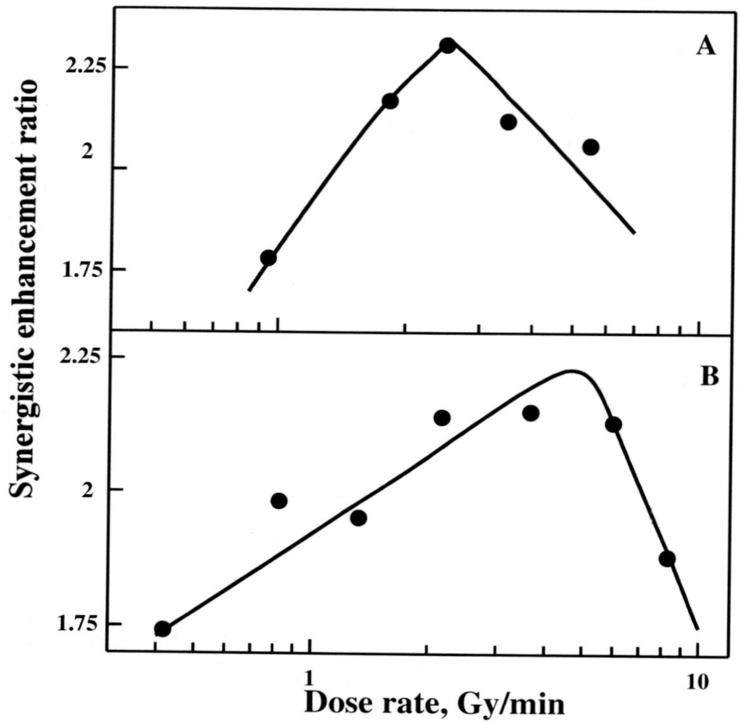

Figure 5.6. The dependence of the synergistic enhancement ratio upon the dose rate for bacterial spores *Bacillus subtilis* exposed to ionizing radiation at 95 (A) and 105°C (B).

The data on the simultaneous thermoradiation action on *Bacillus subtilis* spores [56] were used to estimate the dependence of the synergistic enhancement ratio on the dose rate (Figure 5.6) for two exposure temperatures 95 (Figure 5.6 A) and 105 °C (Figure 5.6B). One can see that for a constant temperature, at which the irradiation occurs, synergy can be observed within a certain dose rate range. Inside this range an optimal dose rate of ionizing radiation may be indicated that maximizes the synergy. As the exposure temperature is reduced, the optimal dose rate decreases and *vice versa*.

To evidence the importance of synergistic effects at low intensity of inactivating agents, we analysed the dependence of synergistic interaction on the intensity of physical factors or on the concentration of chemical agents applied in combination with hyperthermia. Using survival curves data published for simultaneous action of hyperthermia and ionizing radiation on bacteriophage [61], bacterial spores [56], yeast [44; 50] and mammalian cultured cells [6], we were able to calculate the synergistic enhancement ratio for various cell systems and irradiation condition. The results published in the cited paper allowed us to establish the correlation between the dose rate and the exposure temperature, which both provide maximum or other arbitrary levels of synergistic interaction (Figure 5.7). The value of synergism was the highest for bacterial spores (k_{max} = 2.2), diploid yeast cells (k_{max} = 1.7) and was intermediate for bacteriophage (k = 1.3) and cultured mammalian cell (k = 2.2). This was due to the fact that for the last cell systems the highest synergistic effect was not obtained for all dose rates used in the experiments. One can see that linear relationships are found between these values for various cellular objects. This means the general importance of the dose rate of ionizing radiation in the manifestation of synergistic interaction. It can be inferred that the temperature at which ionizing radiation is delivered should be diminished to obtain the maximum or a definite synergistic effect with dose rate decreasing and *vice versa*.

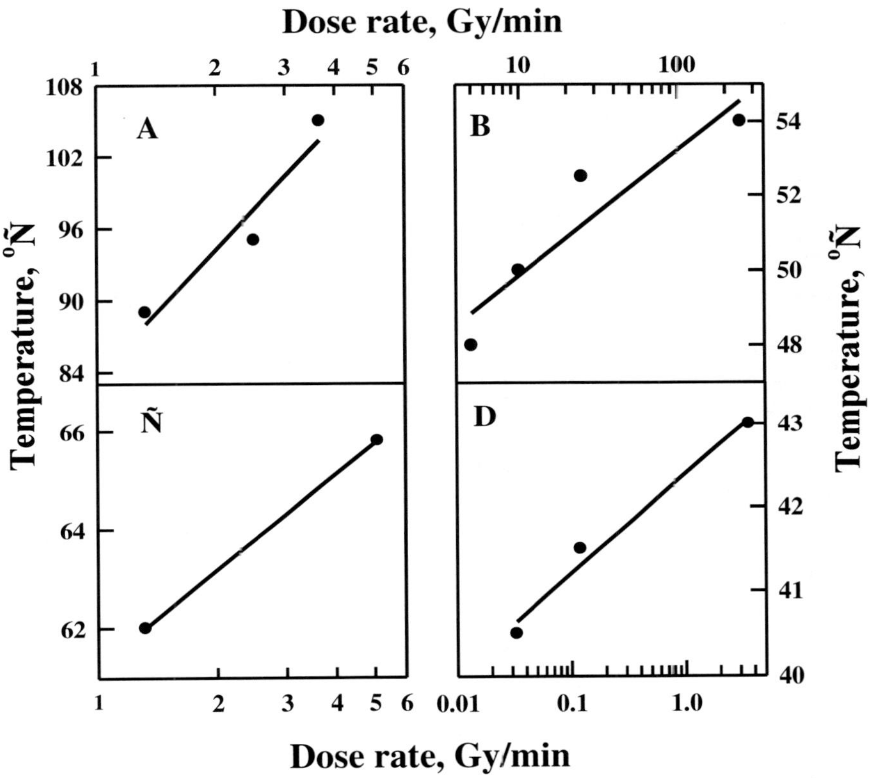

Figure 5.7. Correlation of the dose rate of ionizing radiation and exposure temperature providing the same synergistic interaction under simultaneous thermoradiation action: A, bacterial spores (*Bacillus subtilis*); B, diploid yeast cells (*S. cerevisiae*, strain XS800); C, bacteriophage (T4); D, Chinese hamster cells.

To check the universality of this rule, the data on simultaneous effect of hyperthermia combined with UV light [46; 50] or ultrasound [49; 54] on diploid yeast cells, as well as with tris(1-aziridinyl)-phosphine sulphide (thio-TEPA) [49] and cis-diamminedichloroplatinum (II) (cis-DDP) [21] on cultured mammalian cells were involved. The last two set of data include the relationship between exposure temperature, concentration, and the rate of cell inactivation for chemical agents used in clinical chemotherapy. Hence, they have no direct attitude toward environmental harmful agents and they used here to demonstrate that the concentration of chemical agents is also important for their synergistic interaction with heat. Using data published in the above cited works, we could obtain the relationships between the intensity of physical factors or the concentration of chemical agents with the exposure temperature which both provide the greatest synergy (Figure 5.8).

Here again, circles denote the results of our calculations based on the survival curve data published earlier [55]. The value of the synergistic enhancement ratio was about 1.6 for UV light, 2.5 for ultrasound, 2.5 for thio-TEPA and 3.2 for cis-DDP combined with hyperthermia. In all cases, at a smaller intensity of the physical factor or concentration of the chemical agents, it was required to reduce the acting temperature to preserve the highest synergistic effect.

Thus, several common regularities of synergistic effect display have been demonstrated, which are universal and do not depend on the biological object and test being studied as well as on the agents used in combined actions. These data indicate the need to elaborate a new theoretical conception of the synergy.

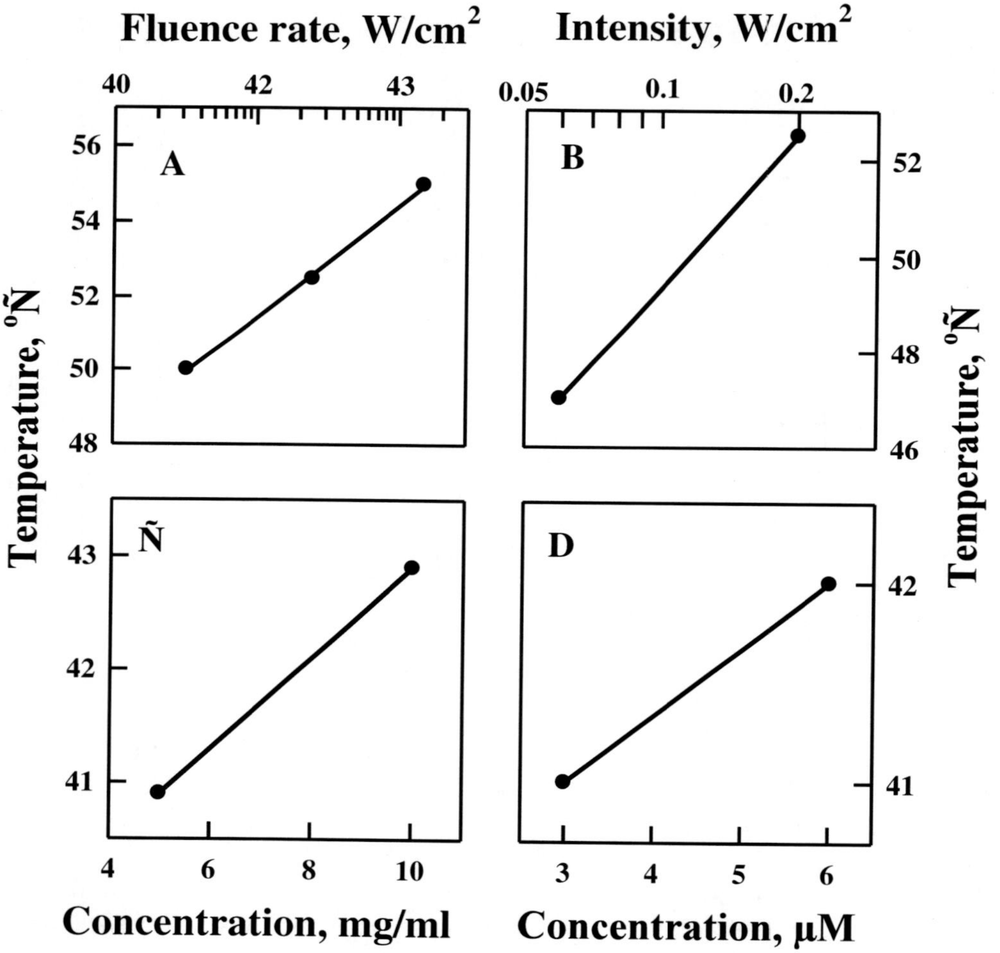

Figure 5.8. Correlation of exposure temperature with UV light fluence rate (A), ultrasound intensity (B), and concentration of tris(1-aziridinyl)-phosphine sulphide (thio-TEPA) (C) and cis-diamminedichloroplatinum (II) (cis-DDP) (D) providing the highest synergistic interaction of these agents with heat.

5.2. Sublesion Manifestation in the Distant Descendants of Yeast Cells Surviving after Irradiation

The inhibition of cell recovery is commonly considered as a reason of synergy of the combined action of ionizing radiation with various agents. However, the inhibition may be preceded via either the damage of the mechanism of the recovery itself or *via* the formation of irreversible damage which can not be repaired at all. It was shown in the previous Chapter that the synergistic interaction is not related with the direct damage to recovery process itself. It is not excluded that synergism could be expected to result from some additional lethal lesions arising from the interaction of sublesions induced by both modalities. The idea of the interaction of damage is applied in radiobiology by many authors. Therefore it might be reasonable to assume that some additional effective damage produced during combined action are responsible for the synergistic interaction. Before presentation of our mathematical model of the synergistic interaction, we consider the experimental data demonstrating the existence of sublesions in yeast cells surviving after irradiation.

It is a well-known fact that single cells among both surviving and inactivating parts of a homogeneous population respond differently to the same dose of irradiation. Inactivated cells produce microcolonies consisting of various numbers of cells (different inactivation forms). Surviving cells produce on solid nutrient media colonies of various sizes and morphology which appear within different time interval after irradiation. The late appearance of clones produced by cell surviving after ionizing radiation exposure is expressed only for diploid and polyploid cells for which sigmoid survival curves after exposure to ionizing radiation are obtained. Each damage is lethal only for haploid cells. In diploid cells, effects of several primary lesions (sublethal lesions) may be summarized and inherited resulting in various mitotic disorders. These disorder effects are the higher the greater number of primary sublethal lesions is produced in the originally irradiated cells [31; 32].

It was shown above (section 1.1) that different manifestation of cell radiation damage can be quantitatively described by the probability model. The model is based on the supposition that clone formation is a probable process and the probability for the successful division P of cell is determined by a number of sublethal damages. In fact, the probability model is a peculiar synthesis of hit-and-target principle and biological stochastic. According to this model, single irradiated cells are damaged in a random fashion in accordance with the hit-principle, as in classical models [2; 33; 59], and the number of damages defines the probability P for the successful division. For independent interaction of radiation damages, the probability for the successful division of a cell with i primary damage (sublethal lesion, hit) may be presented by Eq. (1.1).

It was demonstrated that the reduced probability for the successful division can remain unchanged through the successive division of cells [24; 25]. Hence, the mechanism of clone formation is the well-known "birth-and-death" process. The process operates when an entity, in our case a single cell, either gives rise to progeny like itself (birth), or is removed in some way (death), and these two events occur in a random fashion. Taking into account this approach, it would be of interest to determine various biological responses in the dependence of the primary sublethal lesions.

The following diploid yeast strains were used in this study: *S. ellipsoideus (vini)*, strain Megri 139-B, and *S. cerevisiae*, strain 5*a*3b*α* that is heterozygous for the *ade2* mutation.

Yeast cells were incubated before irradiation for 3–5 days at 30 °C on a complete nutrient agar layer to a stationary phase. Primary colonies, which appeared after irradiation simultaneously with a control and did not differ from the control phenotypically, were identified as stable (normal cell clones). Colonies, which formed slowly growing subclones and/or subclones that differed in morphology, were believed to be unstable clones. We believed that the latter colonies were appearing the later the more of the initial sublesions contained in a parent cell surviving after irradiation and produced the colony.

To determine the content of mitotic recombinants in diploids that are heterozygous for the *ade1* and *ade2* mutations, colonies were replated on nutrient media and the number of white, red and sectored colonies was counted. When estimating culture growth rate, cells of separate clones were grown in liquid nutrient media in tubes under aeration; the number of cells, capable of forming colonies was determined in plating. The growth rate was calculated according to the exponential region of curve reflecting the propagation of cells. The quantitative evaluation of respiratory mutants was performed as was published by others [40], and the content of nonviable cells was determined by detecting, under a microscope, budding and nonbudding cells on the surface of the nutrient agar after growing for one day. Other details have been published [32; 47].

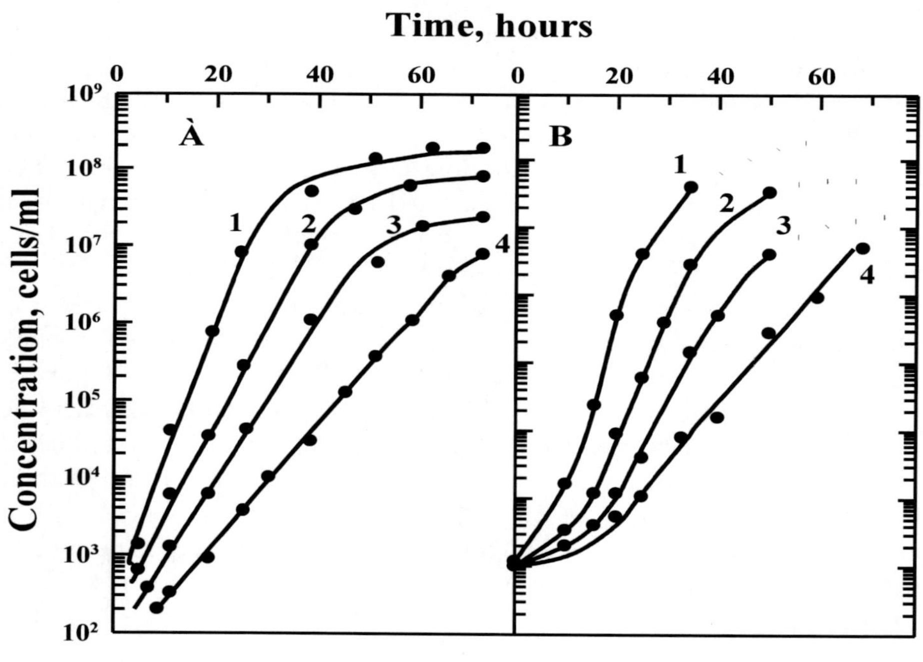

Figure 5.9. Dynamics of the propagation of nonirradiated diploid yeast cells (curves 1) and distant progenies of cells surviving after irradiation and generating unstable colonies (curves 2–4); strains Megry 139-B (A) and 5*a*3b*α* (B).

We determined the growth rate of cells from colonies of various sizes appearing in different time after irradiation. Figure 5.9 shows the growth of control cells in liquid nutrient medium at 30 °C (curves 1). Curves 2–4 represent growth curves of yeast cells from individual colonies, produced by cells surviving after irradiation and having 1–3 primary sublethal lesions (hits), respectively. These curves have much lower slopes (i.e., slower growth rate).

It was shown before [25] that for diploid strains Megry-139B and 5*a*3Bα the probability of refusal of successful division $\alpha = 0.12$. Then in accordance with Eq. (1.1), the following probabilities for the successful division of cells with 0, 1, 2, 3 and 4 sublesions can be obtained: $P_0 = 1$, $P_1 = 0.88$, $P_2 = 0.77$, $P_3 = 0.68$ and $P_4 = 0.60$. It leads to the decrease of clone-formation rate immediately after irradiation and upon the successive plating.

If the reduction of growth rate is related with the decreasing of the probability P_i for the successful division, the specific growth rate of cells μ_i with i sublesion may be given as

$$\mu_i = \mu_c(2P_i - 1), \tag{5.1}$$

where μ_c is the specific growth rate of control cells. Knowing μ_i and μ_c, we calculated P_i for various clones. It would be of interest to test whether clones with various values of sublesions differ in their radiosensitivity, viability, sensitivity to environmental conditions, frequency of recombination and respiratory-deficient mutations.

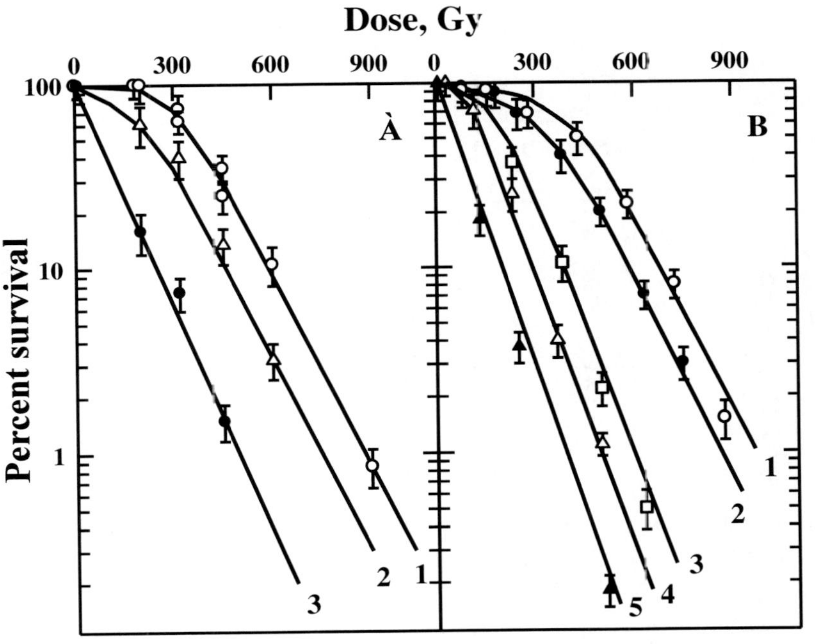

Figure 5.10. Survival curves of diploid *S. cerevisiae* control cells (curves 1) and cells from unstable clones containing various number of primary sublethal lesions (curves 2–5); strains: Megry 139-B, i = 0, 1, 2 for curves 1, 2, 3, respectively (A) and 5*a*3b*α*, i = 0, 1, 2, 3 and 4 for curves 1, 2, 3, 4 and 5, respectively (B).

The cells from clones of *S. cerevisiae* haploid and diploid strains, which were produced after a single irradiation, were irradiated repeatedly. The radiosensitivity of repeatedly irradiated haploids does not differ from that of the original cultures. However, in diploids, cells from some clones that appeared after irradiation showed an increased radiosensitivity. These results were interpreted by mitotic inheritance of sublethal lesions occurring in diploid cells. It was therefore of interest to know whether there was a difference between radiosensitivity of cells from normal clones and from those containing the various numbers of sublethal lesions. Figure 5.10 exhibits the results obtained. It can be seen that cells from unstable clones containing various number of primary sublethal lesions (curves 2–5) have an increased radiosensitivity compared to control cells (curves 1). This is expressed mainly as a decrease in the survival curve shoulder and only in a small increase of its slope. In formal terms, the shoulder decrease means that cells initially had a number of primary sublethal lesions. As can be seen from Figure 5.10, the radiosensitivity of distant progenies of single cell, surviving after irradiation, correlates with the number of primary sublethal lesions. Cells with a great number of primary sublethal lesions are inactivated with a great probability in comparison with those with a lesser number of sublesions, the degree of radiosensitivity being the higher the greater number of sublesions are in the original cell surviving after irradiation.

The appearance among macrocolonies produced by irradiated diploid yeast cells various morphological variants (saltants) may be attributed to the expression of primary sublethal lesions. The content of such kind of cells in colonies produced by survived diploid yeast cells (strain Megry 139-B, ^{60}Co γ-rays, 600 Gy) is summarized in Table 5.2.

Table 5.2. The content of morphologically changed cells (saltants) among colonies produced by irradiated diploid yeast cells

The number of primary sublethal lesions (hits)	The number of replated colonies	The number of colonies produced after replating	The number of morphologically changed colonies	
			Total sum	Percentage
0	28	4763	53	1.1
1	50	2978	496	16.6
2	22	2399	2265	94.4
3	28	3717	3077	82.8

One can see from Table 5.2 that the content of morphologically changed cells with two and three sublethal lesions are significantly greater comparing with that for clones without or with one primary sublethal lesion (hit).

The increased probability of unsuccessful division (refusal) should result in the existence of nonviable cells in clones produced by single cells surviving after irradiation. One can believe that any deviation of cultivation conditions from optimal should result in the increase in the probability of unsuccessful division. Then there is a reasonable base to expect the increased number of nonviable cells in unstable clones. Experimental data concerning the content of cells incapable to proliferation at optimal (30 °C, standard nutrient media) and suboptimal (37 °C, standard nutrient media + 7% NaCl) conditions for distant progenies of diploid yeast cells (strain Megry 139-B, 600 Gy) with various number of primary sublethal radiation lesions are presented in Table 5.3. It could be easily seen from the results presented that the relative yield of nonviable cells in clones produced by surviving diploid yeast cells is increased with the number of inherited primary radiation sublethal lesions. Due to this fact, the growth effectiveness of cells under suboptimal conditions of culture was 76, 46 and 13%, respectively, to clones with 0, 2 and 3 primary sublethal lesions.

Table 5.3. The content of nonviable cells in clones produced by diploid yeast cells survived after exposure to ionizing radiation

The number of primary sublethal lesions (hits)	The number of tested clones	The content of nonviable cells, %	
		Optimal condition	Suboptimal condition
0	28	17 ± 6	24 ± 7
1	50	20 ± 5	44 ± 11
2	22	27 ± 9	54 ± 15
3	28	36 ± 13	87 ± 13

Some unstable clones differ by a high segregation rate of respiratory mutants. Although respiratory mutants are virtually not encountered among primary colonies of irradiated diploid, these mutants can constitute more than 30 % of all subclones in plating of some unstable clones. The data concerning the yield of respiratory mutants in clones with various numbers of primary sublethal lesions are presented in Table 5.4. These experiments were performed with diploid yeast cells (strain Megry 139-B) which were irradiated in the stationary phase of growth (^{60}Co γ-rays, 600 Gy). To identify respiration deficiency in yeast we used the tetrazolium overlag technique [40]. According to this method, cells with normal respiratory ability were colored in red color while clones consisting of respiratory mutants stayed white. One can see that irradiation resulted in an increased content of respiratory mutants in clones produced by irradiated cells. The effect was particularly expressed for clones with a greater number of primary sublethal lesions (hits).

Table 5.4. The content of respiratory mutants in clones produced by surviving diploid yeast cells *S. ellipsoideus* (vini) exposed to ionizing radiation

The number of primary sublethal lesions (hits)	The percentage of clones containing various number of respiratory mutants, %			
	0–2%	>2–10%	>10–50%	>50–100%
0	100	–	–	–
1	80	16	2	2
2	53	12	20	15
3	32	16	28	24

It is well-known fact that potential radiation damages are realized during the first cell division (proliferation). Then it should be expected that distant progenies of cells surviving after a single irradiation and revealing the decreased probability of successful division have to be characterized by the increased yields of various mutation. Moreover, it should be expected that the number of primary sublesions (hits) must correlate with the yield of mutant cells. We tested this expected conclusion in experiments studying mitotic recombination in diploid yeast *S. cerevisiae* (strain 5a3bα) heterozygous in adenine loci. The cells of this strain produced white colonies on a solid nutrient media while a recombinant *ade1* or *ade2* cells produced the whole or sectored red colonies. The recombinant cells were developed during the growth of clones produced by the surviving cells exposed to ionizing radiation (γ-rays of ^{60}Co, 300 Gy). The content of recombinant cells in clones with various number of primary sublethal lesions is presented in Table 5.5. One can see that clones with greater number of such lesions contain a greater number of segregant cells.

Table 5.5. The content of recombinant cells in distant progenies of diploid yeast *S. cerevisiae* (strain 5a3bα) surviving after irradiation

The number of primary sublethal lesions (hits)	The number of investigated clones	The number of colonies emerging after replating	The number of recombinant cells			
			Whole coloration	Sectored coloration	Total number	Percentage
0	13	768	1	2	3	0.4
1	59	6966	181	33	214	3.1
2	34	3695	267	29	296	8.0
3	36	927	158	32	190	20.5

Thus, it was shown in this study that unstable clones were formed among colonies grown from diploid yeast cells surviving after irradiation. Cells from unstable clones exhibited an increased (as compared with the control) content of morphologically changed cells, nonviable cells, respiratory and recombinant mutants. Cells from unstable clones are characterized by the enhanced radiosensitivity. The degree of expression of the foregoing effects was the higher the greater number of primary sublethal lesions was in the originally irradiated cell. The data obtained here can be explained by the hypothesis in accordance with which the primary radiation lesion is not absolutely lethal for the diploid cell. With some probability which depends on the total number of primary sublethal lesions and conditions of culture, they result in the disturbance of cell division. Such a disturbance revealed by cell inactivation (nonviable cells among clones produced by irradiated cells), incorrect distribution of mitochondria between daughter cells (respiratory mutants). As a whole, the data of this section demonstrate the existence of sublesions in yeast cells surviving after irradiation.

5.3. The Basic Assumptions and Development of the Theoretical Model of Synergy under the Simultaneous Action of Agents

It is our aim here to develop a simple mathematical approach to synergistic interaction which would be able to predict several universal patterns of synergistic effect display that do not depend on biological object and test being studied as well as on the agents applied. To gain some more insight into the mode of interaction, we are going to implement the following tasks.

(1) To suggest a unified biophysical conception of synergistic interaction that permits to interpret the reveled general regularities of synergistic interaction.
(2) To formulate a common semi-empirical mathematical model to describe, optimize and predict the synergy.
(3) To find out the condition under which the highest synergistic effect can be achieved.
(4) To prove formula for quantitative estimation of synergistic interaction.

The results of the previous section, pointing out the negligible role of the recovery damage itself in the mechanism of synergistic interaction, strongly invoke the need to elaborate a new theoretical conception of the synergy which, being useful for environmental radiation protection, take into account the regularities revealed. Synergism is defined as the combined interaction of two agents that exceeds the sum of their individual effects. With the use of this definition, it might be reasonable to assume that some additional effective damages, responsible for the synergy, are arisen from the interaction of sublesions induced by both agents and these sublesions are ineffective when each agent is applied separately. We suppose that one sublesion produced by one agent (for example, ionizing radiation) interacts, for concreteness sake, with one sublesion from heat to produce one additional effective damage. This assumption is not crucial since we do not define sublesion nature concretely. It seems also plausible to suppose that the number of sublesions is directly proportional to the number of the effective damages. To follow to our previous publications [28; 49; 51; 54], we use here index 1 for the agent applied together with heat (index 2). Let p_1 and p_2 be the numbers of sublesions that occur for one effective damage and N_1 and N_2 be the mean numbers of the effective damages in a cell produced by these agents. According to these assumptions, the synergistic effect is attributed to the formation of some additional effective damage N_3. Then the synergistic enhancement ratio k may be expressed as

$$k = \frac{N_1 + N_2 + N_3}{N_1 + N_2}. \tag{5.2}$$

A number of additional damages N_3 arising from the interaction of sublesions produced by the two agents may be given as

$$N_3 = \begin{cases} p_1 N_1, & if \quad p_1 N_1 \le p_2 N_2 \\ \\ p_2 N_2, & if \quad p_1 N_1 \ge p_2 N_2 \end{cases}. \tag{5.3}$$

In fact, this equation shows that the additional effective damage is a minimal value from two variable quantities: $p_1 N_1$ and $p_2 N_2$, which are the mean number of sublesions produced by the corresponding agent. Taking into account Eq. (5.3), the expression for synergistic effect (Eq. 5.2) can be written as

$$k = \begin{cases} \dfrac{N_1 + N_2 + p_1 N_1}{N_1 + N_2}, & if \quad p_1 N_1 \le p_2 N_2 \\ \\ \dfrac{N_1 + N_2 + p_2 N_2}{N_1 + N_2}, & if \quad p_1 N_1 \ge p_2 N_2 \end{cases} . \tag{5.4}$$

After simplification we have:

$$k = \begin{cases} 1 + \dfrac{p_1}{1 + N_2 / N_1}, & if \quad p_1 N_1 \le p_2 N_2 \\ \\ 1 + \dfrac{p_2 N_2 / N_1}{1 + N_2 / N_1}, & if \quad p_1 N_1 \ge p_2 N_2 \end{cases} \tag{5.5}$$

This equation has two parameters (p_1 and p_2) to be estimated from experimental data. It is evident from here that the effectiveness of the synergistic interaction will be determined by the least value from the two functions:

$$f_1 = 1 + \frac{p_1}{1 + N_2 / N_1} \tag{5.6}$$

and

$$f_2 = 1 + \frac{p_2 N_2 / N_1}{1 + N_2 / N_1}. \tag{5.7}$$

The relationship between the synergistic enhancement ratio and the ratio of N_2 / N_1 (Eq. 5.5) shows an increase in the synergistic enhancement ratio with N_2 / N_1 to a maximum, followed by a subsequent decrease at higher N_2 / N_1. The absolute value of this maximum as well as the ratio of N_2 / N_1 at which it is attained are obviously of great interest not only in thermobiology but for environmental protection consideration as well. Since f_1 decreases while f_2 increases with increasing N_2 / N_1 (Figure 5.11), the greatest synergistic effect will be obtained when $f_1 = f_2$, i.e.

$$\left[1+\frac{p_1}{1+N_2/N_1}\right]=\left[1+\frac{p_2N_2/N_1}{1+N_2/N_1}\right]. \tag{5.8}$$

After simplification we have

$$N_2/N_1 = p_1/p_2 \tag{5.9}$$

or

$$p_1N_1 = p_2N_2. \tag{5.10}$$

Eqs. (5.9) and (5.10) enable us to predict the condition underwhich the highest synergistic interaction can be achieved. The latter equation suggests that the highest synergistic interaction occurs when both agents produce equal numbers of interacting sublesions. For a particular case $p_1 \cong p_2$, the highest synergistic interaction occurred when the both agents produce almost equal number of effective damages, i.e., $N_1 \cong N_2$.

Combining Eqs. (5.5) and (5.9), one can deduce the value of the highest synergistic enhancement ratio which can be given by

$$k_{\max} = 1+\frac{p_1 \cdot p_2}{p_1 + p_2}. \tag{5.11}$$

It is evident that the value of the highest synergistic effect does not depend on the ratio of N_2/N_1 and is completely determined by the basic parameters of the model.

It is curiously but the model considered does not demand a knowledge of absolute values of N_1 and N_2 but only their dimensionless ratio, N_2/N_1, which can be derived for cell inactivation (the main end point used in this study) from the ratio of slopes of the corresponding curves

$$N_2/N_1 = D_{01}/D_{02} = t_{01}/t_{02}, \tag{5.12}$$

where D_0 (or t_0) being a measure of the dose (or duration of exposure) for survival to be reduced to $1/e$ on the exponential part of survival curve. According to Poisson distribution, the corresponding number of lethal lesions can be calculated [19; 33; 58] as

$$N = -\ln S, \tag{5.13}$$

where S is the surviving fraction.

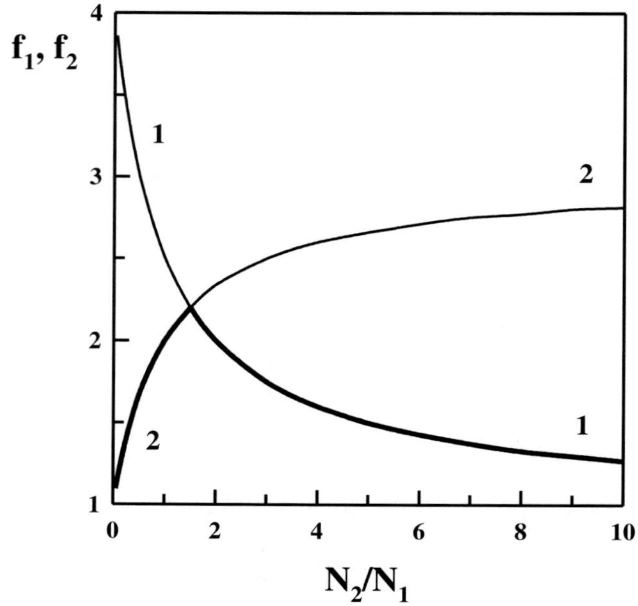

Figure 5.11. The calculated dependency of functions f_1 (curve 1) and f_2 (curve 2) as well as the synergistic enhancement ratio (bold line) on the N_2 / N_1 ratio for the following values of the basic parameter: $p_1 = 3$ and $p_2 = 2$.

Again, suppose that we deal with simultaneous action of ionizing radiation and high temperatures. If the observed biological effect is mainly induced by heat ($p_1N_1 < p_2N_2$) then taking into account Eq. (5.5), the parameter p_1 can be expressed as

$$p_1 = (k_1 - 1)(1 + N_2 / N_1), \tag{5.14}$$

where k_1 is the value of synergistic enhancement ratio observed in experiments performed in this condition. On the contrary, if the observed biological effect is mainly induced by ionizing radiation ($p_1N_1 > p_2N_2$), we have

$$p_2 = (k_2 - 1)(1 + N_1 / N_2), \tag{5.15}$$

where k_2 is the experimental value of the synergistic enhancement ratio observed for the condition $p_2N_2 < p_1N_1$.

It is easily to demonstrate that the model under consideration can predict two N_2 / N_1 ratio, at which equieffective values of the synergistic enhancement ratio (k_i) can be observed. For the case $p_1N_1 < p_2N_2$, we have

$$N_2 / N_1 = \frac{p_1 - k_i + 1}{k_i - 1}, \tag{5.16}$$

while for the case $p_1 N_1 > p_2 N_2$

$$N_2 / N_1 = \frac{k_i - 1}{p_2 - k_i + 1}. \tag{5.17}$$

We can estimate the possible range of the highest synergistic effect in terms of sublesions values. Taking the highest and the least values of p_1 and p_2, represent the Eq. (5.11) as follows

$$k_{max} = 1 + \frac{p_{min}}{1 + p_{min} / p_{max}}. \tag{5.18}$$

From here, we obtain the range of possible variation in the synergistic enhancement ratio:

$$(1 + p_{min} / 2) \le k_{max} \le (1 + p_{min}) \tag{5.19}$$

Consequently, the highest value of synergistic effect is primarily determined by the minimum value from p_1 and p_2 (p_{min}). It follows that even a great number of sublesions from one of the agents do not provide an effective interaction if the other agent creates little sublesions.

Thus, based on Eqs. (5.14) and (5.15) and experimental data for k_1 and k_2, we can estimate the basic model parameters p_1 and p_2 and then predict the value of the synergistic enhancement ratio for any N_1 and N_2 (Eq. 5.5), the highest value of the synergistic enhancement ratio (Eq. 5.11), and the condition (Eq. 5.10) under which it can be achieved. Some examples of theoretically predicted dependencies of the synergistic enhancement ratio on the N_2 / N_1 ratio for various values of the basic parameters p_1 and p_2 are depicted in Figures (5.12) and (5.13).

It can be seen that the synergistic enhancement ratio for arbitrary values of p_1 and p_2 passes through the maximum and its value increases with an increase in p_1 and p_2. For a constant ratio of p_1 / p_2 the location of maximum remains at the same place of abscissa axis (Figure 5.12). The position of the synergy peak shifts to a higher value of N_2 / N_1 with an increase in the ratio p_1 / p_2 (Figure 5.13).

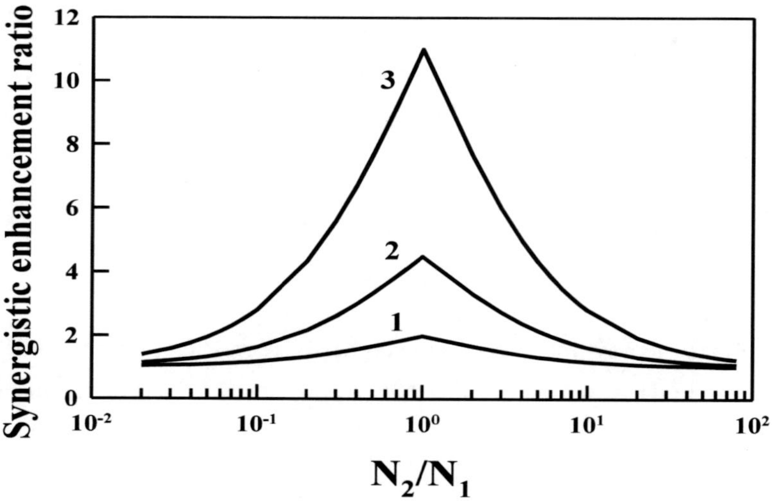

Figure 5.12. Theoretically expected dependences of the synergistic enhancement ratio on the N_2 / N_1 ratio of lethal damages induced by ionizing radiation (N_1) and heat (N_2) for the following values of the basic parameter: $p_1 = p_2$ = 2 (curve 1), 7 (curve 2), 20 (curve 3).

Compare qualitatively the prediction of the considered model with some experimental data described above for simultaneous action of heat and other physical or chemical agents. First of all, the existence of the highest synergistic effect was observed in all our experiments (Figures 5.1–5.6). The increase in the ratio N_2 / N_1 means the increase in the thermal load and the decrease in the damage produced by ionizing radiation. It follows that the increase in the exposure temperature should result at first in an increase in the effectiveness of interaction, after achieving the optimum value the synergistic effect should be decreased with increasing exposure temperature because of the decrease in the whole duration of treatment and the corresponding small number of damage produced by another agents (ionizing radiation). These predictions mean that the synergy should be observed only within a certain temperature range. Actually, this effect was observed in our various findings when the exposure temperature was increased and as a consequence the contribution of ionizing radiation, or another agent combined with heat, in the overall effect of thermoradiation action decreased (Figures 5.1–5.6).

The increase in cellular thermosensitivity may be interpreted as and increase in p_2. Hence the decrease in the ratio of p_1 / p_2 should result in shifting the whole dependence of synergy towards small region of N_2 / N_1. Both theoretical predictions (Figure 5.13) and experimental results Figures 5.1 and 5.4 evidence in favor of this point of view.

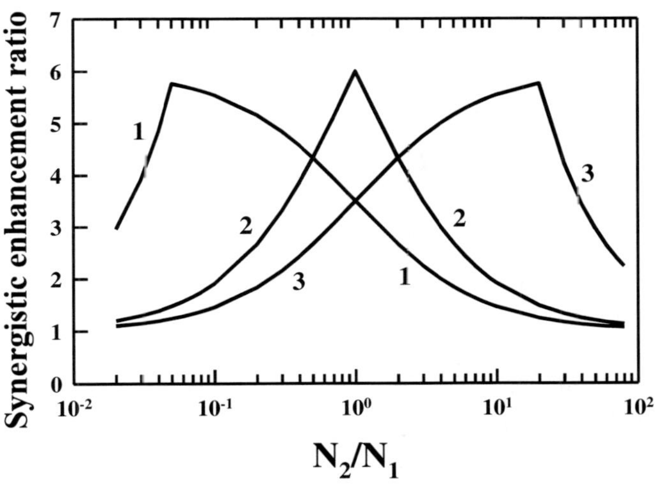

Figure 5.13. Theoretically expected dependences of the synergistic enhancement ratio on the N_2 / N_1 ratio of lethal damages induced by ionizing radiation (N_1) and heat (N_2) for the following values of the basic parameter: $p_1 = 5$ and $p_2 = 100$ (curve 1), $p_1 = p_2 = 10$ (curve 2), $p_1 = 100$ and $p_2 = 5$ (curve 3).

The specific temperature that maximizes the synergistic effect was shown to decrease with decreasing of exposure rate. This effect was demonstrated experimentally for various agents applied with heat Figures 5.6 and 5.7 and also can be qualitatively expected from the model discussed. For example, in accordance with the model, the decrease in the dose rate means that more time needs to achieve the same value of damage produced by ionizing radiation (N_1) and as a consequence, the large number of heat damage should be produced. To preserve the condition for achieving the greatest synergy (Eq. 5.10) we must reduce the exposure temperature. This also means that for a constant temperature at which the irradiation occurs, synergy can be observed within a certain dose rate range and inside this range an optimal intensity of physical agent may be indicated that maximizes the synergy. This effect is also confirmed by experimental findings (Figures 5.5 and 5.6).

5.4. The Dependence of Synergy on the Intensity of Applied Agents: Mathematical Modeling and Interpretation

Now let us analyze the possibility to predict the influence of dose rate on the value of synergy from the position of the model described in the previous section. If the biological effect does not depend on the dose rate, the increase in the dose rate of ionizing radiation in the case of its simultaneous application with heat is accompanied with the decrease in the exposure duration and the mean lethal duration after ionizing radiation applied alone. This condition is implemented for radiobiological studies with bacteria, yeast and mammalian cells. The dose rate dependence is manifested mainly for protracted exposures when the duration of irradiation, for example of yeast cells, is extended for more than 10 hours [52].

The synergistic effect is determined by the ratio of N_2 / N_1 of the effective damage produced by inactivating agents (Eq. 5.5). This ratio can be found from the corresponding slopes of survival curves after action of ionizing radiation and hyperthermia applied separately (Eq. 5.12). In accordance with this equation, the decrease in the mean lethal time of ionizing radiation, t_{01}, that takes place after the increase in the dose rate, had to result in the corresponding decrease in N_2 / N_1 ratio because of smaller exposure time, i.e., due to the decrease in the number of thermal damage N_2 under the unchanged quantity of the effective damage N_1 produced by ionizing radiation. To preserve the condition guaranteeing the highest synergistic effect (Eq. 5.10), we should increase the number of thermal damage N_2 as much as the dose rate increased. The increases in N_2 can only occur due to the increase in the acting temperature. Consequently, the exposure temperature, which should provide the maximum synergistic effect with a larger dose rate, must be higher than that which was corresponded to the same synergy at the initial dose rate of ionizing radiation. Similarly, to keep the greatest synergy after a reduction in dose rate of ionizing radiation, the acting temperature should be diminished.

Let the Eq. (5.10) holds for an arbitrary selected dose rate and some exposure temperature. If the dose rate decreases by a factor of m_1, the number of effective thermal damage increases also by a factor of m_1, i.e., will be equal to $m_1 N_2$ and the condition $p_1 N_1 < p_2 N_2$ will be executed. Then the new value of the synergistic enhancement ratio according to the Eq. (5.5), with keeping in mind Eq. (5.10), is equal to

$$k_1 = 1 + \frac{p_1}{1 + m_1 p_1 / p_2}. \tag{5.20}$$

This expression, taking into account the Eq. (5.11) can be rewritten as

$$k_1 = 1 + \frac{(k_{\max} - 1) \cdot (1 + p_1 / p_2)}{1 + m_1 p_1 / p_2}. \tag{5.21}$$

Eqs. (5.20) and (5.21) show that a decrease in synergy will be greater for larger values of m_1 and the ratio of p_1 / p_2.

If the dose rate increases by a factor of m_2, the number of effective thermal damage decreases also by a factor of m_2, i.e., will be equal to N_2 / m_2 and the condition $p_1 N_1 > p_2 N_2$ will be executed. Then the new value of the synergistic enhancement ratio according to the Eq. (5.5), with keeping in mind Eq. (5.10), is equal to

$$k_2 = 1 + \frac{p_1}{(m_2 + p_1 / p_2)}. \tag{5.22}$$

This expression, taking into account the Eq. (5.11), may be rewritten as

$$k_2 = 1 + \frac{(k_{max} - 1)\cdot(1 + p_1 / p_2)}{(m_2 + p_1 / p_2)}. \tag{5.23}$$

It follows from Eqs. (5.22) and (5.23) that a decrease in synergy will be greater for larger values of m_2 and the ratio of p_1 / p_2. This circumstance can explain why a strong synergistic effect was not observed by Kiefer et al. [27]. In their study the dose rate of ionizing radiation was equal to 480 Gy/min and the maximum temperature used was 50 °C. In our experiments at 50 °C the highest synergistic effect (k =2.5) was observed for the same species of diploid yeast cells at a dose rate of 10 Gy/min. The ratio of the dose rates used was m = 48. Assuming p_1 / p_2 = 2 and the synergistic enhancement ratio was the greatest, we estimated (Eq. 5.23) its value should not be more than 1.09. According to our estimates, in fact, it was 1.05, which corresponds to the assumptions made. In order to achieve the highest synergistic effect, Kiefer et al. [27] should increase the exposure temperature to about 57–58 °C. We presented above equations for the synergistic enhancement ratio for any ratio of acting agents N_2 / N_1 (Eq. 5.5), the condition at which the highest effect can be achieved (Eq. 5.10) and its value (Eq. 5.11). These expressions can be easily modified to include the dose rate of ionizing radiation dD/dt. Let r_1 and r_2 be the ratio of the survival curve slopes after separately exposure to hyperthermia and ionizing radiation for the two cases, when the rate of inactivation by heat is greater than by radiation (r_1) and when the dominant contribution to cell inactivation is provided by ionizing radiation (r_2). Taking into account Eqs. (5.5, 5.14 and 5.15) and that the mean inactivation time after ionizing radiation treatment $t_1 = D_0 /(dD/dt)$, we obtain the expression determining the value of the synergistic enhancement ratio including the dose rate of ionizing radiation:

$$k = \begin{cases} 1 + \dfrac{(k_1 - 1)\cdot(1 + r_2)}{1 + D_0 / t_2(dD/dt)}, & if \quad p_1 N_1 \le p_2 N_2 \\[2ex] 1 + \dfrac{D_0(k_2 - 1)\cdot(1 + r_2)/(dD/dt)t_2}{1 + D_0 / t_2(dD/dt)}, & if \quad p_1 N_1 \ge p_2 N_2 \end{cases}. \tag{5.24}$$

Eq. (5.10) can be rewritten as

$$\frac{D_0}{t_2(dD/dt)} = \frac{(k_1 - 1)\cdot(1 + r_1)}{(k_2 - 1)\cdot(1 + 1/r_2)} \tag{5.25}$$

and the highest synergistic effect according to Eq. (5.11) can be presented by the following expression

$$k_{max} = 1 + \frac{(k_1 - 1)\cdot(1 + r_1)}{1 + \dfrac{(k_1 - 1)\cdot(1 + r_1)}{(k_2 - 1)\cdot(1 + 1/r_2)}}. \tag{5.26}$$

The presented analysis shows that, in principle, the model described here can be applied to interpret and describe the relationship between synergistic effect and the dose rate of ionizing radiation for simultaneous exposure to heat and radiation. Quite similar considerations are also applicable to prove the dependence of synergy on the intensity of any other combination of physical and chemical agents.

5.5. Prediction of Irreversible Component of Radiation Damage after Combined Action

That high temperature can enhance radiation sensitivity of a wide variety of biological systems both *in vitro* and *in vivo* can no longer be doubted. Heating at temperatures above 40 °C are now used to increase the cellular radiosensitivity, the greatest synergistic effect being observed when heat and irradiation are given simultaneously. It has so far generally been assumed that thermal radiosensitization results from the heat-induced inhibition of the repair of radiation damage. This inhibition was revealed in a decrease of both the rate and extent of recovery which can be interpreted as being due to either the impairment of the recovery process itself (inactivation of repair enzymes, for instance) or to the production of more severe irreversible lethal damage which can not be repaired. The latter case was proved above to be realized for the combined treatment of heat and ionizing radiation on yeast and cultured mammalian cells as well as for the exposure of yeast cells to heat and UV light. These novel observations provide insight into the mechanisms underlying the effects of combined exposure and should be taken into account for interpretation and mathematical modeling of synergistic interaction of the combined exposures of various environmental agents. A simple mathematical approach has been described above to predict and optimize the synergistic interaction of various acting agents. The model was found to predict the dependence of synergistic interaction of the ratio of lethal events produced by every agent used, the greatest synergistic effect and the condition at which the maximal interactive effect can be achieved. However, this model does not take into consideration novel data indicating that additional damage responsible for the synergy is irreversible.

As was mentioned, if irradiated cells are held in a liquid or solid non-nutrient media at 30 °C before plating on to a growth medium, their survival is increased reflecting the process of LHR [30; 31; 43]. When samples were plated at different times after the treatment, the number of viable cells increased due to the LHR as a function of time, reaching a plateau after about 2–3 days. The availability of the plateau reflects the existence of the irreversible component of radiation damage which cell is incapable to recover from. In accordance with the definition, the irreversible component can be quantitatively estimated as the ratio of the number of irreversible damage to the total number of the produced damage. The ratio shows the portion of damage which can not be repaired. Each type of the effective damage (N_1, N_2 and N_3) can be characterized by its own value of irreversible component (K_1, K_2 and K_3, respectively). Then for the synergistic interaction of two agents the irreversible component (K) might be presented as

$$K = \frac{K_1 N_1 + K_2 N_2 + K_3 N_3}{N_1 + N_2 + N_3}. \tag{5.27}$$

Taking into account Eq. (5.3) and the basic postulates of the model, Eq. (5.27) can be rewritten as

$$K = \begin{cases} \dfrac{K_1 N_1 + K_2 N_2 + K_3 p_1 N_1}{N_1 + N_2 + p_1 N_1}, & if \quad p_1 N_1 \le p_2 N_2 \\[2ex] \dfrac{K_1 N_1 + K_2 N_2 + K_3 p_2 N_2}{N_1 + N_2 + p_2 N_2}, & if \quad p_1 N_1 \ge p_2 N_2 \end{cases}, \tag{5.28}$$

where again K_1 is the irreversible component after separate exposure to the agent combined with heat, K_2 is the irreversible component after thermal treatment applied alone and K_3 is the irreversible component of additional damages responsible for the synergy. As was shown above, this additional damage is irreversible, i.e., K_3 = 1. For convenience calculation ability, this equation may be given by

$$K = \begin{cases} \dfrac{K_1 + K_2 N_2 / N_1 + K_3 p_1}{1 + N_2 / N_1 + p_1}, & if \quad p_1 N_1 \le p_2 N_2 \\[2ex] \dfrac{K_1 + K_2 N_2 / N_1 + K_3 p_2 N_2 / N_1}{1 + N_2 / N_1 + p_2 N_2 / N_1}, & if \quad p_1 N_1 \ge p_2 N_2 \end{cases}. \tag{5.29}$$

It is plausible to assume that biological damage produced by each acting agent (N_1 and N_2) during combined action can be repaired in the same extent and with the same probability as they were observed after application of these agents individually. One can estimate K_1 and K_2 values from experiments with separate action of each agent used in combination.

Analysis of numerous data has shown that synergistic interaction of ionizing radiation with other environmental chemical and physical agents results in the inhibition of some cellular repair mechanisms. The inhibition was revealed in a decrease of both the rate and extent of recovery. Two hypotheses could be envisaged. Such lesions, resulting from the simultaneous application of the two insults, might not be reparable. Alternatively, the damage might be similar, but the repair enzymes might be modified so that their ability of repair is changed. The synergistic effects of the combined action of heat with ionizing radiation or UV light were interpreted as being due to more severe or more complex damage produced during the combined treatment. The most likely explanation for these results is that heating after irradiation modified the damage so that enzymes could then not deal with the lesions.

The major points to be inferred from the data presented here are that the mathematical model basing on the minimum initial postulates describes in general form the dependence of the synergistic effect on the ratio of the effective damages produced by agents used in combination, predicts the greatest synergistic interaction, condition under which it can be achieved, the dependence of synergy on the intensity of agents used in combination and the portion of the irreversibly damage resulted from the combined exposures.

In summary, it can be concluded that the mechanism of the synergistic interaction of hyperthermia with ionizing radiation or UV light after their simultaneous treatment in yeast cells is not related to the impairment of the recovery capacity itself and it may be attributed to an increased yield of irreversible damage.

CONCLUSION

Thus, the main common regularities of the synergistic interaction obtained in this investigation may be summarized as follows.

1. After simultaneous thermoradiation action for any constant intensity of ionizing radiation or UV light, the synergy can be observed only within a certain temperature range.
2. The temperature range synergistically increasing the effect of radiation is shifted to lower temperature for thermosensitive objects.
3. Inside this range, there is a specific temperature that maximizes the synergistic effect.
4. An increase in exposure rate resulted in an increase of this specific temperature to achieve the greatest synergy and *vice versa*.
5. For a constant temperature at which the irradiation occurs, synergy can be observed within a certain intensities of physical agent range.
6. Inside this range an optimal intensity may be indicated that maximizes the synergy.
7. As the exposure temperature is reduced, the optimal intensity decreases and *vice versa*.
8. The recovery rate after combined action is decelerated due to an increased number of irreversible damage.

9. The probability of recovery is independent of exposure temperature for cells irradiated with ionizing radiation.
10. Chemical inhibitors of cell recovery act through the formation of irreversible damage but not *via* damage the recovery process itself.

The last two rules were formulated on the basis of our results presented in the previous Chapters. These common patterns of synergistic effects display are universal and do not depend on the biological object and test being studied as well as on the agents used in combined actions. These data strongly invoke the need to elaborate a new theoretical conception of the synergy which, being useful for environmental radiation protection, took into account the regularities revealed. A great number of experimental data obtained in this area stress the need for a mathematical approach in efforts to optimize and predict the interaction of environmental agents.

The remarkable result of this Chapter concerns a simple semi-empirical mathematical model that describes the synergistic effect of combined action of two different agents. The model is based on the assumption that synergism is conditioned by additional effective damage arising from the interaction of subiesions induced by both agents. The model is not concerned with the molecular nature of subiesions, and the mechanisms of their interaction remain to be elucidated. Moreover, they may be unknown or different for various combination of chemical and physical factor. These subiesions are noneffective when each agent is applied alone. The idea of lesion interaction is now widely used in radiobiology [8; 9; 18; 22; 26; 29; 34-36; 39; 48; 49; 60; 63]. In the proposed model, the synergistic interaction is given by $\min\{p_1N_1;p_2N_2\}$. This means that one sublesion caused by irradiation interacts with one sublesion caused by heat. This process is assumed to proceed until the very last sublesion of the less frequent type is used up. However, it does not mean that the model requires some kind of long-range interaction across the whole cell or at least across the whole cell nucleus. To estimate the basic parameters p_1 and p_2 we have used the experimental values of the synergistic enhancement ratio k_1 and k_2. It follows that the model takes into consideration only the actual interaction determining the synergistic effect. The model is very simple and requires the estimation of only two unknown parameters, p_1 and p_2, which are assumed to be entirely independent of experimental conditions (temperature, dose rate, and exposure time). This fact may contribute to differences in experimental and predicted results. For example, some of our investigations of yeast cell responses to heat applied alone or combined with γ-rays were depicted in an Arrhenius plot [45]. These curves for heat treatment alone show a distinct break of the inactivation constant at 52.5 °C for diploid *Saccharomyces* yeast cells. This suggests that hyperthermic killing above and below this temperature is mediated by different mechanisms. It was shown [50] that the values of mean number of sublesions p_2 that arise for one lethal heat event may be somewhat different below and above 52.5 °C.

The model predicts the dependence of synergistic interaction on the ratio N_2 / N_1 of effective damage produced by every agent applied (Eq. 5.5), the greatest value of the synergistic effect (Eq. 5.11) as well as the condition under which it can be achieved (Eq. 5.10). In spite of the approximation used in this simplified model, it is evident from the data presented that a quantitative agreement appears to exist between theoretical and experimental results. The degree of synergistic interaction was found to be dependent on the ratio of lethal damage N_2 / N_1 induced by the two agents applied. The synergistic interaction is not observed at any N_2 / N_1 ratios. The effectiveness of synergistic interaction appears to decline with a deviation of the ratio N_2 / N_1 from optimal value. The model can predict the irreversible component of radiation damage after combined action of agents.

The most interesting result, predicted by the model, is the conclusion that for a lower intensity of physical agents or a lower concentration of chemicals a lower temperature must be used to provide the greatest synergy. Actually, any decrease in the intensity of physical agents would result in an increase of the duration of thermoradiation action to achieve the same absorbed dose. Therefore, the number of thermal sublesion will also be increased resulting in the disruption of the condition at which the highest synergy should be observed (Eq. 5.10). Hence, to preserve an optimal N_2 / N_1 ratio with any decrease in the dose rate (or the intensity of other agents) the exposure temperature should be decreased. Herewith the following general rule should be executed: the lower the intensity of one of the agents employed, the lower intensity of the another agent should be used to maximize the synergistic interaction. Several examples of such relationships have been presented above. Exactly such dependence is of the greatest interest for environmental problems and assessment of the potential significance of mutual enhancement of harmful action of agents really occurring in the biosphere. Because of the fundamental importance of this issue, in the next Chapter we will analyze our extensive own experimental data and those published by other authors, in terms of the synergy dependence on the dose rate of ionizing radiation or the intensity of other agents used under simultaneous combined exposure. It can be concluded on this basis that for a long duration of interaction, which is important for problems of radiation protection, low intensities of deleterious environmental factors may, in principle, synergistically interact with each other or with environmental heat.

There are the next principal issues to be drawn concisely from the results described here.

1. Some common regularities of synergistic effect display have been revealed, which are universal and do not depend on the biological object and test being studied as well as on the interactive agents. These data strongly invoke the need to elaborate a new theoretical conception of the synergy which, being useful for environmental radiation protection, took into account the regularities revealed.
2. Cells from unstable clones exhibited an enhanced radiosensitivity, increased content of morphologically changed cells, nonviable cells, respiratory and recombinant mutants. The degree of expression of the foregoing effects was the higher the greater number of primary sublethal lesions was in the originally irradiated cell. These data indicate the possibility of sublesion preservation in the distant descendants of cells surviving after irradiation.
3. A unified biophysical conception for synergistic interaction mechanism was suggested. In accordance with this conception, the synergistic effect of combined action of two agents is conditioned by the additional effective damage arising from the interaction of subiesions induced by both agents. These sublesions are considered as non-effective after each agent being applied separately.

4. A simple semi-empirical mathematical model of synergy for simultaneous action of two interacting agents was developed. In the proposed model, one sublesion from one agent interacts with a sublesion from another agent to produce an additional effective damage. This process is assumed to proceed until the very last sublesion of the less frequent type is used up. The model takes into consideration the actual interaction determining the synergistic effect since the basic parameters of the model have been determined from experimental results.
5. The model predicts the bell-shaped dependence of synergistic enhancement ratio on the ratio of the effective damage produced by both agents, on the exposure temperature applied simultaneously with ionizing radiation, UV light, ultrasound or chemical compounds. A qualitative agreement between the model predictions and experimental results was demonstrated for various biological objects and tests.
6. The dependence of synergy on the intensity of applied agents was predicted. Herewith the following general rule is executed: the lower the intensity of one of the agents employed, the lower intensity of the another agent should be used to maximize the synergistic interaction. Exactly such dependence is of the greatest interest for environmental problems and assessment of the potential significance of mutual enhancement of harmful action of agents really occurring in the biosphere.

REFERENCES

[1] Ager, DD; Haynes, RH. Mathematical description of the interactions between cellular inactivating agents. *Radiat. Res.*, 1987; 110, 129-141.

[2] Alper, T. *Cellular Radiobiology*. Cambridge University Press, 1979.

[3] Baverstock, K F; Stather, JW. *Low Dose Radiation: Biological Bases of Risk Assessment*. Edition. London: Taylor & Francis, 1989.

[4] Brannen, JP. A temperature- and dose rate-dependent model for the kinetics of cellular response to ionizing radiation. *Radiat. Res.*, 1975; 62(3), 379-387.

[5] Ben-Hur, E. Mechanisms of the synergistic interaction between hyperthermia and radiation in cultured mammalian cells. *J. Radiat. Res.,* 1976; 17, 92-98.

[6] Ben-Hur, E; Elkind, MM; Bronk, BV. Thermally enhanced radio-response of cultured Chinese hamster cells – Inhibition of repair of sublethal damage and enhancement of lethal damage. *Radiat. Res.*, 1974; 58, 38-51.

[7] Burlakova, EB. *Low Doses of Radiation: Are They Dangerous?* Edition. New York: Nova Science Publishers, Inc., 2000.

[8] Cebulska-Wasilewska, A; Leenhouts, HP; Chadwick, KH. Synergism between EMS and X-rays for the induction of somatic mutations in *Tradescantia. Int. J. Radiat. Biol.*, 1981; 40, 163-173.

[9] Chadwick, KH; Leenhouts, HP. *The Molecular Theory of Radiation Biology*. Berlin, Heidelberg, New York: Springer-Verlag, 1981.

[10] Dethlefsen, LA; Dewey, WC. *Cancer Therapy by Hyperthermia, Drugs, and Radiation*. Edition. Washington, D.C.: U.S. Dept. of Health and Human Services, National Cancer Institute, 1982.

[11] Dikomey, E. Differential cytotoxic effects on hyperthermia below and above 43 degree C alone or combined with X irradiation. *Radiat. Res.*, 1981; 88(3), 489-501.

[12] Dineva, SB; Abramov, VI; Shevchenko, VA. The genetic effects of treatment of Arabidopsis thaliana seeds by the sodium lead of chronic irradiatied population. *Genetics*, 1993; 29, 1914-1920.

[13] Dugan, VL. A kinetic analysis of spore inactivation in composite heat and gamma radiation environment. *Space Life Sci.*, 1971; 2, 498-505.

[14] Gerner, EW; Leith, JT; Boone, ML. Mammalian cell survival response following irradiation with 4 MeV X-rays or accelerated helium ions combined with hyperthermia. *Radiology*, 1976; 119, 715-720.

[15] Goodhead, DT. Track structure considerations in low dose and low dose rate effects of ionizing radiation. *Adv. Radiat. Biol.*, 1992; 16, 7-44.

[16] Hahn, GM. *Hyperthermia and Cancer*. New York: Plenum Press, 1982.

[17] Hall, EJ. Radiation dose-rate: A factor of importance in radiobiology and radiotherapy. *Br. J. Radiol.*, 1972; 45, 81-97.

[18] Harder, D; Virsik-Peuckert, RP; Bartels, ER. Theory of intratrack pairwise lesion interaction. *Radiat. Prot. Dosimetry*, 1994; 52, 13-16.

[19] Haynes, RH. The interpretation of microbial inactivation and recovery phenomena. *Radiat. Res.,* 1966; Suppl. 6, 1-29.

[20] Jain, VK; Pohlit, W. *Byocybernetic of Cancer. Optimizimg Cancer Treatment with Ionizing Radiation*. Banglore: INSDOC, 1986.

[21] Johnson, HA; Pavelec, M. Thermal enhancement of thio-TEPA cytotoxicity. *J. Natl. Cancer Inst.*, 1973; 50, 903-908.

[22] Jung, H. A generalized concept for cell killing by heat. *Radiat. Res.*, 1986; 106(1), 56-72.

[23] Kappos, A; Pohlit, W. A cybernetic model for radiation reactions in living cells. I. Sparsely ionizing radiation, stationary cells. *Int. J. Radiat. Biol.*, 1972; 22, 51-65.

[24] Kapultcevich, YuG. *Quantitative Regularities of Cell Radiation Injury*. Moscow: Atomizdat, 1978.

[25] Kapultsevich, YuG; Petin, VG. Probability model for cell responses to irradiation. *Studia Biophysica*, 1977; 62, 151-165.

[26] Kellerer, AM; Rossi, HH. A generalized formulation of dual radiation action. *Radiat. Res.*, 1978; 75, 471-488.

[27] Kiefer, J; Kraft-Weyrather, W; Hlawica, MM. Cellular radiation effects and hyperthermia. Influence of exposure temperatures on survival of diploid yeast irradiated under oxygenated and hypoxic conditions. *Int. J. Radiat. Biol.*, 1976; 30, 293-300.

[28] Kim, JK; Petin, VG. Theoretical conception of synergistic interactions. *Korean J. Environ. Biol.,* 2002; 20, 277-286.

[29] Kim, JK; Petin, VG; Zhurakovskaya, GP. Exposure rate as a determinant of synergistic interaction of heat combined with ionizing or ultraviolet radiations in cell killing. *J. Radiat. Res.*, 2001; 42, 361-365.

[30] Korogodin, VI. Some regularities of the post-irradiation changes in resting yeast cells. *Biofizika,* 1958; 3, 703-710.

[31] Korogodin, VI. The study of post-irradiation recovery of yeast: the premolecular period. *Mutat. Res.*, 1993; 289, 17-26.

[32] Korogodin, VI; Bliznik, KM; Kapul'tcevich, YuG et al. Cascade mutagenesis: regularities and mechanisms. In: Korogodina VL, Cigna AA, Durante M editors. *Modern Problems of Genetics, Radiobiology, Radioecology and Evolution*. Dubna: JINR, 2007; 419-447.

[33] Lea, DE. *Action of Radiation on Living Cells*. Cambridge University Press, 1946.

[34] Leenhouts, HP; Chadwick, KH. An analysis of synergistic sensitization. *Brit. J. Cancer*, 1978; 37, Suppl. III, 198-201.

[35] Leenhouts, HP; Sijsma, MJ; Cebulska-Wasilewska, A et al. The combined effect of DBE and X-rays on the induction of somatic mutations in Tradescantia. *Int. J. Radiat. Biol.*, 1986; 49, 109-119.

[36] Little, JB. Low-dose radiation effects: interaction and synergism. *Health Phys.*, 1990; 59, 49-55.

[37] Loshek, DD; Orr, JS; Solomonidis, E. Interaction of hyperthermia and radiation. *Br. J. Radiol.*, 1977; 50, 893-901.

[38] Menezes, S; Costa, JA. Methylene blue inhibits polymerase 1 enzyme and sensities *Escherihia coli* bacteria to X-rays. *Int. J. Radiat. Biol.*, 2000; 76(9), 1289-1294.

[39] Murthy, MS; Deorukhakar, VV; Rao, BS. Hyperthermic inactivation of diploid yeast and the interaction of damage caused by hyperthermia and ionizing radiation. *Int. J. Radiat. Biol.*, 1979; 35, 333-341.

[40] Ogur, H; John, R; Nagai, S. Tetrazolium overlag technique for population studies of respiration deficiency in yeast. *Science*, 1957; 125, 928-931.

[41] Overgaard, J. Editor. *Hyperthermic Oncology*. London, Philadelphia: Taylor & Francis, 1984; 1.

[42] Overgaard, J. Editor. *Hyperthermic Oncology*. London, Philadelphia: Taylor & Francis, 1985; 2.

[43] Patrick, MH; Haynes, RH. Dark recovery phenomena in yeast. II. Conditions that modify the recovery process. *Radiat. Res.*, 1964; 23, 564-579.

[44] Petin, VG; Berdnikova, IP. Effect of elevated temperatures on the radiation sensitivity of yeast cells of different species. *Radiat. and Environ. Biophys.*, 1979; 16, 49-61.

[45] Petin, VG; Berdnikova, IP. Responses of yeast cells to heat applied alone or combined with gamma-rays. *Int. J. Radiat. Biol.*, 1981; 39, 281-290.

[46] Petin, VG; Dergacheva, IP; Romanenko, AG et al. A new concept of optimization and prediction of synergism effects during the combined action of chemical and physical factors of the environment. *Mendeleev Chemistry J.*, 1997; 41(3), 129-140.

[47] Petin, VG; Kim, JK; Kritsky, RO et al. Mathematical description, optimization and prediction of synergistic interaction of fluoride and xylitol. *Chemosphere*, 2008; 72, 844–849.

[48] Petin, VG; Kim, JK; Zhurakovskaya GP et al. Some general regularities of synergistic interaction of hyperthermia with various physical and chemical inactivating agents. *Int. J. Hyperthermia*, 2002; 18, 40-49.

[49] Petin, VG; Kim, JK; Zhurakovskaya; GP et al. Mathematical description of synergistic interaction of UV light and hyperthermia for yeast cells. *J. Photochem. Photobiol. B: Biology*, 2000; 55, 74-79.

[50] Petin, VG; Komarov, VP. *Quantitative Description of Radiosensitivity Modification*. Moscow: Energoatomizdat, 1989.

[51] Petin, VG; Komarov VP. Mathematical description of synergistic interaction of hyperthermia and ionizing radiation. *Mathem. Biosci.*, 1997; 146, 115-130.

[52] Petin, VG; Pohlit, W. Influence of dose rate on survival and recovery of yeast cells. *Radiobiologiy*a, 1969; 9, 492-499.

[53] Petin, VG; Zhurakovskaya, GP. The peculiarities of the interaction of radiation and hyperthermia in *Saccharomyces cerevisiae* irradiated with various dose rates. *Yeast*, 1995; 11(6), 549-554.

[54] Petin, VG; Zhurakovskaya; GP; Komarova, LN. Mathematical description of combined action of ultrasound and hyperthermia on yeast cells. *Ultrasonics*, 1999; 37, 79-83.

[55] Petin, VG; Zhurakovskaya, GP; Komarova, LN. *Radiobiological Basis of Synergistic Interaction in Biosphere*. Moscow: GEOS, 2012.

[56] Reynolds, MC; Brannen, JP. Thermal enhancement of radiosterilization. In: *Radiation Preservation of Food*. Vienna: International Atomic Energy Agency, 1973; 165-176.

[57] Scott, BR. Theoretical models for estimating dose effect relationship after combined exposure to cytotoxicants. *Bull. Math. Biol.*, 1983; 45, 323-345.

[58] Timofeeff-Resovsky, NW; Zimmer, KG. *Das Trefferprinzip in der Biologie*. Stuttgart: Hierzel Verlag S., 1947.

[59] Timofeeff-Ressovsky, NV; Ivanov, VI; Korogodin, VI. *Die Anwendung des Trefferprinzip in der Strahlengbiologie*. Vienna: VEB Gustav Fischer Verlag, 1972.

[60] Tobias, CA. The repair-misrepair model in radiobiology: comparison to other models. *Radiat. Res.*, 1985; Suppl.8, 77-95.

[61] Trujillo, R; Dugan, VL. Synergistic inactivation of viruses by heat and ionizing radiation. *Biophys. J.*, 1972; 12, 92-113.

[62] Urano, M; Kahn, J; Majima, H et al. The cytotoxic effect of cis-diamminedichloroplatinum (II) on culture Chinese hamster ovary cells at elevated temperatures: Arrhenius plot analysis. *Int. J. Hyperthermia*, 1990; 6, 581-590.

[63] Zaider, M; Rossi, HH. The synergistic effects of different radiations. *Radiat. Res.*, 1980; 83, 732-739.

Chapter 6

QUANTITATIVE PROGNOSIS OF SYNERGISTIC EFFECTS AND RECOVERY FOR VARIOUS BIOLOGICAL OBJECTS

ABSTRACT

A mathematical model of synergistic interaction was applied to describe, predict and optimize the synergistic interaction of simultaneous action of ionizing radiation and heat to the simple microbiological objects: viruses, bacterial spores, streptococci, bacterial and yeast cells. In order to apply this model quantitatively to the experimental data, two basic parameters of the model were estimated from real experiments. They were calculated for such experimental condition, when one of the agents applied induces more sublesions than another. The basic observation made during the course of this investigation was that the synergistic interaction of both modalities occurs only within a certain temperature range and there was a specific temperature that maximizes the synergy. Any deviation of the exposure temperature from the optimal value results in a decrease of the synergistic enhancement ratio. Theoretically expected synergistic enhancement ratios agree fairly well with experimental findings. The predicted dependence of synergy on the intensity of applied agents was confirmed for bacteriophage, bacterial spores and yeast cells simultaneous treated with ionizing radiation and heat. The crucial role of fluence rate of UV light and the intensity of ultrasound after their combined application with heat was demonstrated for yeast cells. The correlation between the optimal temperature that maximized the synergy and the intensity of applied agents was elucidated: this temperature shifted towards lower temperature values with a decrease in the intensity. Just such a response should be expected according to the model considered. These data imply that the synergistic interaction may be important at low intensities of harmful agents really observed in biosphere. The model was shown to predict the irreversible component of radiation damage for yeast cells after simultaneous action of hyperthermia with ionizing radiation or UV light. The peculiarities of the sequential thermoradiation action have been analyzed experimentally and theoretically by the mathematical model.

INTRODUCTION

Many physical and chemical environmental factors can synergistically interact with each other. The possibility to predict the synergistic effect and its highest value appears to be important for the establishment of safe level of agents after combined exposures. The conceptual idea and mathematical model developed in the previous Chapter need to be verified experimentally. This Chapter will deal with the describing of experimental data by this model.

The effect of ionizing radiation on bacterial, plant and animal viruses has been extensively investigated. Viruses constitute a certain life forms that can penetrate into living cells and multiply only inside these cells. The inactivation of viruses by ionizing radiation has been shown to be strongly dependent on temperature. The utilization of this synergistic interaction in medicine enables to reduce the concentration of medication or the dose of physical agent to provide the same medical effect or under the same value of drug to reach more effective response [7; 14; 40; 66]. In particular, both fluoride and xylitol are known to be widely used as cariostatic agents and the synergistic effect of fluoride and xylitol on glycolysis by mutans sreptococci was established [36; 57; 60]. The authors pointed out that percent inhibition by combination of fluoride and xylitol was larger than the sum of percent inhibition by fluoride alone and percent inhibition by xylitol alone. Inactivation of *Bacillus subtilis* spores are widely used in hygiene practice as a test for verification of the effectiveness of sterilization, including spacecraft during launch into space. It has been shown that the desired level of sterilization may be achieved at lower temperatures and exposure durations when simultaneous thermoradiation action is applied [13; 55; 56]. However, the authors of the cited papers did not make an attempt to estimate this effect quantitatively. Our first task should be related with the application of synergistic model for thermoradiation and other inactivating agents to viruses, *Bacillus subtilis* spore, *Streptococcus* and bacteria *E. coli* cells.

There are reports indicating the dependence of synergy on the dose rate of ionizing radiation for viruses [66], bacterial spores [13; 55] and mammalian cells [3, 16; 33]. However, these studies were not identified general patterns of synergy dependence on intensity of agents applied. In particular, in none of these and other works the bell-shaped dependence of synergy on dose rate (Figures 5.5 and 5.6) was observed. Moreover, in many reviews and monographs devoted to the analysis of the combined action of hyperthermia and ionizing radiation [18; 31; 64] neither there was information concerning the dependence of synergy on dose rate or approved that it can be observed only after high dose and dose rates. On the contrary, Kuzin [32] asserted that high dose rate and temperature reveal an additive interaction while synergistic effect can be obtained as both these parameters reduced. This conclusion was confirmed by the data showing that the synergistic effect of the simultaneous action of heat and ionizing radiation on diploid yeast cells was observed at 10 Gy/min [48] and was almost absent at 480 Gy/min [29]. However, in the last paper the range of exposure temperatures used in experiments was too small and the duration of heat action was not enough to provide the number of thermal sublesions needed for the greatest synergistic interaction. It was therefore of interest to analyze the synergy of thermoradiation action in a wide range of exposure temperatures and dose rates.

UV light is a part of the solar electromagnetic radiation and it is usually classified into three parts according to the wavelength: UV-A (320–390 nm); UV-B (280–320 nm) and UV-C (below 280 nm) [63]. The increased background of UV-radiation attributed to the depletion of ozone layer in the stratosphere is a potential hazard in the environment. Several lines of evidence show that UV light exhibits synergistic interaction between UV light of different energies [68], with ionizing radiation [20; 23; 38], various chemical agents [1; 22; 62] and hyperthermia [25; 37; 67]. But these data have been obtained mainly at relatively high fluence rates. Fluence rate is known to constitute a very important parameter in photobiology. While it is generally accepted that lowering the UV light fluence rate results in a decrease of the cell killing or mutagenesis efficiency per unit dose [24], the matter is still unclear with regards to the synergistic interaction of UV light and another environmental agent. All the above mentioned results were obtained for the inactivation of various unicellular organisms. There have been no studies for the similar effects on the genetic level, though the evaluation of the genetic risks is necessary for any potentially hazardous environmental factors. Moreover, UV radiation is considered the most important initiator and co-factor for carcinogenesis of human skin exposed to UV radiation [61]. Moreover, UV exposure of skin is often encountered at hot ambient temperatures. Therefore, the elucidation of new fundamental aspects of the simultaneous action of UV radiation and heat is an actual goal. The problem of synergistic interaction of UV light with other agents is important in view of heating effect in photoradiation therapy, which may provide synergistic contribution to the effectiveness of photochemotherapy in the treatment of cancer in experimental and clinical investigations. Indeed, synergistic interaction between the photodynmamic effect and heat has been observed in some experimental models [41; 69]. Thus, our next purpose is to establish whether the UV fluence rate alters the synergistic interaction between hyperthermia and UV light for inactivating and mitotic recombinogenic effects in yeast. The role of ultrasound intensity in its synergistic interaction with heat is not studied in detail. To demonstrate the common role of agent intensity in synergy, it is of interest to establish whether there is any influence of ultrasound intensity on the synergistic interaction of ultrasound and heat.

Among other prognostications resulted from our mathematical model of synergy deserve consideration the quantitative estimation of irreversible component of radiation damage after simultaneous action of heat with ionizing or UV radiations. Besides, the model has not been applied to predict the degree of interaction after the sequential action of modalities. Extension of the model to the sequential treatment of heat and ionizing radiation and experimental evidence of the consequences from this variance of the model seems to be of interest. These data are expected to be valuable to provide additional information on the interaction of ionizing radiation with heat and other environmental agents.

Therefore, the main aims of this Chapter may be formulated as follows.

1. To describe mathematically the synergistic responses of viruses, *Bacillus subtilis* spores, *Streptococcus* and bacteria *Escherichia coli.*
2. To analyze the peculiarities of the interaction of ionizing radiation and heat in yeast cells irradiated with various dose rates.
3. To investigate the fluence rate significance in the synergistic interaction of UV light and heat for survival and recombination of yeast cells.
4. To predict yeast cell recovery after the simultaneous action of hyperthermia combined with ionizing radiation or UV light.
5. To estimate the role of ultrasound intensity in synergistic interaction of ultrasound and hyperthermia.
6. To investigate the peculiarities of the sequential thermoradiation action on yeast cells after sharply differing dose rates and exposure temperature.

6.1. Synergistic Responses of Viruses, *Bacillus subtilis* Spore, *Streptococcus* and Bacteria *Escherichia coli*

Thermoradiation inactivation of viruses. The experimental protocol described by Trujillo and Dugan [66] consisted of exposing a T4 bacteriophage suspension (approximately $5 \cdot 10^6$ phage/ml) to heat alone, ionizing radiation (^{60}Co γ-rays, dose rates 1.32 and 5.1 Gy/min) at ambient temperature, or a composite environment of heat and gamma radiation. Two dose rates were investigated in order to determine the dose rate dependency of the rate of phage inactivation in the composite environment. A constant temperature bath provided a temperature control of ±0.1 °C at the desired temperature. Simultaneous heat and radiation exposure was achieved by placing the water bath, containing the samples, inside gamma irradiation facility. Various elevated temperatures (55–66 °C) were supported only in the process of irradiation. The T4 phage inactivation in all set of experiments was exponential. The basic observation made during the course of experiments was that T4 bacteriophage was inactivated in a synergistic manner by the simultaneous application of heat and ionizing radiation. This phenomena is illustrated by the data inserted in Table 6.1 where the mean inactivated times after separate action of ionizing radiation (t_1) and high temperature (t_2) are listed. These parameters represent duration of the corresponding exposures after which phage survival reduces by a factor of e=2.72.

The values of t_1 and t_2 were evaluated by us on the basis of averaged results published by others [14; 66]. Basing on these data, we calculated the expected duration of the simultaneous action of heat and radiation for independent (additive) interaction of these modalities (t_3) by the expression

$$t_3 = \frac{t_1 \cdot t_2}{t_1 + t_2}. \qquad (6.1)$$

The calculation results are also summarized for different exposure regimens in Table 6.1. This table also includes the values of mean lethal durations of exposure estimated in experiments after the simultaneous treatment with ionizing radiation and hyperthermia (t_4). One can see the synergistic interaction of the applied agents. In other words, the term synergistic interaction may be used for the survival data obtained on simultaneous exposure of T4 phage to heat and ionizing radiation when $t_4 < t_3$. Experimental values of the synergistic enhancement ratio listed in Table 6.1 were determine by

$$k_{\exp} = t_{indep} / t_{\exp} = t_3 / t_4. \qquad (6.2)$$

One can see that the synergistic interaction of both modalities occurs only within a certain temperature range and there is a specific temperature (60 °C and 5.1 Gy/min) that maximizes the synergy. Any deviation of the temperature from the optimal value results in a decrease of the synergistic enhancement ratio. The temperature range for 1.32 Gy/min was not large enough to judge strongly this conclusion.

Table 6.1. Parameters characterizing the simultaneous thermoradiation phage T4 inactivation

T, °C	Dose rate, Gy/min	t_1, min	t_2, min	t_3, min	t_4, Min	N_2/N_1	$k_{\exp}$	k_{theor}
23	1.32	353	–	353	353	0	1.00	1.00
62	1.32	353	76	62	49	4.7	1.29	1.40
64	1.32	353	27	25	22.4	13.2	1.12	1.16
23	5.1	87	–	87.0	87.0	0	1.00	1.00
55	5.1	87	328	69.0	60.0	0.27	1.15	1.34
60	5.1	87	110	48.6	28.3	0.79	1.72	1.72
64	5.1	87	26.8	20.5	13.3	3.25	1.54	1.34
66	5.1	87	9.6	8.6	6.0	9.1	1.43	1.23

It is of interest to test the applicability of the mathematical model described in the previous Chapter to these results. The ratio of damage produced by hyperthermia and ionizing radiation (N_2/N_1) was calculated by Eq. (5.12). Theoretically expected synergistic enhancement ratio (k_{theor}) was estimated by Eq. (5.5). The solid line in Figure 6.1 is the prediction from our model and circles denote experimental values of k. Predicted values of k were estimated for both dose rates used in the original papers at p_1 = 2.3 and p_2 = 1.6. These values were established by means of Eqs. (5.14) and (5.15) and the corresponding values of k derived from experiments of irradiation (dose rate 5.1 Gy/min) at 64 and 60 °C, respectively. We believed that in the first case one can use Eq. (5.14) since $p_1N_1 < p_2N_2$. On the contrary, in the second case $p_1N_1 > p_2N_2$ and Eq. (5.15) is valid. Theoretically expected synergistic enhancement ratio (k_{theor}) is also included in Table 1. The statistical error in determining the radiobiological parameters including the mean lethal dose is typically within 5–10%. To estimate quantitatively the synergistic interaction the ratio of the mean lethal doses is used. Hence, for such a case the error can increase to 10–20%. With this feature, the comparison of experimental and theoretically predicted synergistic enhancement ratios (Table 6.1, Figure 6.1) agree fairly well with each other. The remarkable aspect of Figure 6.1 is that there appears to be a sharp dependence of synergistic effect on the ratio N_2/N_1, the greatest value of synergy (k_{max} = 2) expected at N_2/N_1 = 1.40. The synergistic effect appears to decline with any deviation of the ratio N_2/N_1 from the optimal value.

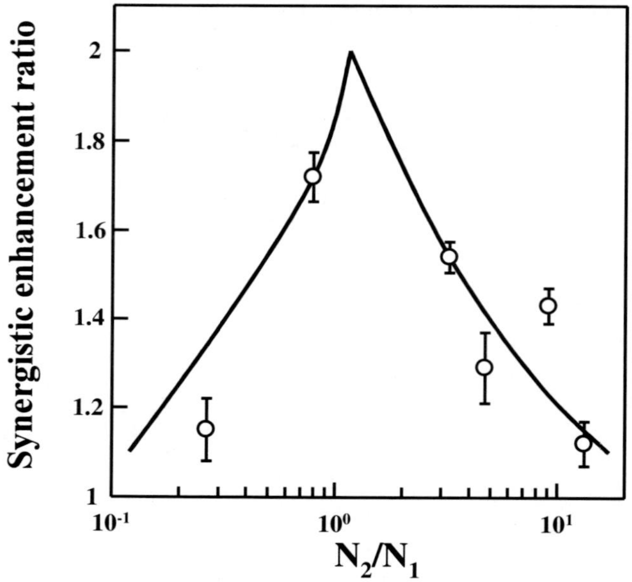

Figure 6.1. Radiation synergistic enhancement ratio by hyperthermia (k) of T4 bacteriophage as a function of ratio N_2 / N_1. Solid line, theoretically predicted dependence; circles, experimentally obtained data.

According to the model discussed, a synergistic effect should depend on the dose rate of ionizing radiation, it being shown that with a decrease in a dose rate the acting temperature should also decrease to achieve the highest synergistic effect. Unfortunately, insufficient set of temperatures at which the irradiation with ionizing radiation at a dose rate of 1.32 Gy/min was carried out did not allow to establish the temperature at which the maximum synergistic effect could be obtained.

Thermoradiation inactivation of bacterial spores. Bacteria are microorganisms with prokaryotic (no nuclear membrane) type of cell structure. Some bacteria form spores – a unique resting forms, whose main function is to maintain the bacteria in adverse environmental conditions.

Extensive experimental data for thermal enhancement of *B. subtilis* var. *niger* spores radiosterilization have been published [4; 14; 55; 56]. All survival cures were exponential. By using the averaged results of these papers, we calculated experimental values of the synergistic enhancement ratio (Eq. 6.2). As can be seen from Figure 5.1B, for a given dose rate (1.33 Gy/min), there was a specific temperature that maximizes the synergistic effect. Similarly, the dependencies of the synergistic effect on dose rate at two fixed temperatures (95 and 105 °C) were obtained (Figure 5.6). It is evident that for a given temperatures, there is a specific dose rate that maximizes the synergistic interaction. Any deviation of temperature or dose rate from the optimal values results in a decrease in synergy. Just such responses would be expected qualitatively according to the model considered. Hence, it is of interest to analyze the model's ability to describe quantitatively the synergistic interaction of simultaneous thermoradiation action on bacterial spore inactivation.

Table 6.2. Parameters determining *Bacillus subtilis* var. *niger* spores inactivation by simultaneous thermoradiation action

T, °C	Dose rate, Gy/min	t_1, min	t_2, min	t_3, min	t_4, Min	N_2/N_1	k_{exp}	k_{theor}
60	1.33	255	3120	236	156	0.082	1.51	1.34
85	1.33	255	520	171	68	0.49	2.51	2.46
90	1.33	255	415	158	78	0.61	2.03	2.35
95	0.95	403	322	182	102	1.21	1.78	1.98
95	1.83	209	332	128	59	0.63	2.17	2.33
95	2.50	135	332	96	41.5	0.41	2.31	2.31
95	3.50	110	332	82.6	39	0.33	2.12	2.12
95	5.50	70	332	57.8	28	0.21	2.06	1.78
100	1.37	250	186	107	62	1.34	1.73	1.93
105	0.42	816	104	92	53	7.8	1.74	1.25
105	0.83	408	104	83	42	3.9	1.98	1.44
105	1.33	255	104	74	39	2.45	1.90	1.63
105	2.17	151	104	62	29	1.45	2.14	1.89
105	3.67	93	104	49	22.8	0.89	2.15	2.15
105	6.00	57	104	37	17.4	0.55	2.13	2.40
105	8.33	41	104	29.4	15.6	0.39	1.88	2.27

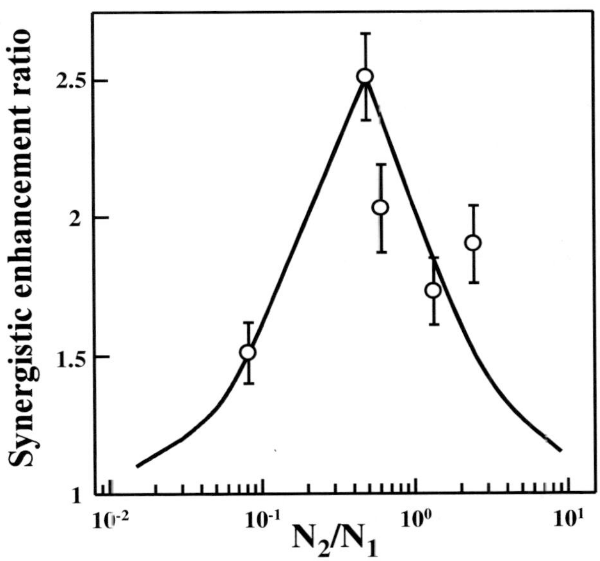

Figure 6.2. Radiation synergistic enhancement ratio by hyperthermia (k) of *Bacillus subtilis* var. *niger* spores inactivation by simultaneous thermoradiation action as a function of ratio N_2/N_1. Solid line, theoretically predicted dependence: circles, experimentally obtained data.

Table 6.2 and Figure 6.2 compare experimental and theoretically predicted values of the synergistic enhancement ratio. The value of p_1 = 2.17 was obtained by substituting values of

k_1 and N_2 / N_1 into Eq. (5.14) for thermoradiation inactivation of spores at 105 °C with a dose rate of 3.67 Gy/min, and p_2 = 4.51 by Eq. (5.15) with 2.5 Gy/min. At the first case, $p_1N_1 < p_2N_2$ and at the second one, the reverse inequality holds. As can be seen from the data presented in Table 6.2 and Figure 6.2, the predicted values of the synergistic enhancement ratio agree fairly well with experimental results. It follows from the model (Eqs. 5.11 and 5.12) that the highest synergistic effect, k_{max} = 2.48, should be expected at N_2 / N_1 = 0.48. It is also in a good agreement with experimental data.

Simultaneous action of fluoride and xylitol on Streptococcus. The purpose of this study was to adjust a simple mathematical model to describe, optimize and predict a synergistic interaction between fluoride and xylitol on acid production by mutans streptococci. The prospects of providing a reasonably useful application of this phenomenon seem to lie in the possibility to predict the highest synergistic effect and condition under which it can be obtained. Therefore, it would be of interest to accommodate this model for the description, optimization and prognosis of the synergistic effects between two chemical agents.

In accordance with the model which is supposed to be adjusted to describe and predict the interaction of two chemical agents, the synergistic enhancement ratio (k) can be defined as the ratio of the effective damages produced by the combined action of these agents, to the sum of the effective damages produced separately (Eq. 5.2). As well as this model is planned to be tested for synergistic interaction of fluoride and xylitol on glycolysis, we designate an amount of some hypothetical effective damage produced by fluoride through N_1 and that produced by xylitol as N_2. Let us suppose that p_1 and p_2 are the numbers of sublesions that occur per a single effective damage induced by fluoride and xylitol, respectively. We assume that one sublesion resulting from one individual agent reacts with one sublesion produced by another agent, thus producing an additional effective damage. Then, the synergistic enhancement ratio, k , may be given by Eq. (5.5) and is determined by a minimal value from two variable quantities, p_1N_1 and p_2N_2, which ascertains the number of additional effective damage arising due to the interaction between sublesions produced by fluoride (p_1N_1) and xylitol (p_2N_2). An actual amount of real damage and sublesion is thought to be difficult to determine quantitatively, in particular because of their molecular nature still should be elucidated. However Eq. (5.5) operates with non-dimensional ratio of the effective damages (N_2 / N_1) produced by the agents under considerations which can be estimated from experimental results. The extent of the inhibition of acid production by oral bacteria may be supposed to be in the direct proportional to some hypothetical effective damages responsible for inactivating of the cariogenic bacteria. In accordance with this supposition, the extent of inhibition of acid production (% control) produced by fluoride (F) and xylitol (X) may be presented as

$$F = cN_1 \quad (6.3)$$

and

$$X = cN_2. \tag{6.4}$$

Here c is a constant of proportionality. Then the expected additive effect can be calculated as

$$(F + X)_{add} = c(N_1 + N_2), \tag{6.5}$$

while the experimentally observed results may be given as

$$(F + X)_{\exp} = c(N_1 + N_2 + N_3). \tag{6.6}$$

The ratio of N_2 / N_1 can be determined on the basis of experimental data as

$$N_2 / N_1 = X / F. \tag{6.7}$$

Then the experimental value of the synergistic enhancement ratio can be as

$$k = \frac{(F + X)_{\exp}}{(F + X)_{add}}. \tag{6.8}$$

Using real experimental results and just reduced equations, we can estimate the basic model parameters p_1 and p_2 (Eqs. 5.14 and 5.15, respectively). Then the synergistic enhancement ratio for any ratio of N_2 / N_1, its greatest value, and the condition under which this is yielded can be derived from Eqs. (5.5), (5.11) and (5.10), respectively. All these parameters are very important for environmental toxicology and risk assessment. To test the validity of the model for chemical agents, one needs a great number of experimental data, obtained after action of two toxic agents applied alone and combined together.

The extensive quantitative study of synergistic inhibition by combination of fluoride and xylitol on glycolysis by mutans streptococci clearly showed [36] that the acid production from glucose by both bacterial strains tested (*Streptococus mutans* NCTC10449 and *Streptococus sobrinus* 6715) was inhibited in the presence of fluoride. The decrease of the acid production was dose-dependent and the effect was synergistically enhanced after simultaneous action of fluoride and xylitol. We used these results to evaluate some model parameters and experimental values of the synergistic enhancement ratio for various conditions of the combined action of fluoride and xylitol on the acid production by *S. mutans* (Table 6.3) and *S. sobrinus* (Table 6.4). Both Tables contain the following data: fluoride concentration, pH, the extent of inhibition of acid production (% from the control) produced by fluoride (F) and xylitol (X) applied individually, the expected additive effect

$(F+X)_{add}$ (Eq. 6.5), the actually observed inhibition in experiment $(F+X)_{exp}$ (Eq. 6.6), the ratio N_2/N_1 (Eq. 6.7), and the experimentally observed synergistic enhancement ratio (Eq. 6.8). These results clearly indicate that for a constant xylitol concentration an optimal concentration of fluoride may exist providing the highest synergistic effect. Just such a dependence is expected qualitatively from the mathematical model presented above. Hence, it would be of interest to apply the described mathematical model to these results.

Table 6.3. The dependence of some model parameters and experimental values of the synergistic enhancement ratio for the combined effect of fluoride and xylitol on the acid production by the cells of *S. mutans* NCTC19449 on the condition of treatment

Fluoride, mM	pH	Fluoride $F=cN_1$	Xylitol $X=cN_2$	$(F+X)_{add}$	$(F+X)_{exp}$	N_2/N_1	k_{exp}
0	7.0	0	9	9	9	–	–
0.8	7.0	3	0	12	20	3.0	1.7
1.3*	7.0	15	9	24	50	0.6	2.1
1.6	7.0	22	9	31	59	0.41	1.9
2.0*	7.0	38	9	47	73	0.24	1.55
2.3	7.0	56	9	65	87	0.16	1.34
0	5.5	0	9	9	9	–	–
0.1	5.5	2	9	11	15	4.5	1.36
0.15*	5.5	4	9	13	29	2.25	2.23
0.20	5.5	6	9	15	42	1.5	2.8
0.25*	5.5	11	9	20	63	0.82	3.15
0.30	5.5	17	9	26	80	0.53	3.1
0.35*	5.5	50	9	59	90	0.18	1.52
0.40	5.5	75	9	84	98	0.12	1.17

* Data obtained by extrapolation.

In order to apply this model quantitatively to the experimental data, it is necessary to define two basic parameters p_1 and p_2. Each of these should be calculated for such experimental condition, when one of the agents applied induces more sublesions than another (Eqs. 5.14 and 5.15). Thus estimated set of these parameters is collected in Table 6.5. Having used them, the synergistic enhancement ratio for any ratio of N_2/N_1, its greatest value, and the condition under which this is yielded can be derived from equations presented above (Figure 6.3).

In Figure 6.3, the synergistic enhancement ratio k of a simultaneous action of fluoride and xylitol on the inhibition of acid production by mutans streptococci from glucose is shown as a function of the ratio of effective damages produced by these agents (N_2/N_1). Some of the extrapolation data presented in Tables 6.3 and 6.4 as well as in Figure 6.3 were estimated by us with a great accuracy using only the original experimental data having a very little scattering. The errors of extrapolated points have been taken as a middle value between errors of neighbor experimental points. One can see that the theoretically predicted interaction (solid lines) corresponds fairly well with the experimental data (circles). It can be inferred that the synergy occurred only within a definite ratio of the effects (N_2 / N_1) produced by each agent. It is also apparent that there is an optimal ratio of N_2 / N_1 which results in the greatest synergistic interaction. Table 6.5 includes the highest synergistic effect that should be expected (Eq. 5.11) and the ratio of N_2 / N_1, at which it can be achieved (Eq. 5.12). Figure 6.3 demonstrates that a good agreement exists between these predicted values and the experimental data. It means that the model presented here can be used for mathematical description, optimization and prediction of synergistic interaction of fluoride and xylitol on glycolysis.

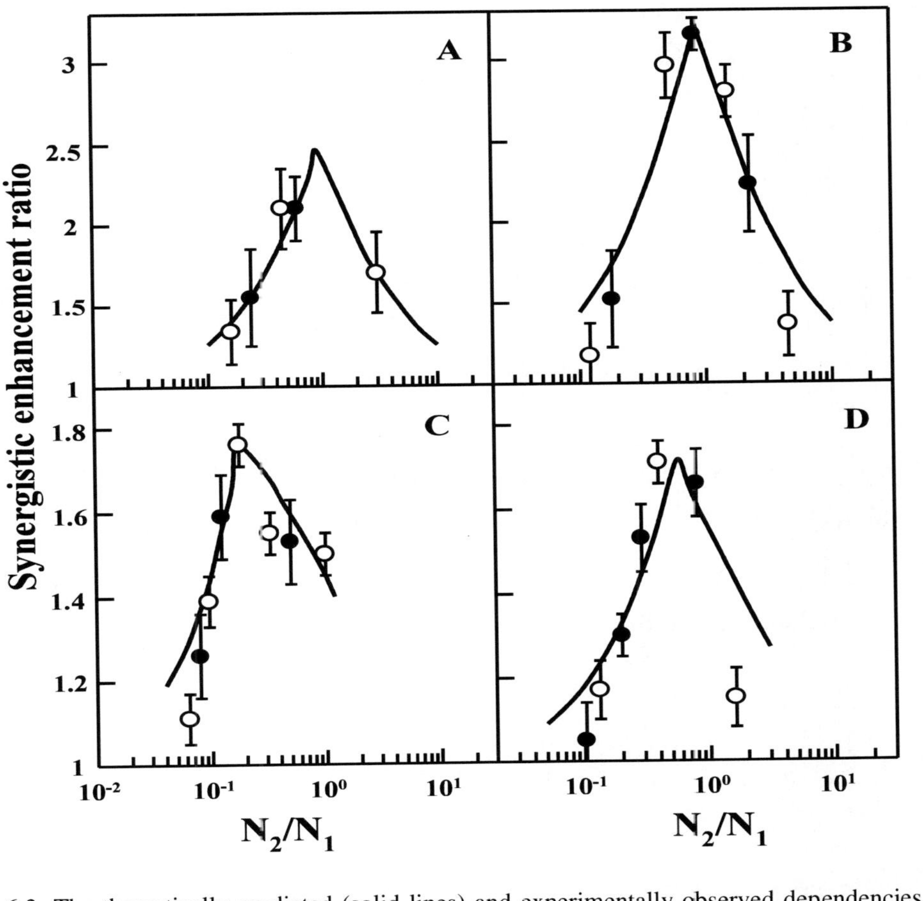

Figure 6.3. The theoretically predicted (solid lines) and experimentally observed dependencies of the synergistic enhancement ratio on the ratio of the effective damages produced by 60 mM xylitol (N_2) and various concentration of fluoride (N_1) after their combined action on the acid production for 10 min from 10 mM glucose by the cells of *S. mutans* 10449 (A, pH 7.0; B, pH 5.5) and *S. sobrinus* 6715 (C, pH 7.0; D, pH 5.5). Closed circles, experimental data; open circles, the data estimated by the extrapolation.

Table 6.4. The dependence of some model parameters and experimental values of the synergistic enhancement ratio for the combined effect of fluoride and xylitol on the acid production by the cells of *S. mutans* NCTC19449 on the condition of treatment

Fluoride, мМ	pH	Fluoride $F = cN_1$	Xylitol $X = cN_2$	$(F+X)_{add}$	$(F+X)_{exp}$	N_2 / N_1	k_{exp}
0	7.0	0	5	5	5	–	–
0.8	7.0	5	5	10	15	1	1.5
1.25*	7.0	10	5	15	23	0.5	1.53
1.6	7.0	15	5	20	31	0.33	1.55
2.4	7.0	28	5	33	58	0.18	1.76
2.9*	7.0	40	5	45	70	0.125	1.56
3.25	7.0	52	5	57	79	0.096	1.39
3.5	7.0	60	5	65	82	0.083	1.26
4.0	7.0	78	5	83	92	0.064	1.11
0	5.5	0	8	8	8	–	–
0.1	5.5	5	8	13	15	1.6	1.15
0.15*	5.5	10	8	18	30	0.8	1.67
0.2	5.5	20	8	28	48	0.4	1.71
0.225*	5.5	28	8	36	55	0.29	1.53
0.25*	5.5	40	8	48	62	0.2	1.29
0.3	5.5	60	8	68	80	0.133	1.18
0.35*	5.5	78	8	86	90	0.103	1.05

* Data obtained by extrapolation.

Table 6.5. The dependence of the basic model parameters (p_1 and p_2), the predicted greatest values of the synergistic enhancement ratio (k_{max}) and the ratio of $(N_2 / N_1)_{max}$ at which it can be achieved after the combined effect of fluoride and xylitol on the acid production by the cells of *Streptococus mutans* and *Streptococus sobrinus* on the condition of treatment

Strain	pH	p_1	p_2	$(N_2 / N_1)_{max}$	k_{max}
Streptococcus mutans NCTC10449	7.0	2.8	2.9	0.96	2.42
Streptococcus mutans NCTC10449	5.5	4.1	4.8	0.85	3.21
Streptococcus sobrinus 6715	7.0	0.9	5.0	0.158	1.77
Streptococcus sobrinus 6715	5.5	1.1	2.0	0.55	1.71

Inactivation of bacterial cells. In our experiments, *Escherichia coli* bacterial cells (strain B/r) were subjected to the simultaneous action of hyperthermia and zinc sulfate $ZnSO_4$ (0.01 M) belonging to heavy metals pollutants. Exponential survival curves have been obtained in the dependence of exposure duration for various high temperatures (37–60 °C) after simultaneous action of both modalities and after exposure to each factor separately. An example of these curves was presented above (Figure 3.1D). It is obvious the synergistic interaction of these factors. A complete set of the similar survival curves after various regimens of exposures have been obtained. Having used these data, we calculated the dependence of the synergistic enhancement ratio on the action temperature (Figure 6.4A). As should be expected from our model of synergy, such dependence has a bell shape, i.e., there is an optimal temperature which maximizes the synergistic interaction and some deviation of the acting temperature from an optimal one results in the decrease in the effectiveness of synergistic interaction. It was therefore of interest to verify the applicability of this model to describe quantitatively the experimental results shown in Figure 6.4A.

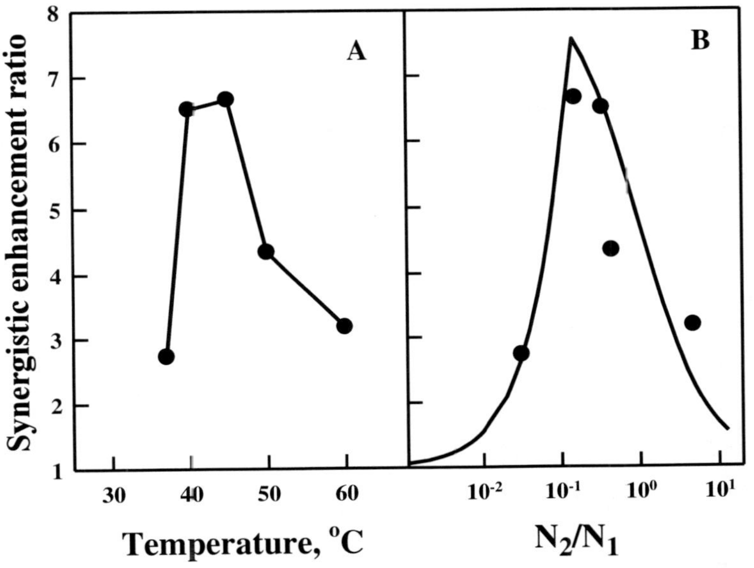

Figure 6.4. The dependence of the synergistic enhancement ratio of simultaneous action of hyperthermia and $ZnSO_4$ (0.01 M) on *Escherichia coli* cell (strain B/r) inactivation on acting temperature (A) and on the ratio N_2/N_1 (B). Panel B: closed circles denote experimental data, solid curve, theoretically predicted relationship.

The number of damage produced by $ZnSO_4$ (N_1) and high temperature (N_2) was determined by means of Eq. (3.3). Using these data for the same exposure durations the ratios N_2/N_1 were calculated. This ratio corresponded to the ratio survival curve slopes after the separate action of hyperthermia and chemical agent (Eq. 5.12). The value of the parameter $p_1 = 7.5$ was obtained by Eq. (5.14) and experimental values of the synergistic enhancement ratio $k = 6.5$, wherein $N_2/N_1 = 0.34$ and exposure temperature was 40 °C. The value of $p_2 = 4.5$ was estimated by Eq. (5.15) for $k = 2.7$ and $N_2/N_1 = 0.034$ and exposure temperature of 37 °C. In the first case $p_1N_1 < p_2N_2$ and in the second case $p_1N_1 > p_2N_2$. Theoretically expected dependence of the synergistic enhancement ratio of simultaneous action of hyperthermia and $ZnSO_4$ on *E. coli* cell inactivation on the ratio

N_2/N_1 Eq. (5.5) is presented in Figure 6.4B by solid curves, closed circles denote experimental data. A good correspondence of these data to each other is evident. It follows from Eqs. (5.11) and (5.10) that the greatest synergistic effect k_{max} = 7.6 should be expected at N_2/N_1 = 0.18. These expected values correspond rather well to the experimental data (Figure 6.4B).

The results presented here show that the developed mathematical model fairly well describes and predicts synergistic interaction of simultaneous action of ionizing radiation and heat on viruses and bacterial spore inactivation, xylitol and fluoride on acid production by *Streptococcus* as well as hyperthermia and zinc sulfate on inactivation of *E. coli* cells.

6.2. The Peculiarities of the Interaction of Ionizing Radiation and heat in Yeast Cells Irradiated with Various Dose Rate

A wild-type diploid yeast of *S. cerevisiae* (strain XS800) in stationary phase of growth was used in these experiments. Aliquots of 10^6 cells/ml were placed in a glass tube and exposed to graded doses of ionizing radiation. The electron beam from a 25 MeV pulsed linear accelerator was used in these experiments. The pulse duration at half peak output was 2.7 μs and the pulse repetition rate was at 50 Hz, the average dose rate being 5, 10, 25 and 250 Gy/min as determined by ferrous sulphate dosimetry [15]. The cells were irradiated under ambient conditions at different temperatures, which were adjusted within ±0.1 °C. The following heating method was used: 0.1 ml of cell suspension at room temperature (about 10^6 cells) was put into 2.4 ml of sterile water preheated to the required temperature in a water bath. For the simultaneous thermoradiation treatment, the time interval between inoculation of the cells into the preheated water and the beginning of radiation exposure was about 0.3 min, which was significantly less than the total treatment time. At the end of the treatment, the samples were rapidly cooled to room temperature, further diluted and plated on nutrient agar to assess colony-forming ability after 5–7 days incubation at 30 °C.

Figure 6.5 provides examples of the basic experimental data observed in this investigation. Three types of survival curve were obtained for every temperature and dose rate investigated: heat alone (curve 1), radiation at room temperature (curve 2), and composite heat and radiation (curve 4). Curve 3 represents a hypothetical curve which would be obtained if inactivation by combined heat and radiation were completely additive. Pooled results of three to five experiments were used to calculate parameters and their statistical errors.

Only exponential survival curves were obtained after hyperthermia was applied alone. Heat sensitivity of cells was progressively increased with temperature. After radiation treatment at room temperature, a typical sigmoid survival curve was observed. The yeast radiation sensitivity appeared to be independent of dose rate over the investigated range. Similar data have been published previously for X-ray-irradiated yeast cells [28; 51]. The data in Figure 6.5 indicate that hyperthermia in conjunction with radiation can both decrease the shoulder and increase the slope. Similar survival curves were obtained for the higher dose rate: 10, 25 and 250 Gy/min (Figures 6.6, 6.7 and 6.8, respectively).

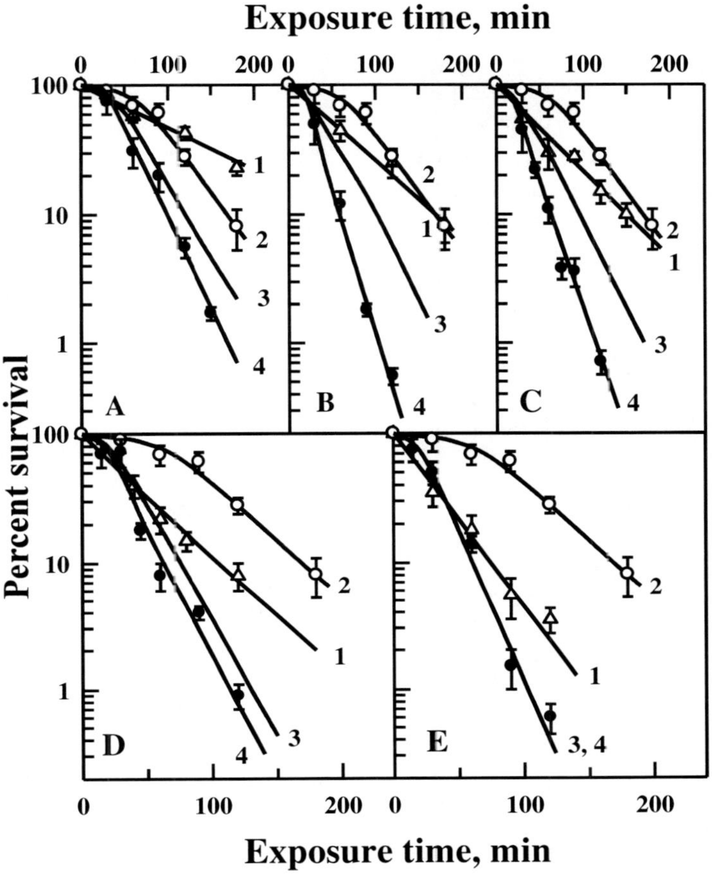

Figure 6.5. Survival curves of *S. cerevisiae* diploid yeast cells (strain XS800) after simultaneous application of hyperthermia (A, 47 °C; B, 48 °C; C, 49 °C; D, 50 °C; E, 51 °C) and ionizing radiation (25 MeV electrons, 5 Gy/min): 1, heat treatment alone; 2, electron irradiation at room temperature (19 °C); 3, calculated curves for independent action of ionizing radiation and high temperatures; 4, experimental curves after simultaneous thermoradiation action. Error bars represent the standard error of the mean.

To quantitate synergistic effect we used the synergistic enhancement ratio (k), defined as the ratio of the calculated radiation dose (assuming an additive effect of radiation and hyperthermia) to that observed from the experimental survival curve for the simultaneous action of electron radiation and hyperthermia at a fixed level of survival, i.e., for 10% survival $k = D_2 / D_1$ (Figure 3.2, Eq. 3.5) or as the ratio of the corresponding D_0 or exposure duration. It is worth noting that the ratio of D_0 values corresponds to absolute sensitivity variation, while the ratio of D_{10} values express the relative sensitivity, taking into account the sensitization due to the decrease in the shoulder width and the decrease in the D_0 values.

Results of our experiments carried out with yeast cells exposed to ionizing radiation at various temperatures and dose rates are summarized in Figure 6.9A. In this figure, the dependence of the synergistic enhancement ratio (k) upon irradiation temperature and dose rate is depicted.

It is evident from the data given in this figure that, for the simultaneous thermoradiation action, both temperature and dose rate affect the synergy. As can be seen, for a given dose rate there is a specific temperature that maximized the synergistic effect. Any deviation of temperature from optimal value results in a decrease of synergism. This implies that the synergistic interaction between hyperthermia and ionizing radiation is observed only within a certain temperature range. For temperature below this range, cell killing was mainly determined by ionizing radiation, while for temperatures above this range, cell killing was caused by heat. These data put in order the contradictory of experimental data published in literature. Due to bell-shaped dependence of the synergistic effect on temperature, the synergistic interaction may increase, decrease or stay unchanged with a decrease or an increase in dose rate of ionizing radiation.

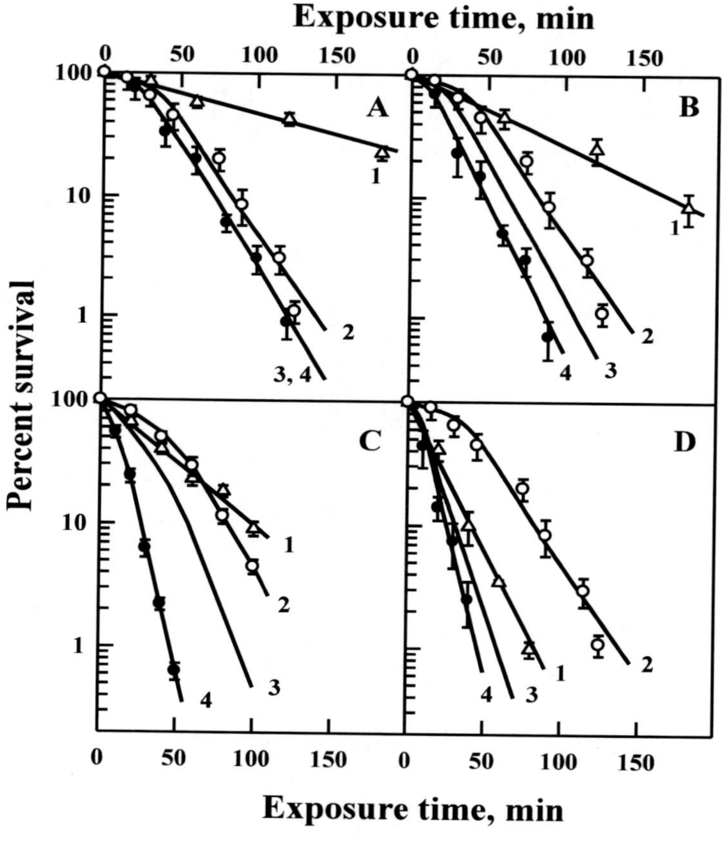

Figure 6.6. Survival curves of *S. cerevisiae* diploid yeast cells (strain XS800) after simultaneous application of hyperthermia (A, 45 °C; B, 47.5 °C; C, 50 °C; D, 52.5 °C) and ionizing radiation (25 MeV electrons, 10 Gy/min): 1, heat treatment alone; 2, electron irradiation at room temperature (19 °C); 3, calculated curves for independent action of ionizing radiation and high temperatures; 4, experimental curves after simultaneous thermoradiation action. Error bars represent the standard error of the mean.

From these findings, it can be inferred that the temperature at which radiation is delivered should be reduced to obtain maximum synergistic effect with decreasing dose rate (Figure 6.9B). Such dependence indicate a potential significance of synergistic interaction at low dose rates existing in biosphere. One can also conclude that for any fixed temperature there should be a specific dose rate that maximized the synergistic effect and any deviation of dose rate from optimum would result in a decrease of synergy. The data of this section are in qualitative agreement with the predictions expecting on the basis of mathematical model presented above. A quantitative description of the results is also published [52].

In summary, the main conclusion is that, under simultaneous thermal treatment and irradiation, both temperature and dose rate are important for a synergistic effect. Optimum inactivation may be achieved under either a fixed temperature with variation of dose rate, or *vice versa*.

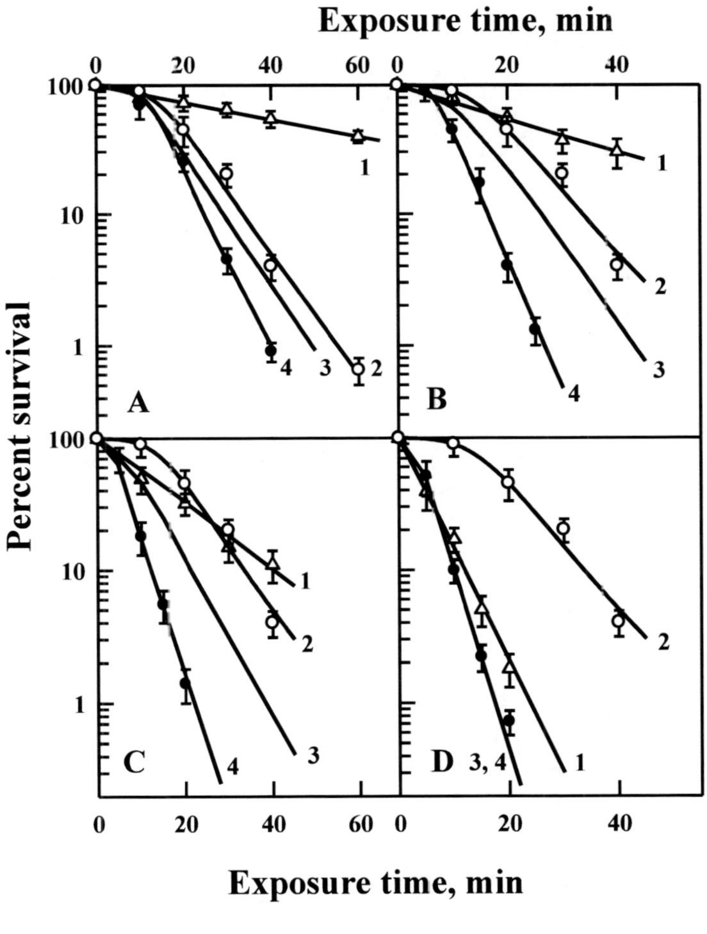

Figure 6.7. Survival curves of *S. cerevisiae* diploid yeast cells (strain XS800) after simultaneous application of hyperthermia (A, 49 °C; B, 51 °C; C, 52.5 °C; D, 54 °C) and ionizing radiation (25 MeV electrons, 25 Gy/min): 1, heat treatment alone; 2, electron irradiation at room temperature (19 °C); 3, calculated curves for independent action of ionizing radiation and high temperatures; 4, experimental curves after simultaneous thermoradiation action. Error bars represent the standard error of the mean.

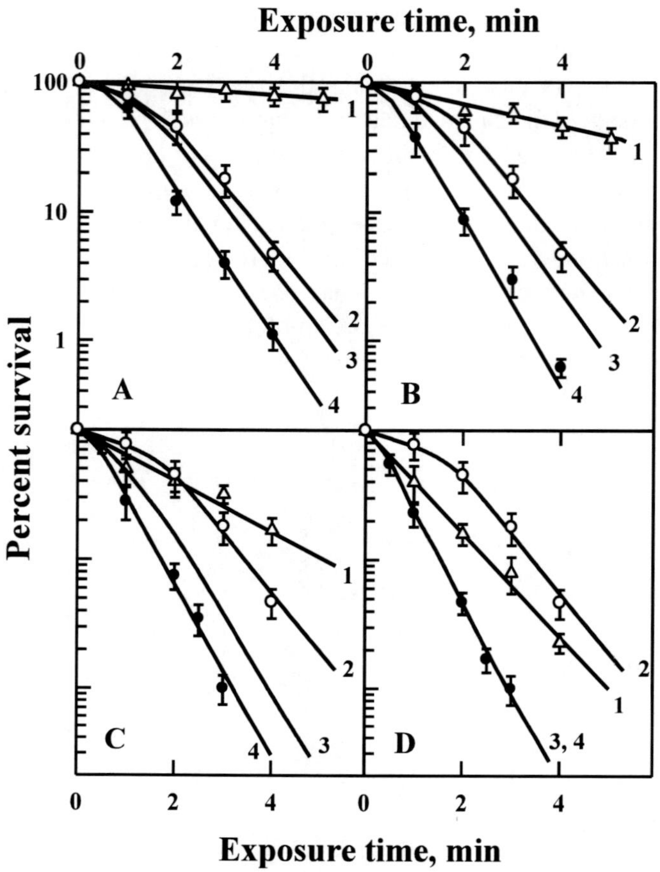

Figure 6.8. Survival curves of *S. cerevisiae* diploid yeast cells (strain XS800) after simultaneous application of hyperthermia (A, 52.5 °C; B, 54 °C; C, 55.5 °C; D, 57.5 °C) and ionizing radiation (25 MeV electrons, 250 Gy/min): 1, heat treatment alone; 2, electron irradiation at room temperature (19 °C); 3, calculated curves for independent action of ionizing radiation and high temperatures; 4, experimental curves after simultaneous thermoradiation action. Error bars represent the standard error of the mean.

6.3. The Fluence Rate Determines the Synergistic Interaction of UV Light and Heat

The purpose of this section is to investigate the effect of the UV intensity on the synergistic and/or antagonistic interaction of the simultaneous application of UV light and heat in *S. cerevisiae.* Wild-type diploid yeast of *S. cerevisiae* (strains XS800 and T1) as well as homozygous UV-sensitive mutant (*rad*18/*rad*18, strain XS1924) were used in these experiments. Before irradiation, the cells were incubated for 3–5 days at 30 ºC on a full nutrient agar layer. The cells were irradiated with germicidal lamps that emitted predominately UV light of wavelength 254 nm (far-UV radiation) at fluences of 0.033, 0.25 and 1.5 W/m^2 under ambient conditions at different temperatures (45–57.5 °C), which were adjusted within ±0.2 °C. The source of far-UV light at 254 nm was a low pressure mercury vapor lamp with an integral filter that provided more than 95% of the emitted radiation at 254 nm. Variation of the intensity was achieved by means of calibrated wire nets. The fluence rates at 254 nm were measured using a calibrated General Electrical germicidal meter.

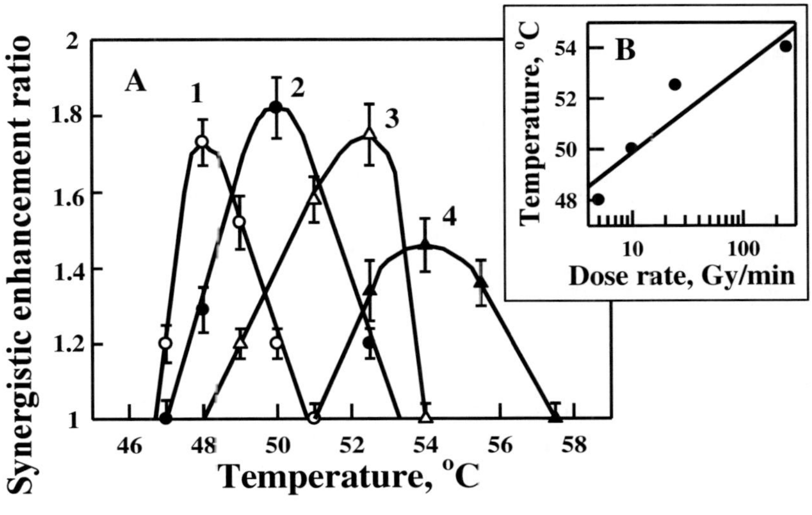

Figure 6.9. The results of simultaneous thermoradiation action on *S. cerevisiae* yeast cell inactivation (strain XS800). A, the dependence of the synergistic enhancement ratio upon exposure temperature and dose rate (1 – 5, 2 – 10, 3 – 25, 4 – 250 Gy/min). Error bars show interexperiment standard errors. B, the relationship between the dose rate and acting temperature providing the highest synergistic interaction under simultaneous thermoradiation action.

The following heating method was used: 0.1 ml of cell suspension at room temperature (about 1.5 x 10^6 cells) was placed into 1.4 ml of sterile water preheated to a required temperature in a water bath. For the simultaneous thermal and UV light treatment, the time interval between introduction of the cells into the preheated water and the beginning of exposure was about 0.1 min, which was significantly less than the total treatment time. At the end of the treatment, the samples were rapidly cooled to room temperature. To avoid photoreactivation, UV exposure, dilution and other procedures were performed under red ambient light, while post-irradiation incubation was carried out under dark conditions.

In all our combined experiments, both modalities were applied simultaneously. This means that the durations of the UV light irradiation and heat exposure were identical, and might be determined as the ratio of the UV dose (J/m^2) to the fluence rate (W/m^2). The duration of heat incubation alone at different temperatures was varied to reduce cell survival to about 10%.

Figure 6.10 exhibits typical examples of the basic experimental data observed in this investigation. The same set of survival curves were obtained in this study for the different intensities of UV light and acting temperatures for *S. cerevisiae* diploid yeast cells, strains XS800 (Figures 6.11–6.13) and T1 (Figures 6.14–6.15).

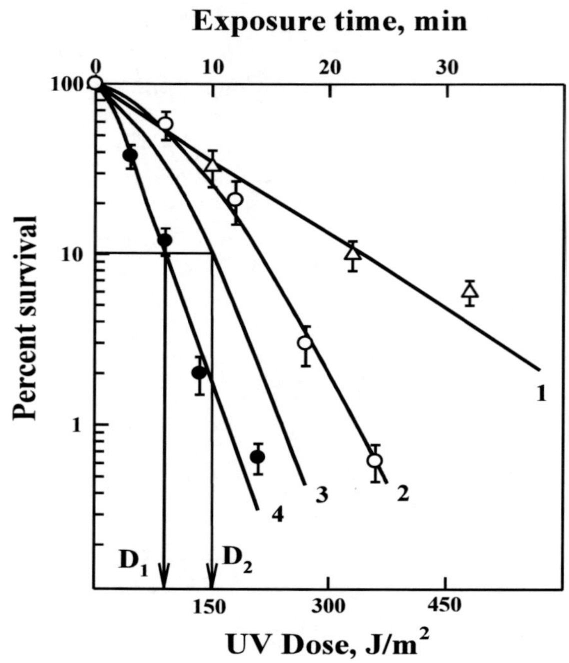

Figure 6.10. Survival curves of *S. cerevisiae* diploid yeast cells, strains XS800: 1, heat treatment (52.5 °C) alone; 2, UV light (254 nm) at 0.25 W/m^2 and room temperature 20 °C; 3, calculated curve for independent action of UV light (0.25 W/m^2) and high temperature (52.5 °C); 4, experimental curve after simultaneous action of these modalities. The error bars represent the standard error of the mean.

These data indicate that, in conjunction with UV light, hyperthermia can decrease the shoulder and increase the slope. Only exponential survival curves were obtained after hyperthermia was applied alone. The heat sensitivity of the cell was progressively increased with increasing temperature, as judged by the D_0 values. After UV light exposure at room temperature, a typical sigmoid survival curve was observed. The yeast UV light sensitivity appeared to be slightly dependent on the intensity over the range investigated. To quantify the synergistic effect, we used the synergistic enhancement ratio k, defined as the ratio of the calculated UV light dose (assuming independent effect of UV light and hyperthermia) to the UV light dose observed from the experimental survival curve for the simultaneous action of UV light and hyperthermia at a fixed level of survival, i.e., for 10% survival, $k = D_2 / D_1$ (Figure 6.10).

Pooled results of 3–5 experiments were used to calculate the mean number of these parameters and their statistical errors. This parameter is depicted in Figure 6.16 as functions of the exposure temperature and fluence rate. It is evident from the data given in Figure 6.16 that, for the simultaneous action of heat and UV light, both the temperature and fluence rate affect the synergy. As can be seen, for a given intensity, there is a specific temperature that maximizes the synergistic effect. This implies that the synergistic interaction between hyperthermia and UV light, like to simultaneous action of ionizing radiation and high temperature, is observed only within a certain temperature range. For temperatures below this range, killing of cells mainly determined by UV light exposure, while, for temperatures above this range, killing of cells was caused by heat. Any deviation of the temperature from the optimal value results in a decrease in synergism. From Figure 6.17, the antagonistic interaction (k <1) of UV light and relatively low temperature damage can also be observed. The antagonistic interaction of UV light and heat may be related to the induction of some reparation systems, but this problem is beyond the main scope of this study. It is clear from the data presented that the fluence rate is a very important parameter and may serve as a determinant for synergy. Figures 6.16B and 6.16D show the correlation of the fluence rate and exposure temperature, which both provide maximum synergistic interaction for simultaneous treatment with heat and UV light. Closed circles denote our experimental results. One can see that a linear relationship is found between these values. From these findings, it can be inferred that the temperature at which UV light is delivered should be decreased to obtain the maximum synergistic effect with decreasing fluence rate.

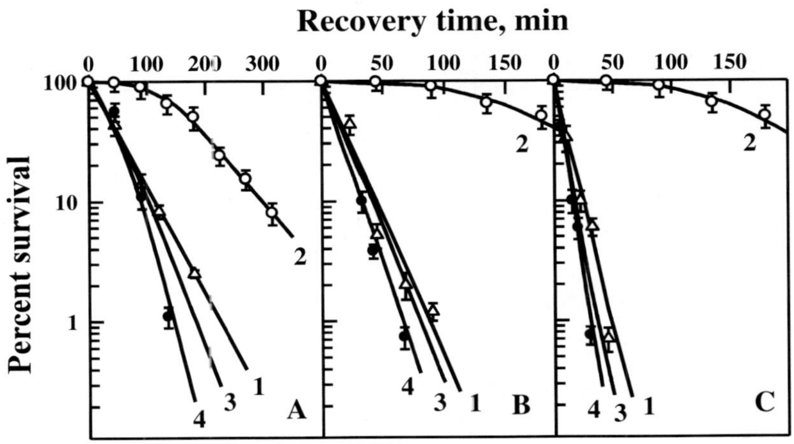

Figure 6.11. Survival curves of *S. cerevisiae* diploid yeast cells (strain XS800): curves 1, heat treatment alone (A, 47.5, B, 50.0, C, 52.5 °C); curves 2, UV light (254 nm) irradiation alone at room temperature and 0.033 W/m^2; curves 3, calculated curves for independent action of UV light and high temperature; curves 4, experimental curves after simultaneous action of both agents.

One can also see, for any fixed temperature, there should be a specific intensity that maximizes the synergistic effect, and any deviation of the UV light intensity from the optimal value would result in a decrease in synergy. These results seem to be related to the fact that, if the fluence rate increases, then the lethal UV dose is delivered in a short time, so that the duration of heat incubation decreases and could explain the higher temperature that should be applied to the cells. For the UV-sensitive *S. cerevisiae* mutant XS1924, which presumably lacks certain dark reactivation enzymes [44], the survival curves were exponential both after heat treatment alone and after UV exposure at different temperatures. It is worth noting that, under the conditions of these experiments, the UV-sensitive strain is about three times as sensitive to UV light (254 nm) as is its wild-type parent strain XS800.

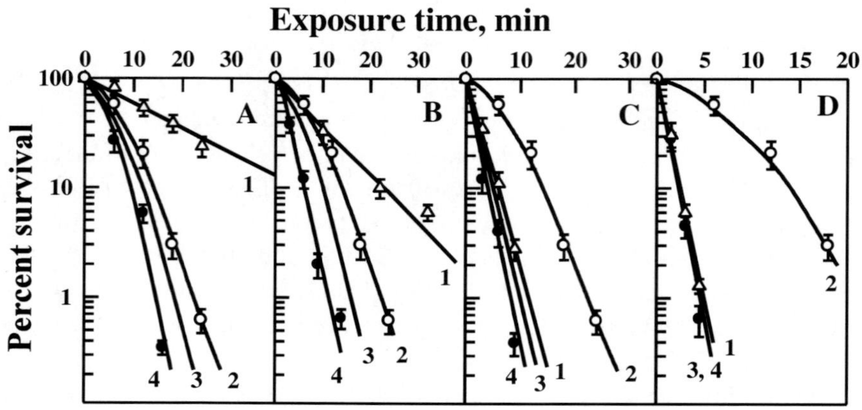

Figure 6.12. Survival curves of *S. cerevisiae* diploid yeast cells (strain XS800): curves 1, heat treatment alone (A, 50, B, 52.5, C, 55, D, 57.5 °C); curves 2, UV light (254 nm) irradiation alone at room temperature and 0.25 W/m^2; curves 3, calculated curves for independent action of UV light and high temperature; curves 4, experimental curves after simultaneous action of both agents.

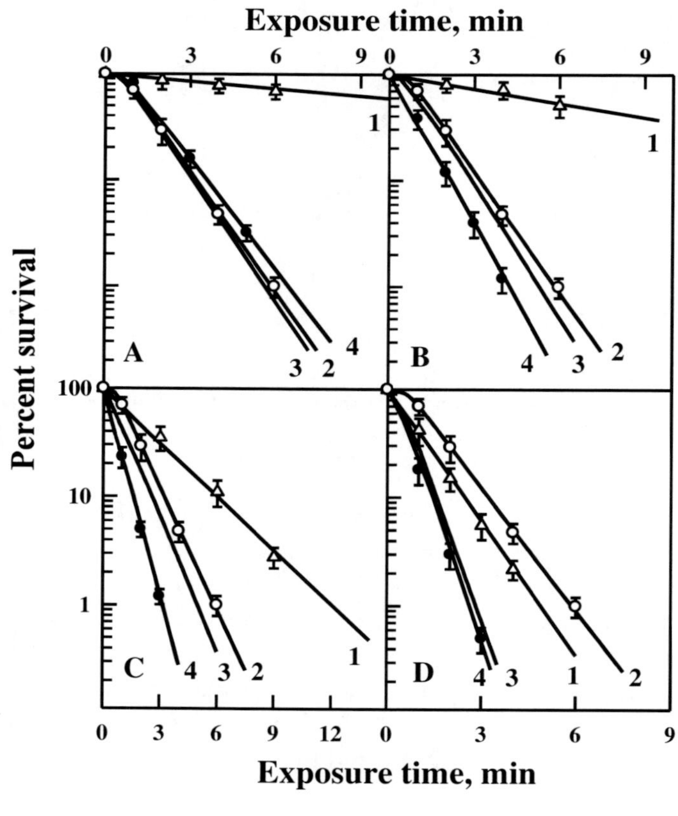

Figure 6.13. Survival curves of *S. cerevisiae* diploid yeast cells (strain XS800): curves 1, heat treatment alone (A, 50, B, 52.5, C, 55, D, 57.5 °C); curves 2, UV light (254 nm) irradiation alone at room temperature and 1.5 W/m^2; curves 3, calculated curves for independent action of UV light and high temperature; curves 4, experimental curves after simultaneous action of both agents.

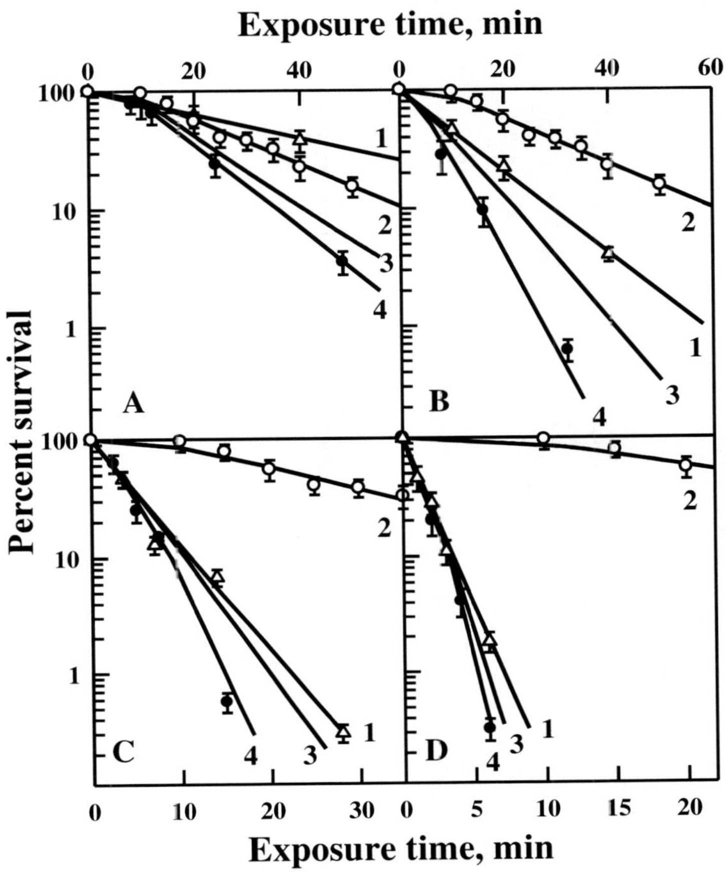

Figure 6.14. Survival curves of *S. cerevisiae* diploid yeast cells (strain T1): curves 1, heat treatment alone (A, 47.5, B, 50, C, 52.5, D, 55 °C); curves 2, UV light (254 nm) irradiation alone at room temperature and 0.15 W/m^2; curves 3, calculated curves for independent action of UV light and high temperature; curves 4, experimental curves after simultaneous action of both agents.

Table 6.6. Effect of hyperthermia on UV light survival curve parameters of XS1924 (*rad*18/*rad*18) diploid yeast cells*

Fluence rate, W/m^2	Parameters	Temperature, °C				
		20	50	52.5	55	57.5
0	D_0, min	–	16.0±1.4	6.0±0.5	2.0±0.1	0.5±0.1
0.25	D_0, J/m^2	17.8±1.5	16.3±1.6	15.5±1.2	11.0±1.3	7.3±0.8
	D_0 (additive), J/m^2	17.8	16.0	14.7	11.2	7.7
	k	–	0.98±0.05	0.95±0.08	1.02±0.05	1.05±0.06

* All values include the standard deviations of the estimates. Inter-experimental errors are included for the synergistic enhancement ratio.

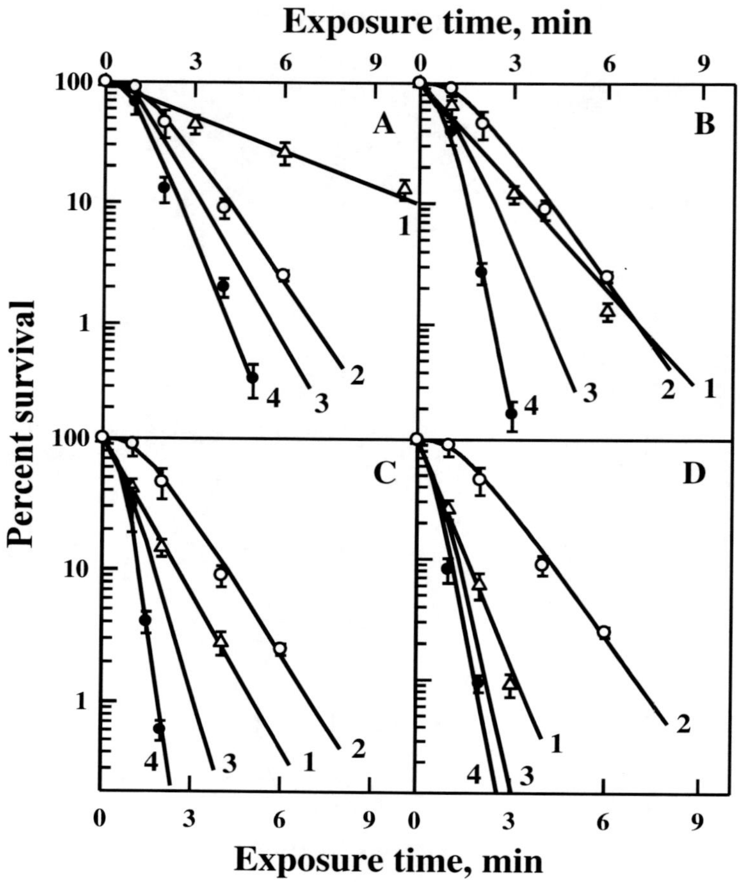

Figure 6.15. Survival curves of *S. cerevisiae* diploid yeast cells (strain T1): curves 1, heat treatment alone (A, 50, B, 52.5, C, 55, D, 57.5 °C); curves 2, UV light (254 nm) irradiation alone at room temperature and 1.5 W/m^2; curves 3, calculated curves for independent action of UV light and high temperature; curves 4, experimental curves after simultaneous action of both agents.

Also, there appears to be no interaction between UV light and high temperatures in the UV-sensitive strain – the synergistic enhancement ratio is close to unity (Table 6.6). Because there was no synergistic or antagonistic interaction between UV light and heat for the UV-sensitive mutant exposed to the single fluence rate (0.25 W/m^2) within a wide temperature range, we believed that there was no need to extend these experiments for other fluence rates.

To estimate mitotic recombination, wild-type diploid yeast of *S. cerevisiae* (strains T1) was used. Strain T1 has been specifically designed for the screening of the recombinogenic effects [73]. This strain is heterozygous for the gene ADE2 (*ade2-192/ade2-45*), which makes it possible to detect the recombinogenic events [46; 73]. The alleles differ in their extent of colony pigmentation and complement each other so that diploid is prototroph and produces white colonies. Mutation of *ade2-192* locus produces red colonies while the *ade2-45* produces pink colonies. The induction of mitotic crossing-over was measured by determining twin-sectored red/pink colonies together with other monochrome or sectored colonies. The strain is highly sensitive to various damaging agents of chemical and physical origin [73]. The cells were grown before irradiation up to stationary stage of growth for 3–5 days at 30 °C on a complete growth medium of the following composition (in water): peptone, 20 g/l; yeast extract, 1 g/l; agar, 25 g/l; glucose, 20 g/l. The budding cell component in the experimental culture was found to be less than 2%. In order to determine mitotic recombination and cell survival after treatments with UV and hyperthermia the following complete nutrient medium was used: glucose 20, agar 20, yeast extract 1, KH_2PO_4 2, $MgSO_4$ 1, $(NH_4)_2SO_4$ 1 g/l of water. The salts in the complete medium were used instead of peptone for more clear expression of recombinants [46].

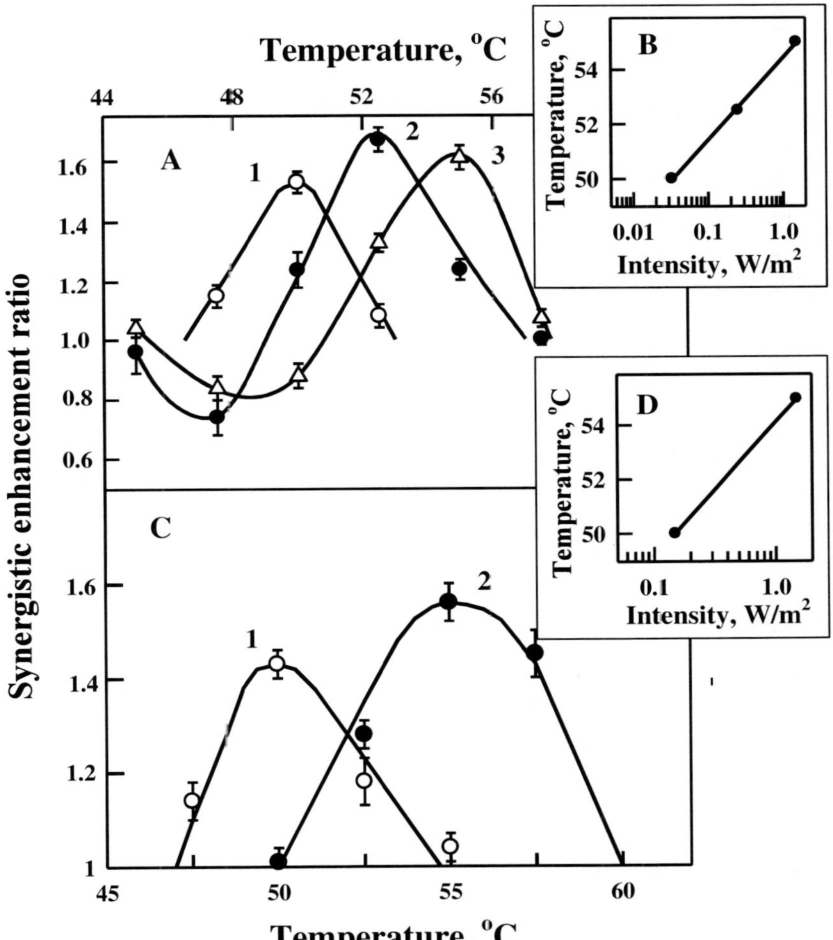

Figure 6.16. The results of a simultaneous action of UV light (254 nm) and high temperature on the inactivation of diploid *S. cerevisiae* wild-type yeast cells, strains XS800 (A and B) and T1 (C and D). A: the dependence of the synergistic enhancement ratio on the exposure temperature and UV light fluence rate (curve 1, 0.033 W/m^2; curve 2, 0.25 W/m^2; curve 3, 1.5 W/m^2). B: the relationship between exposure temperature and UV light fluence rate providing the highest synergistic enhancement ratio. C: the dependence of the synergistic enhancement ratio on the exposure temperature and UV light fluence rate (curve 1, 0.15 W/m^2; curve 2, 1.5 W/m^2). D: the relationship between exposure temperature and UV light fluence rate providing the highest synergistic enhancement ratio.

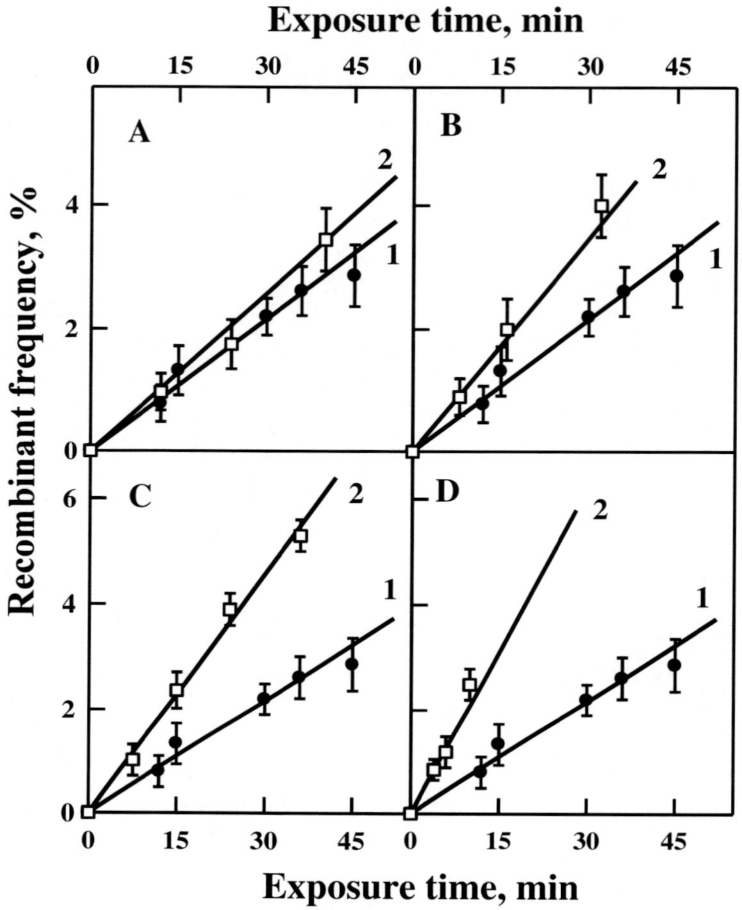

Figure 6.17. The recombinant frequency of yeast cells (*S. cerevisiae*, strain T1) after exposure with UV light (254 nm) alone (0.15 W/m^2, – curves 1) and simultaneous action of UV light and hyperthermia – curves 2 (A, 47.5, B, 50.0, C, 52.5, D, 55.0 °C).

The experimental protocol used was as follows: the interaction of two agents was tested at two intensity levels. As a start, increasing doses of UV irradiation at low dose rate were combined with hyperthermia treatments at different temperatures ranging from 47.5 to 55 °C and the yield of recombinants in the surviving population was assessed. After that, at a 10-fold higher dose rate of UV irradiation, the effects of simultaneous hyperthermia treatments at temperature ranging from 52.5 to 60 °C were studied for mitotic recombination. Mitotic crossing-over results in the formation of twin-sectored colonies with red and pink sectors. The occurrence of recombination estimated includes, besides these twin-sectored colonies, other sector or complete monochrome colony coloration. A recombination of cells was determined by counting of colonies on the 10th day after irradiation. In fact, the induction of mitotic recombination was determined by counting the twin spots, i.e., real crossing-over events together with some very rare mutations in the ADE2 locus. The possible loss of chromosomes was not taken into account and the frequency of recombination was expressed as a percentage of the aberrant colonies to the total number of colonies produced after the exposure. From

1000 to 2000 colonies were scored to determine the recombinant frequency for each experimental point. The survival response was determined by colony counts on the 5th day of incubation. Each data point represents an average value from two to five independent experiments pooled together. Error bars in all figures show inter-experimental errors. No recombination was detected in unexposed cells or cells treated at high temperature alone (over 10,000 colonies were scanned). For the frequency of recombination, linear regression calculation was used to fit the experimental point.

Results of action of UV light and hyperthermia on recombination and inactivation of yeast cells have been obtained for two fluence rates and the corresponding temperature ranges: 0.15 W/m^2 and 47.5–55° C; 1.5 W/m^2 and 52.5–60°C. Figure 6.17 summarizes our experimental results obtained for the recombinant frequency of yeast cells as a function of UV exposure time applied alone (curve 1) at fluence rate of 0.15 W/m^2 and combined with heat (curves 2: A, 47.5; B, 50.0; C, 52.5; D, 55.0 °C). A similar set of data was obtained at a ten-fold higher fluence rate of UV light and at higher temperatures: A: 52.5, B: 55.0, C: 57.5, D: 60.0 °C (Figure 6.18).

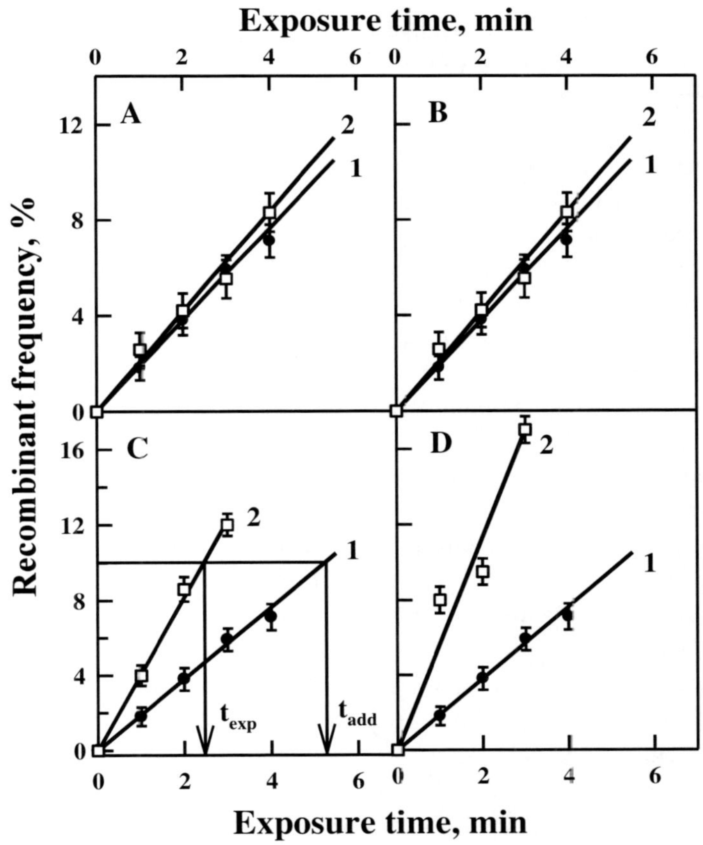

Figure 6.18. The recombinant frequency of yeast cells (*S. cerevisiae*, strain T1) after exposure with UV light (254 nm) alone (1.5 W/m^2. curve 1) and simultaneous action of UV light and hyperthermia, curve 2 (A, 52.5; B, 55.0; C, 57.5; D, 60.0 °C).

It is of importance that hyperthermia alone has no effects on mitotic recombination in these experiments. Because of that any increase in recombinants observed after UV in the presence of hyperthermia can be considered as synergistic effect. Thus, the two blocks of data show that the agents used in this investigation and applied together can result in synergistic effect to UV irradiation by hyperthermia when a number of recombinant cells induced under simultaneous action of both modalities exceed the sum of the effects produced by every agent applied separately.

The synergistic enhancement ratio (k) after simultaneous action of UV light and hyperthermia on the yield of mitotic recombination was calculated by means of

$$k = t_{add} / t_{\exp}, \tag{6.9}$$

i.e., by the ratio of the durations for additive interaction of the agents to that observed in experiments after simultaneous applications of both agents. The example of t_{add} and $t_{\exp}$ estimation is shown in Figure 6.18C.

Figure 6.19 presents the dependence of the synergistic enhancement ratio on the temperature at which the irradiation occurred in the case of a simultaneous action of UV radiation and hyperthermia on the frequency of the recombinants in the diploid yeast cells. Curve 1 shows this dependence for UV radiation delivered at 0.15 W/m^2, while curve 2 shows this dependence for UV fluence rate of 1.5 W/m^2. These data have been calculated based on a whole set of dose-effect curves. It is seen that for a fixed intensity a rise in temperature is accompanied by an increase of the synergistic enhancement ratio over the used range of temperatures. It is also evident that a decrease in the intensity of UV exposure results in the necessity to decrease the temperature at which irradiation occurs to provide the same value of the synergistic enhancement ratio.

For the recombinogenic test, in opposite to cell inactivation, the synergistic enhancement ratio did not pass through a maximum value and was linearly related with the temperature. It can be explained by the fact that the mitotic crossing-over frequency would have increased to a peak at moderate exposure doses and then decreased rapidly [46]. That is why in our investigation we tested only relatively low doses of UV exposure within which the frequency of the recombination and the synergistic enhancement ratio were linear functions of the UV dose and exposure temperature. For the recombinant yield the temperature range synergistically increasing the effect of UV radiation was shifted to lower temperatures with a decrease in the fluence rate (Figure 6.19).

The data presented here allow plotting the recombinant frequency as a function of cell survival (Figure 6.20). It demonstrates two specific examples for the two different fluence rates and for identical temperature treatment (52.5 °C). One can see that hyperthermia modified recombinant expression, the recombinant frequency for the same survival being smaller after combined action in comparison with that produced by UV irradiation alone. The frequency of recombinants at a given survival level (10%) is 2.2-fold higher at a high fluence rate of UV than that obtained at a low fluence rate (Figure 6.20B, curve 1 compared to Figure 6.20A, curve 1).

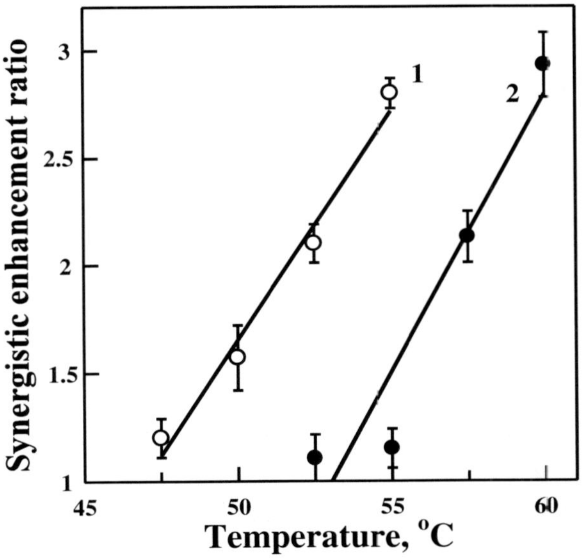

Figure 6.19. The dependence of the synergistic enhancement ratio on the temperature at which the UV irradiation (254 nm) occurred on the frequency of the recombinants in the diploid yeast cells (*S. cerevisiae*, strain T1). Curve 1 exhibits this dependence for UV radiation delivered at 0.15 W/m^2. Curve 2 presents this dependence for UV fluence rate of 1.5 W/m^2.

In combination with hyperthermia 52.5 °C, the recombinant frequency is (at a given survival) five-fold lower in case of low fluence rate UV irradiation + hyperthermia than in case of high fluence rate UV irradiation + hyperthermia. Similarly, at a given survival level (2%) the recombinant frequency is approximately four-fold higher in the case of high fluence rate UV irradiation + hyperthermia than in the case of low fluence rate UV irradiation + hyperthermia suggesting that low fluence rate irradiation even in combination with hyperthermia is much less genotoxic than high fluence rate UV irradiation in combination with hyperthermia.

To interpret the results obtained in this study, a mathematical model described above may be engaged. According to the model, additional recombinogenic damage would result from the interaction of some pre-recombinogenic sublesions, which are ineffective when each agent is applied separately. It is worth noting that the nature of the supposed sublesions and the mechanism of their interaction may be different for the lethal and recombinogenic effects and still remain to be elucidated. Let N_1 and N_2 be the numbers of recombinogenic events induced by the UV light and high temperature, respectively. We believe that even for the case when $N_2 < 1$ (ineffectiveness of hyperthermia to induce mitotic recombination) there are some sublesions produced by hyperthermia to provide the synergistic effects observed. Then the following predictions of the synergy modification should be expected. First, an increase in the synergistic enhancement ratio with an increase in the exposure temperature for any constant exposure time and fluence rate of UV light (the increasing of N_2 at the constant of N_1) should be observed. Just such a dependence of the synergistic enhancement ratio as a function of the exposure temperature was observed in this investigation (Figure 6.19).

Secondly, to maintain the same synergistic effectiveness with a decreasing UV light intensity the temperature of the hyperthermia treatment should be lowered to decrease the amount of thermal lesions (N_2). Exactly such results were observed in this study (Figure 6.19).

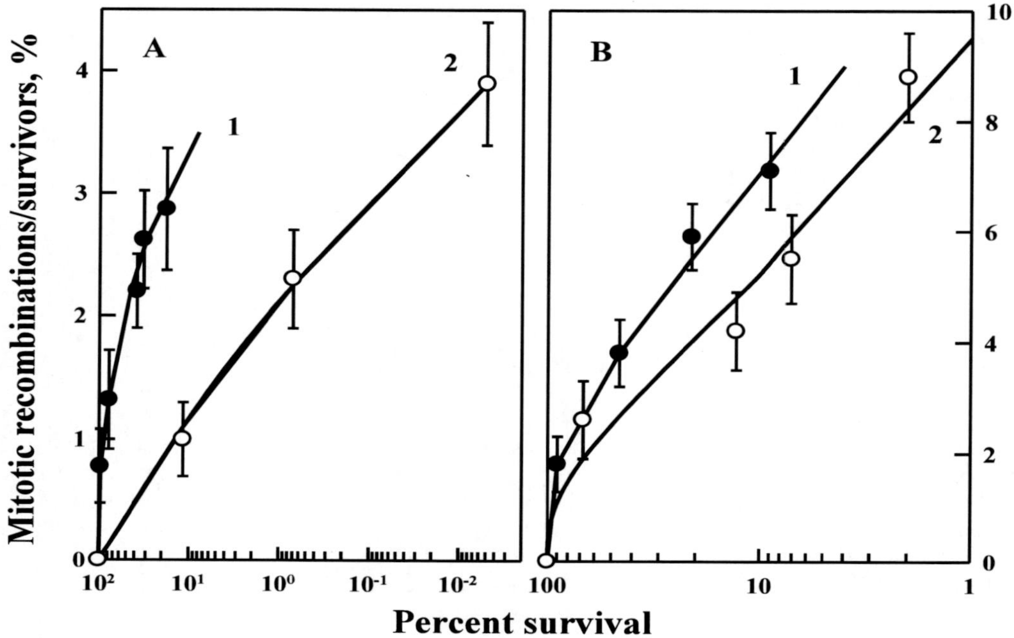

Figure 6.20. The dependence of the recombinant frequency on percent survival of *S. cerevisiae* diploid yeast cells (strain T1) exposed to UV light (254 nm) alone (curves 1) and combined with hyperthermia (curves 2). A, 0.15 W/m², 50 °C; B, 1.5 W/m², 52.5 °C.

6.4. Prognosis of Yeast Cell Recovery after the Simultaneous Action of Hyperthermia with Ionizing Radiation or UV Light

It has so far generally been assumed that thermal radiosensitization results from the heat-induced inhibition of the repair of radiation damage. This inhibition was revealed in a decrease of both the rate and extent of recovery which can be interpreted as being due to either the impairment of the recovery process itself (inactivation of repair enzymes, for instance) or to the production of more severe irreversible lethal damage which can not be repaired. As was shown above, the latter case is realized for the combined treatment of heat and ionizing radiation on yeast and cultured mammalian cells as well as for the exposure of yeast cells to heat and UV light. These novel observations provide insight into the mechanisms underlying the effects of combined exposure and were taken into account for interpretation and mathematical modeling of synergistic interaction of the combined exposures of various environmental agents (Chapter 5). The aim of this study is to test the validity of the model to predict yeast cell recovery after the simultaneous action of hyperthermia with ionizing radiation or UV light.

Figure 6.21 exhibits the results of experiments when wild-type diploid yeast cells *S. ellipsoideus* (*vini*), strain Megri 139-B, were subjected to simultaneous treatments with ionizing radiation (^{60}Co γ-ray, 10 Gy/min) at the following temperatures: 20, 40, 45, 50 and 55 °C. To assess the percent of the cell survival by colony forming ability, the exposed cells were plated on nutrient agar immediately after irradiation (curves 1, open circles) and after 3 days LHR (curves 2, filled circles). It appears that in diploid yeast cells thermally enhanced radiosensitivity may be due to both the inhibition of repair from potentially lethal damage and/or the enhancement of expression of lethal damage. The reduction of the survival curve shoulder suggests that hyperthermia also interferes with repair of sublethal damage. The increase in the cell radiosensitivity is accompanied with the production of irreversible damage from which cell is incapable to recover. It means the increase in the irreversible component of radiation damage K, which is determined as a value inversely proportional to the dose-modifying factor that is as the ratio of the D_o after delayed and immediate plating. In our calculations, this value is close to the ratio of the corresponding doses decreasing cell survival to 1%, i.e., $K = D_1 / D_2$ (Figure 6.21A). The calculation showed that the fraction the irreversible component was equal to 0.43, 0.49, 0.76, 0.85 and 0.94 for 20, 40, 45, 50 and 55 °C, respectively (Table 6.7). Data for 52.5 °C are also included in this Table. It is apparent that the cell ability to recover radiation damage is inversely proportional to the temperature at which the exposure was delivered.

Table 6.7. The irreversible component of radiation damage (K) and the synergistic enhancement ratio (k) after simultaneous action of ionizing radiation (γ-ray of ^{60}Co, 10 Gy/min) and hyperthermia on diploid yeast cells of *S. ellipsoideus* (*vini*) (strain Megri 139-B)

Temperature, °C	20	40	45	50	52.5	55
N_2/N_1	0	0.1	0.18	0.55	1.4	7.2
K, experiment	0.43±0.03	0.49±0.04	0.76±0.05	0.85±0.06	0.92±0.07	0.94±0.05
K predicted for:						
$K_3 = 1.0$	0.43	0.57	0.65	0.81	0.89	0.95
$K_3 = 0.5$	0.43	0.48	0.51	0.56	0.61	0.81
$K_3 = 0.0$	0.43	0.39	0.37	0.33	0.33	0.71
k, experiment	1.00	1.00	1.25	1.85	2.25	1.00
k, theory	1.00	1.00	1.40	1.80	2.20	1.35

Yeast cells exposed to ionizing radiation at 55 °C almost completely fail to show recovery after delayed plating. It would be of tempting to test the ability of the model considered above to predict the extent of the recovery after the combined action of heat and ionizing radiation. We can reasonably believe that the irreversible component induced by ionizing radiation alone K_1 = 0.43 (irradiation at room temperature). Since in all our experiments with yeast cells no LHR was observed after hyperthermia alone, we consider that for this agent K_2 = 1.0. It means that all thermal damages are irreversible.

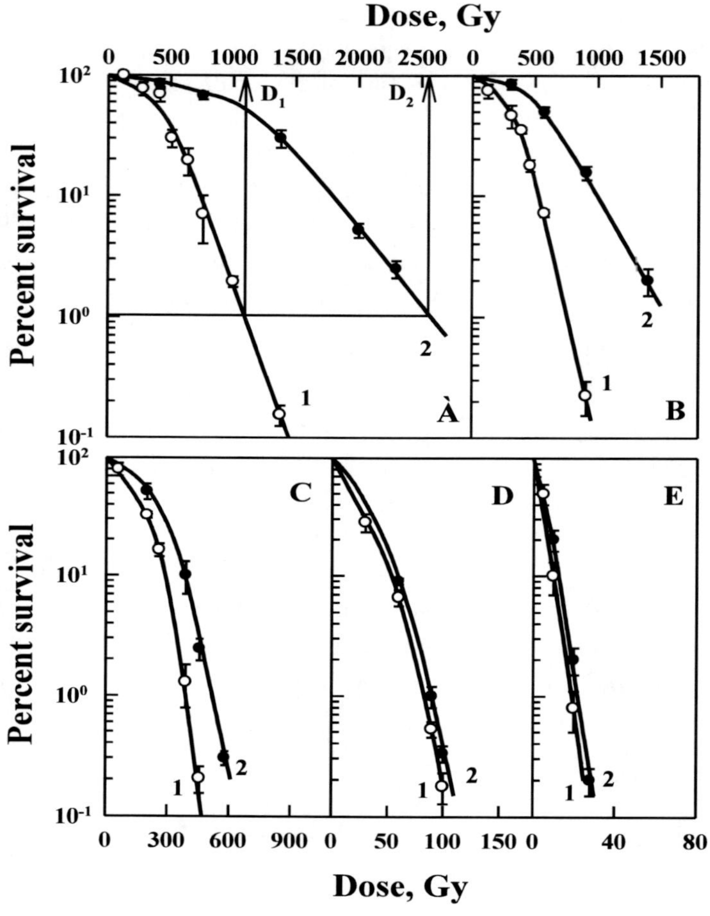

Figure 6.21. Survival curves of diploid yeast cells *S. ellipsoideus* (*vini*) (strain Megri 139-B) irradiated with ^{60}Co γ-rays (10 Gy/min) at different temperatures: 20 (A), 40 (B), 45 (C), 50 (D) and 55 °C (E). Cells were plated on nutrient media to assess colony forming ability just after irradiation (curves 1) and after 3 days of the LHR (curves 2). Pooled results of three experiments are presented. Error bars show inter-experimental standard errors.

It is of interest to verify the possibility of predicting the irreversible component of radiation damage (K) after the simultaneous action of ionizing radiation and hyperthermia (Eq. 5.29). The basic parameters of the model (p_1 and p_2) may be found from experiments. Figure 6.22 shows survival curves for diploid yeast cells *S. ellipsoideus* (*vini*) (strain Megri 139-B) after simultaneous action of ionizing radiation and high temperature (A, 50 °C, B, 52.5 °C).

The experimental value of the synergistic enhancement ratio is defined by $k = D_{theor} / D_{\exp}$ (Figure 6.22A). Such set of survival curves has been obtained for other conditions of thermoradiation action, they are not presented here. At the first case (50 °C) the inequality $p_1N_1 > p_2N_2$ holds and k = 1.9. Using this value and N_2 / N_1 = 0.55 estimated by Eq. (5.12), we have (Eq. 5.15) p_2 = 2.54. On the contrary, for temperature 52.5 °C the inequality $p_1N_1 < p_2N_2$ is correct, k = 2.3 and N_2 / N_1 = 1.36. In accordance with Eq. (5.14) we obtain p_1 = 3.07. Using these results and Eq. (5.29), we calculated the irreversible component K expected after simultaneous action of ionizing radiation and hyperthermia for 3 various values of the irreversible component of damage responsible for the synergy (K_3).The results of calculation together with experimental data of the proportion of irreversible damage after simultaneous action of heat and ionizing radiation are collected in Table 6.7.

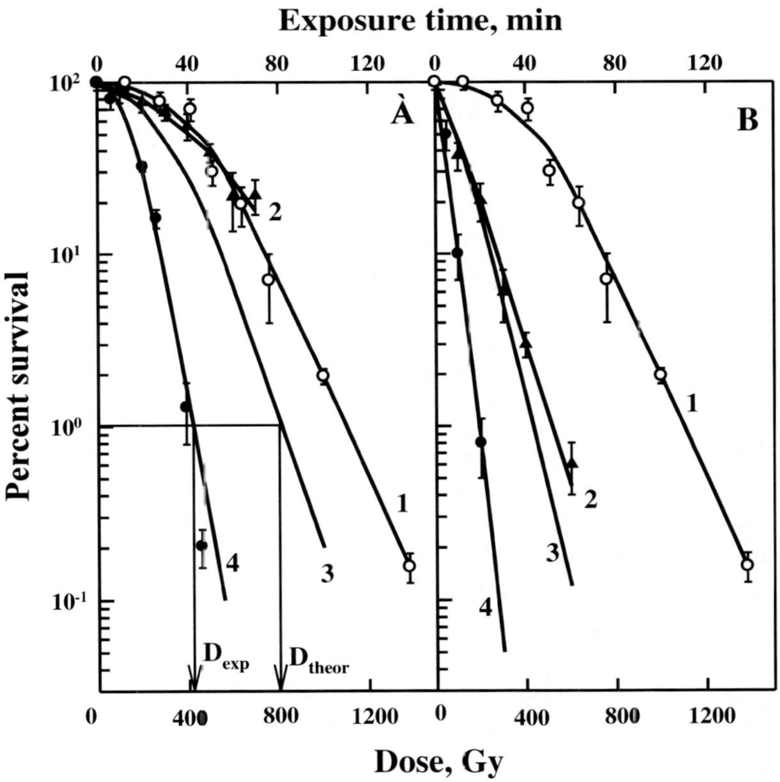

Figure 6.22. Survival curves for diploid yeast cells *S. ellipsoideus* (*vini*) (strain Megri 139-B) after simultaneous action of ionizing radiation and high temperature (A, 50 °C; B, 52.5 °C). Cells were exposed to ionizing radiation (^{60}Co, 10 Gy/min) alone (curves 1), heat alone (curves 2) and simultaneous action of these agents (curves 4). Curves 3, theoretically expected effect for independent action of both agents.

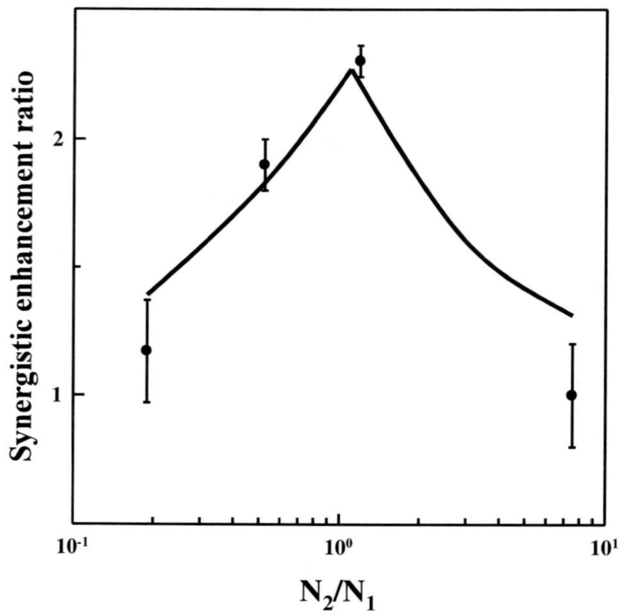

Figure 6.23. The dependence of the synergistic enhancement ratio on the ratio of damage (N_2/N_1) produced by simultaneous action of ^{60}Co γ-rays and heat for diploid yeast *S. cerevisiae* (strain Megri 139-B): points, experimental data; solid line, theoretically expected data.

It can be seen that the values predicted on the basis of Eq. 5.29 for K_3 = 1.0 show the best correspondence with experimental findings. On the contrary, the values predicted for K_3 = 0.5 and K_3 = 0 are not consistent with experimental data. This Table and Figure 6.23 include experimentally obtained and theoretically predicted (Eq. 5.5) values of the synergistic enhancement ratio k. It can also be seen well correspondence of these values.

Similar calculations of the irreversible component K have been fulfilled for diploid yeast cells exposed to ultraviolet light (254 nm, fluence rate 1.5 W/m^2) at different temperatures. Survival curves were obtained for diploid yeast cells *S. cerevisiae* (strain XS800) irradiated with UV light at different temperatures: 20, 53, 54, 55, 56 and 57 °C (Figure 6.24, panels A, B, C, D, E and F, respectively). The increase in cell UV sensitivity is accompanied with the production of irreversible damage which cell is incapable to recover from. It means the increase in the irreversible component K. The calculation showed that the fraction the irreversible component was equal to 0.56, 0.60, 0.69, 0.78, 0.92 and 1.00 for 20, 53, 54, 55, 56 and 57 °C, respectively (Table 6.8). It is evident that the cell ability to recover radiation damage is inversely proportional to the temperature at which the exposure was delivered. Yeast cells exposed to UV light at 57 °C completely fail to show the recovery after delayed plating. It would be of tempting to test the ability of the model considered to predict the extent of the recovery after the combined action of heat and UV light (Eq. 5.29) as well as the synergistic enhancement ratio (Eq. 5.5). We can reasonably believe that the irreversible component induced by UV light alone K_1 = 0.56 (room temperature). Since in all our experiments with yeast cells no LHR was observed after hyperthermia alone, we consider again that $K_2 = 1.0$. It means that all thermal damages are irreversible.

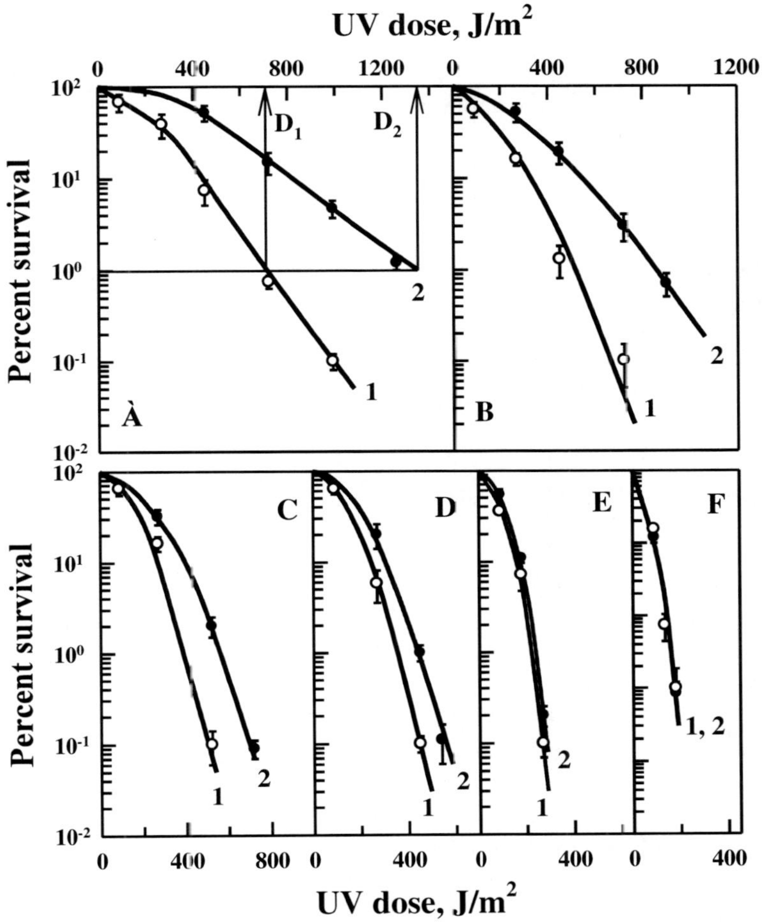

Figure 6.24. Survival curves of diploid yeast cells *S. cerevisiae* (strain XS800) irradiated with UV light (254 nm, 1.5 W/m^2) at different temperatures: 20, 53, 54, 55, 56 and 57 °C (A, B, C, D, E and F, respectively. Cells were plated on nutrient media to assess colony forming ability immediately after irradiation (curves 1) and after 3 days of liquid holding recovery (curves 2). Pooled results of three experiments are presented. Error bars show inter-experimental standard errors.

The basic parameters of the model (p_1 and p_2) were determined from experiments. Figure 6.25 shows survival curves for diploid yeast cells *S. cerevisiae* (strain XS800) after simultaneous action of UV light and high temperature (A, 54 °C, B, 56 °C). The experimental value of the synergistic enhancement ratio is defined by $k = t_2 / t_1$ (Figure 6.25A).

Such set of survival curves has been obtained for other conditions of simultaneous action of heat and UV light, which are not presented here. Using these results and Eq. (5.29), we calculated the irreversible component K expected after simultaneous action of UV light and hyperthermia for 3 various values of K_3 – irreversible portion of damage responsible for the synergy. For simultaneous action of UV light at 54 °C the inequality $p_1N_1 > p_2N_2$ holds and k = 1.5. Using this value and N_2/N_1 = 0.18 estimated by Eq. (5.12): p_2 = 3.5. On the contrary, for temperature 56 °C the inequality $p_1N_1 < p_2N_2$ is correct, k = 1.7 and N_2/N_1 = 1.38. In accordance with Eq. (5.14) we obtain p_1 = 1.5. Using these results and Eq. (5.29), we calculated the irreversible component K expected after simultaneous action of UV light and hyperthermia for 3 various values of the irreversible part of damage responsible for the synergy (K_3).

Table 6.8. The irreversible component K and the synergistic enhancement ratio (k) after simultaneous action of ultraviolet radiation (254 nm) and hyperthermia on diploid yeast cells *S. cerevisiae* (strain XS800)

Temperature, °C	20	53	54	55	56	57
N_2/N_1	0	0.13.	0.18	0.2	1.00	4.10
K, experiment	0.56±0.04	0.60±0.06	0.69±0.06	0.77±0.05	0.92±0.07	1.00±0.05
K predicted for:						
K_3 = 1.0	0.56	0.72	0.76	0.79	0.88	0.95
K_3 = 0.5	0.56	0.57	0.58	0.59	0.69	0.86
K_3 = 0.0	0.56	0.43	0.42	0.38	0.49	0.77
k, experiment	1.00	1.29	1.48	1.70	1.8	1.15
k, theory	1.00	1.34	1.48	1.60	1.8	1.28

The results of calculation together with experimental data of the proportion of irreversible damage after simultaneous action of heat and ionizing radiation are collected in Table 6.8. It can be seen that the values predicted on the basis of Eq. (5.29) for K_3 = 1.0 show the best correspondence with experimental findings. On the contrary, the values predicted for K_3 = 0.5 and K_3 = 0 are not consistent with experimental data. This Table also includes experimentally obtained and theoretically predicted (Eq. 5.5) values of the synergistic enhancement ratio k. It can also be seen well correspondence of these values. According to the model, the highest synergism k_{max} = 2.06 (Eq. 5.11) at N_2/N_1 = 0.44 (Figure 6.26).

In the model presented here sublesions are supposed to be non-effective after each agent taken alone while the additional damage arising from the interaction and responsible for the synergy is irreversible which the cells are incapable to recover from. The main novel contribution of this model concerns the possibility to predict the irreversible component of radiation damage after the combined exposures of different environmental agents.

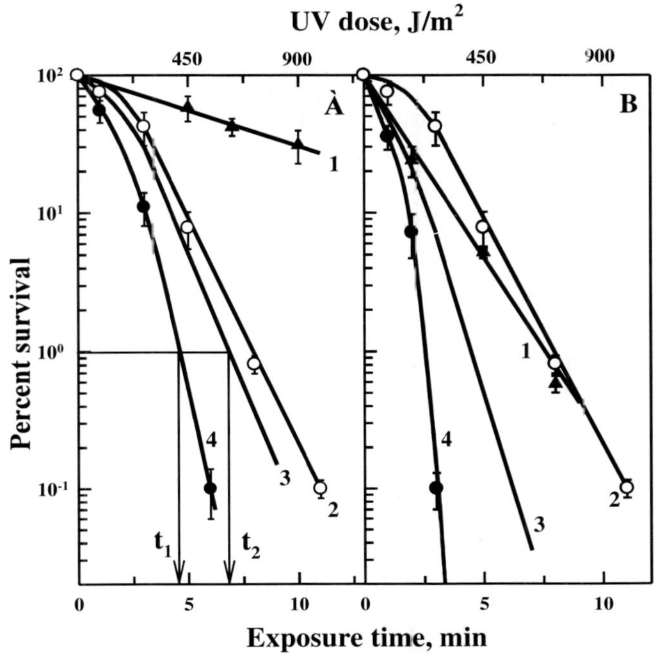

Figure 6.25. Survival curves for diploid yeast cells *S. cerevisiae* (strain XS800) after simultaneous action of UV light (254 nm, 1.5 W/m^2) and high temperature (A, 54 °C; B, 56 °C). Cells were exposed to heat alone (curves 1), UV light at room temperature (curves 2) and simultaneous action of these agents (curves 4). Curves 3, theoretically expected effect for independent action of both agents.

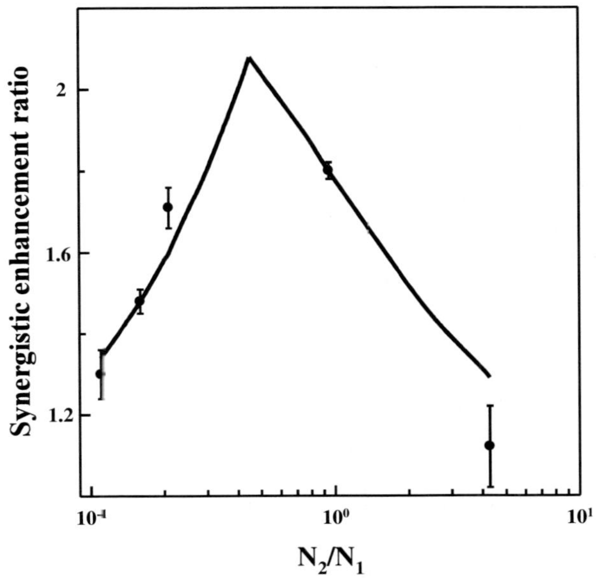

Figure 6.26. The dependence of the synergistic enhancement ratio on the ratio of damage (N_2/N_1) produced by simultaneous action of UV light (254 nm) and heat for diploid yeast *S. cerevisiae* (strain XS800): points, experimental data; solid line, theoretically expected data.

To test the approach, the results of experimental research of diploid yeasts cells survival after simultaneous action of heat combined with ^{60}Co ionizing radiation or 254 nm UV light have been used here. It was shown that the cell ability to the LHR decreased with an increase in the temperature, at which the exposure was occurred. A good correspondence between experimental results and model prediction was demonstrated only if the initial supposition about the irreversible character of additional damage holds. Otherwise, the values of irreversible component predicted on the basis of the model developed were not consistent with experimental data.

In summary, it can be concluded that the mechanism of the synergistic interaction of hyperthermia with ionizing radiation or UV light after their simultaneous treatment in yeast cells is not related to the impairment of the recovery capacity itself and it may be attributed to an increased yield of irreversible damage.

6.5. Ultrasound Intensity and Its Role in Synergistic Interaction of Ultrasound and Hyperthermia

As has been shown, a physical factor like ultrasound (US) may enhance the response to ionizing radiation of cultured mammalian cells [39; 65] and the bacterium *E. coli* [10]. It would be of interest to study the combined action of ultrasound and ionizing radiation for yeast cells with different ability to recover radiation damage. We demonstrated the significance of dose rate of ionizing radiation and UV light fluence rate in synergistic effect of these agents with heat. Hence, it will be of interest to estimate the role of ultrasound intensity in synergistic interaction of ultrasound and hyperthermia.

The following strains of the yeast, *S. cerevisiae*, were used for these experiments both diploid (XS800) and haploid (S288C) cells of wild-type, its X-ray-sensitive mutant strains (*rad*51/*rad*51: XS806; *rad*51: XS774-4d). The gamma-irradiation of the samples in a glass tube achieved by exposure of 7 ml sample in a 500 Ci ^{60}Co facility at a dose-rate of 10 Gy/min, as measured by a Siemens ionization Chamber. Sonication was achieved by lowering the sterilized probe into a 44 kHz ultrasonic unit (continuous mode) at 19 °C. The diameter of the ultrasonic transducer was 15 mm. The absorbed heat was partly removed by cooling with water. A semiconductor thermister was inserted to a glass tube to measure the temperature rise during sonication. This was found to be only about 10 °C for the maximum intensity and exposure time used in these experiments. Since the cells can easily tolerate temperatures of 30–40 °C a temperature rise of 10 °C seems unlikely to be significant. The ultrasonic dose rate was calculated by noting the initial rate of increase of the sample temperatures (without cooling) and computing the rate of energy deposition. In most experiments the ultrasonic energy output of the transducer was 40 W/cm^2.

The lethal effect of ultrasound on yeast cells has received relatively little attention. So before the study of yeast cell responses to combined action of ultrasound and ionizing radiation it would be worth to consider the lethal effects of ultrasound alone on yeast cells. First of all, we have investigated the dependence of the cell inactivation on a number of cells in the sample. Different cell concentrations (10^3–10^9 cells/ml) have been sonicated with the same intensity and duration. The results are summarized in Table 6.9. It can be easily seen that the effect is not dependent on cell concentration in a wide range, excepting the highest values. So in all subsequent experiments we have used concentration 10^6 cells/ml.

In Figure 6.27 are plotted the survival curves of diploid yeast cells of *S. cerevisiae* (strain XS800) exposed to US at different exposure rates (40, 30, and 6 W/cm^2). Lower power rates were inaccessible because of technical reasons. The shape of the survival curves was exponential only at small doses and was different from that obtained with ionizing radiations. From Figure 6.27 it can be seen that there was no exposure rate effect in our experiments. Similar data have been obtained with cultured mammalian cells by Martins et al. [39]. In all the following experiments the exposure rate was 40 W/cm^2.

Table 6.9. The influence of cell concentration on the survival of diploid yeasts *S. cerevisiae,* strain XS800, exposed to 44 kHz frequency ultrasound (2 minutes at exposure rate 40 W/cm^2)

Cell concentration, cells/ml	10^3	10^4	10^5	10^6	10^7	10^8	10^9
Survival, %	13.1	11.7	12.6	11.8	11.5	11.4	28.3

Survival of XS800 cells exposed to ultrasound at stationary (curve 1) and exponential (curve 2) stages of growth is demonstrated in Figure 6.28. Inactivation was found to follow almost a single-hit law. The yeast cell culture at stationary phase of growth consists mainly of G-1 phase cells (~95%); the asynchronous exponentially growing population contains about 80–90% of budded cells. Examination of Figure 6.28 reveals that budded cells exhibit a strikingly enhanced sensitivity towards ultrasound exposure when compared to resting cells. On the contrary, when the yeast cell population is irradiated with ionizing radiation, cells at exponential phase are most resistant – their survival curves has a large resistant tail [2; 54].

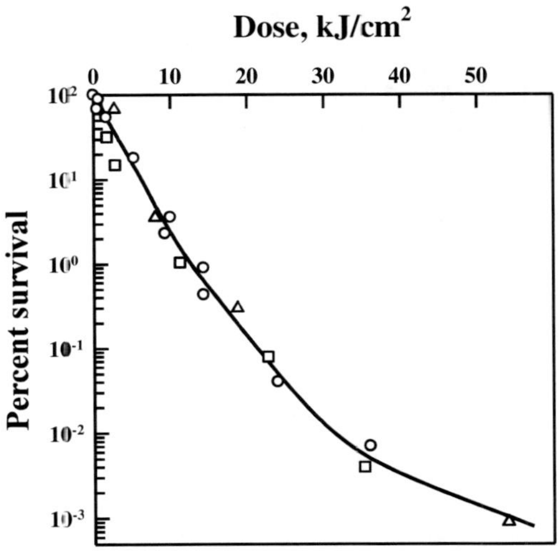

Figure 6.27. Survival of *S. cerevisiae* diploid yeast cells, strain XS800, exposed to 44 kHz frequency ultrasound at different exposure rates: circles, 40 W/cm^2, triangles 30 W/cm^2, squares, 6 W/cm^2.

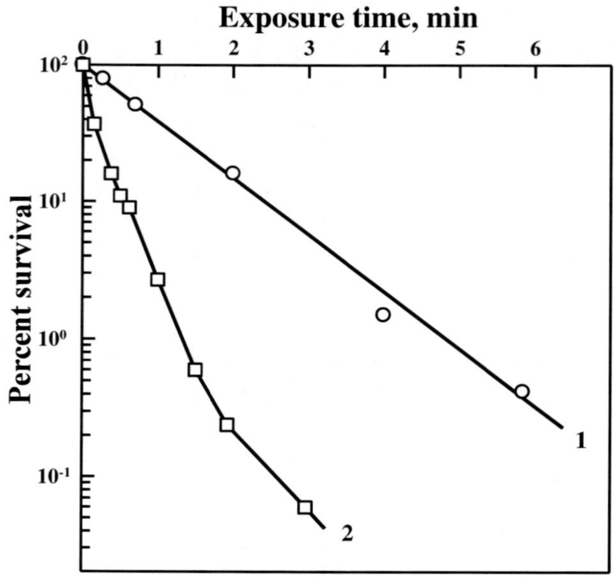

Figure 6.28. Survival of *S. cerevisiae* diploid yeast cells, strain XS800, exposed to 44 kHz frequency ultrasound at stationary (curve 1) and exponential (curve 2) stages of growth.

This would seem to indicate that different cell sites are responsible for cell inactivation due to sonication and ionizing radiation. A similar conclusion has been made for mammalian cells [39]. If the main cell sensitive targets for ultrasound differ greatly from those for ionizing radiation, being, perhaps, membranes in the first [6; 8] and nuclei in the second case, it would seem tempting to compare the interaction between the two kinds of damage after combined action of these agents.

Figure 6.29 shows survival curves of XS800 cells exposed in stationary phase of growth after combined action: ultrasound followed by γ-rays (Figure 6.29A) and γ-rays followed by ultrasound (Figure 6.29B). To estimate quantitatively the synergistic effect of ultrasound and γ-rays we have used the ultrasonic enhancement ratio (UER) which is defined as the ratio of the mean lethal dose D_0 only after the second modality to that after combined action of both modalities. The values of this parameter calculated from the data depicted in Figures 6.29, 6.30, 6.31 are collected in Table 6.10. It follows from the data plotted in Figure 6.29A and listed in Table 6.10 that although the shape of the curves was sigmoid after (US + γ-rays)-treatment, there is a tendency for the shoulder to be smaller and for the slope to be steeper with increasing dose of presonication. The extent of the enhanced effectiveness of combined treatment depends on the order of application, being more pronounced after (γ-rays + US)-treatment (Figure 6.29B, Table 6.10). This enhanced sensitivity is associated with the lack of an initial shoulder on the survival curves and with an increase in the slope. An enhanced effect is obtained even when a threshold radiation dose is applied (curve 2, Figure 6.29B).

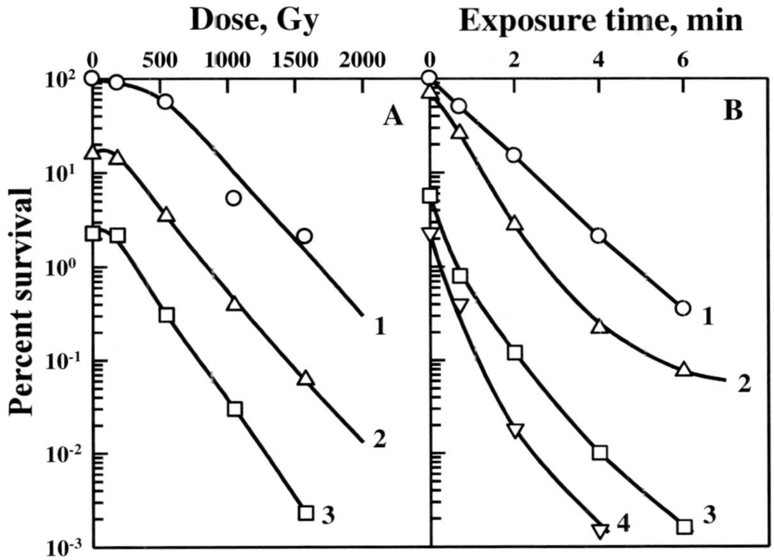

Figure 6.29. Survival of diploid yeast cells *S. cerevisiae* (strain XS800) after combined action of ultrasound 44 kHz and ionizing radiation. A: curve 1 – γ-rays alone; curves 2, 3 – (US + γ-rays)-treatment. B: curve 1 – US alone; curves 2, 3, 4 – (γ-rays + US)-treatment.

Table 6.10. The dependence of the ultrasonic enhancement ratio (UER) on the cell survival after application of the first modality and the order of application of both modalities to yeast cells *S. cerevisiae* of different strains

Strains, Genotype	Survival, % US alone	UER (US + γ-rays)-treatment	Survival, % γ-rays alone	UER (γ-rays + US)-treatment
XS800 RAD/RAD	100.0	1.0	100.0	1.0
	15.0	1.1	90.0	2.0
	2.2	1.4	6.0	2.2
			2.4	2.7
XS806 *rad*51/*rad*51	100.0	1.0	100.0	1.0
	20.0	1.0	14.0	1.0
	0.6	1.0	0.1	1.0
S288C RAD	100.0	1.0	100.0	1.0
	7.0	1.0	5.3	2.3
XS774-4d *rad*51	100.0	1.0	100.0	1.0
	20.0	1.0	14.0	1.0

Nakai and Matsumoto [45] reported that *rad51* mutant cells, both haploid and diploid, in stationary phase of growth were much more sensitive to X-rays and killed almost linearly. The mutant diploid cells were found in our experiments to be sensitive to both ionizing radiation and sonication alone (compare curves 1 in Figures 6.29 and 6.30). For this strain we have not observed the increase in the slope of survival curves after combined action independently of the sequence of application (Figure 6.30, Table 6.10).

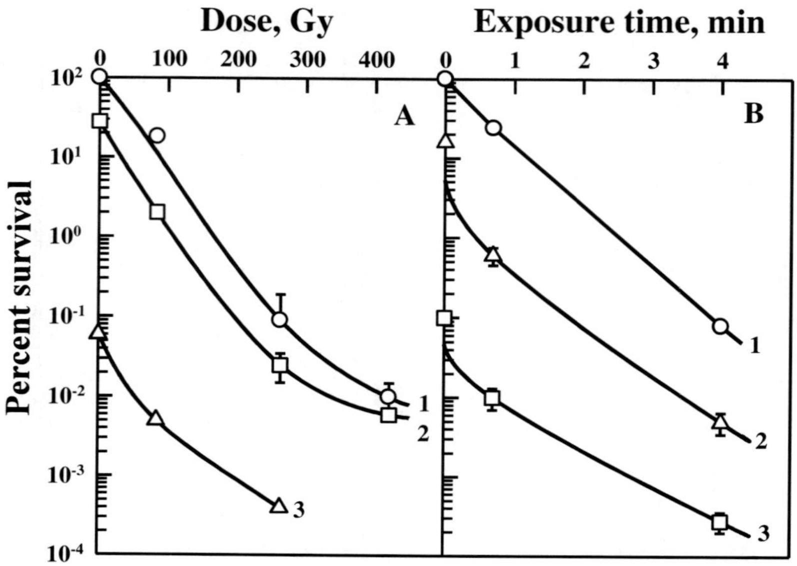

Figure 6.30. Survival of radiosensitive mutant diploid yeast *S. cerevisiae* (strain XS806) after combined action of ultrasound and ionizing radiation. A: curve 1 – γ-rays alone; curves 2, 3 – (US + γ-rays)-treatment. B: curve 1 – US alone; curves 2, 3 – (γ-rays + US)-treatment.

It has been described above that the combined action of elevated temperatures and ionizing radiation had more pronounced effect on diploid yeast cells than on haploid ones. Therefore, it may be worthwhile to fulfill similar investigation for combined action of ultrasound and γ-rays. Figure 6.31A shows survival curves of haploid yeast *S. cerevisiae* of wild-type exposed to γ-rays alone (curve 1) and both (US + γ-rays)- and (γ-rays + US)-treatments (curves 2 and 3, respectively). It was found that the haploid strain is only slightly more sensitive to inactivation by ultrasound irradiation compared to diploid wild-type. Slopes of the curves plotted in Figure 6.31A shows clearly that the effect was only additive for (US + γ-rays)-treatment and synergistic for (γ-rays + US)-exposure (Table 6.10). Figure 6.31B exhibits similar data for *rad51* haploid mutant. No synergistic effect was registered for this strain independently on the order of treatments (Figure 6.31B, Table 6.10).

The main conclusion justified by the above findings on the combined action of ultrasound and ionizing radiation concerns the orders of irradiation to obtain the synergistic interaction of both modalities. The effect of combined action of ultrasound and γ-rays on yeast cells was shown to vary widely with respect to the sequence of application. Presonication of yeast cells did not change the radiosensitivities of all strains studied except diploid of wild-type for which the slope of the survival curves was slightly increased. Post-sonication of irradiated yeast cells was more effective for all strains except haploid *rad51* cells. For (γ-rays + US)-treatment synergism was due to both increase in the slope and disappearing of the shoulder of the survival curves.

It is known that the shoulder is connected with the cell's ability to repair sublethal damage. It is imperative, therefore, to assume that (γ-rays + US)-treatment block this repairing way. However, very similar enhancement was observed for XS806 and S288C cells which are characterized by exponential survival curves. Hence, whether such radiosensitization was related to repair processes has not yet been demonstrated. It is quite plausible that the synergistic interaction of combined (γ-rays + US)-regimen may be due to interference with the repair of sub-lethal damage and/or with enhanced expression of lethal damage. It is also known that the slope of the exponential part of survival curve is in relation to the cell's ability to recover the potentially lethal lesions. The increasing of the slope was observed in our experiments only for wild-type cells capable of recovery and was absent for *rad51* mutants defective on some postradiation recovery processes.

It seems likely that the synergistic effect expressed by the change of the slope may be connected either with the damage of recovery systems or with the production of damages that the cell is incapable of recovery. However, the liquid-holding recovery processes was not damaged after combined action [50]. The above data are most easily explained by assuming that the recovery processes responsible for the manifestation of synergistic effect are performed more quickly and at earlier steps in comparison with the LHR processes. It appears that ultrasonically enhanced radiosensitivity may be due, in part, to inhibition of repair of potentially lethal damage and/or enhancement of expression of lethal damage. On the other hand, it is conceivable that some other reasons may lead to the synergistic effect of ionizing radiation and ultrasound. As far back as 1948, Conger [9] has suggested that the mechanical vibrations caused by ultrasound might prevent the rejoining of the chromosome breaks produced by X-rays. A third alternative also deserves mention. As was pointed out by Martins et al. [39], if ultrasound causes membrane damage and ionizing radiation causes nuclear damage, the synergism may result from an interaction of the two types of damage. These possibilities need not be mutually exclusive, and it is quite possible that other mechanisms could operate in place of, or along with, these, depending on organism.

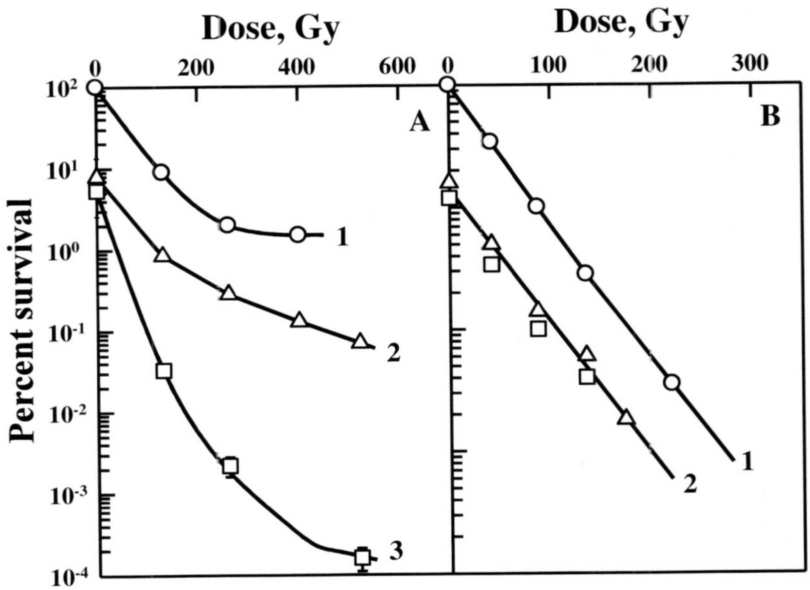

Figure 6.31. Survival of *S. cerevisiae* haploid strains of wild-type (A, strain S288C) and radiosensitive mutant (B, strain XS774-4d) after combined action of ultrasound and ionizing radiation. Curves 1, γ-rays alone; curves 2, (US + γ-rays)-treatment; curves 3, (γ-rays + US)-treatment.

It has been found that there was a non-thermal and non-cavitational component to the cell killing resulting from the combined heat and ultrasound treatment [17]. It is not excluded that this effect may be resulted from the synergistic interaction of these modalities. It is a well known fact that ultrasound is used as a technique for the induction of local heating [19]. In this case the synergistic interaction of heat and ultrasound is also very important. Taking these data as a whole, one may conclude that the main advantage of hyperthermia appears to display in the synergistic interaction of damages produced by heat and ultrasound or other modalities. A combination of agents is said to show the synergistic interaction when the magnitude of its effect is greater than expected, expectation being based on assuming the effects of the individual agents to be independent [35; 68].

To describe the synergistic effects a simple mathematical model has been described above. The model predicts the dependence of synergistic interaction on the ratio of lethal events produced by every agent, the greatest synergistic interaction, and conditions under which the maximal interactive effect can be achieved. Of special interest is the model conclusion that under simultaneous action the less intensity of one of the agent applied the less intensity of another factor has to be used to obtain the highest value of synergism. Several lines of evidence confirm this conclusion. In experiments with yeast cells it was shown above that the synergistic effect of a simultaneous action of ionizing radiation or UV light with hyperthermia was a function of dose rate. It was demonstrated that, with decreasing dose rate to obtain maximum synergistic effect, the temperature at which radiation is delivered should be diminished. Similar data are unknown for ultrasound combined with hyperthermia. It is therefore of interest to establish whether there is any influence of ultrasound intensity on the synergistic interaction of ultrasound and heat. In the work reported here we describe the role of intensity in synergistic interaction of ultrasound and hyperthermia for yeast cells.

Wild-type diploid yeast of *S. cerevisiae* (strain XS800) was used in these experiments. Aliquots of 10^6 cells/ml were placed in a metal vessel and exposed simultaneously to graded time intervals of ultrasound and hyperthermia. Continuous mode of sonication was achieved by a 20 kHz ultrasonic unit (Fisher sonic dismembrator Model 300). The ultrasonic dose rates were 0.053 and 0.195 W/cm^2. The ultrasonic intensity was measured by calorimetric method. To estimate the effect induced by heating alone, the following method was used: 0.1 ml of cell suspension (10^8 cells/ml) at room temperature was put into 9.9 ml of sterile water preheated to the required temperature by a circulating water. This preheated water was placed into a metal vessel which was situated together with the transducer. The same vessel and procedure of cell suspension preparation were used for ultrasonic irradiation. To determine the effect induced by ultrasound alone, the absorbed ultrasound heat was completely removed by cooling with water. To evaluate the effect of simultaneous action of hyperthermia and ultrasound the same heating method was performed simultaneously with sonication. The required temperature was adjusted before experiments with cell suspension and controlled within ±0.2 °C by water circulating around the vessel both for heating applied alone or combined with ultrasound.

Figures 6.32 and 6.33 provide experimental data obtained in our experiments for two ultrasound intensities: 0.053 W/cm^2 (Figure 6.32) and 0.195 W/cm^2 (Figure 6.33) and various temperatures at which sonication was delivered. As usual in combined experiments, three types of survival curves were obtained for any temperature and ultrasound intensity: heat treatment alone, ultrasound action that produces no measurable temperature rise owing cooling, and composite simultaneous heat and ultrasound exposure. All survival curves obtained in these experiments were linear, i.e., the survival decreased exponentially with exposure time increasing. Linear regression calculation based on one-target single-hit mathematical model was used to fit the survival curves. Each data point represents an average survival for three to six Petri dishes, each containing from 50–200 clones. Pooled results from three to five experiments were used to calculate the mean number of this parameter and its statistical errors.

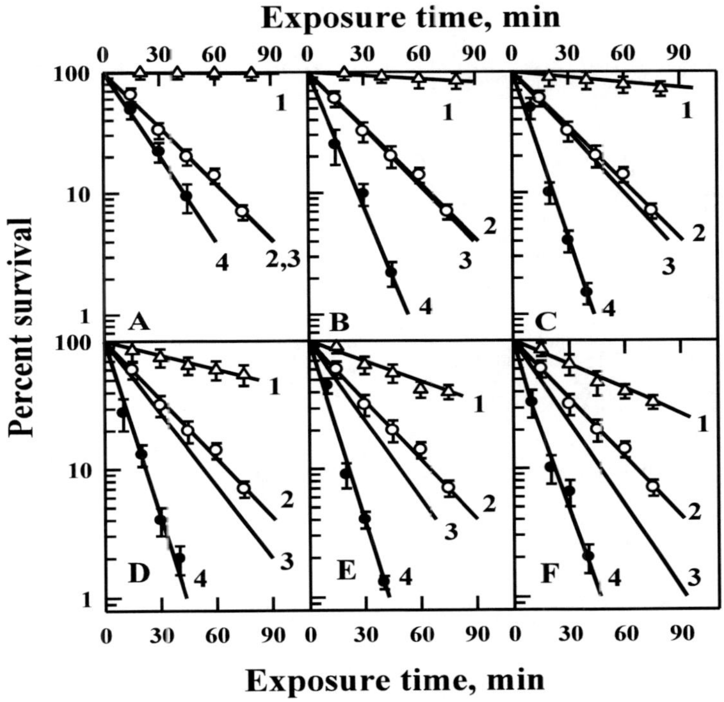

Figure 6.32. Survival curves of wild-type diploid yeast cells of *S. cerevisiae* (strain XS800) after exposure to hyperthermia (curves 1: A, 44 °C, B, 45 °C, C, 46 °C, D, 47 °C, E, 48 °C, F, 49 °C), ultrasound treatment (0.053 W/cm^2) under room temperature (17 °C, curves 2), theoretical curves after combined action of these agents expected for additive interaction of hyperthermia and ultrasound (curves 3), experimental survival curves after simultaneous action of these agents (curves 4). Error bars represent the standard error of the mean.

One can see that for most results experimental survival curves after combined action of both agents are lower than theoretical curves expected for additive sum of the effects induced by every modality. It follows that ultrasound and hyperthermia interact in a synergistic manner. All survival curves may be quantitatively described by the only parameter: D_0, the mean lethal exposure duration that reduces the cell survival to 37%. The values of D_0 derived from all experimental data are summarized in Table 6.11. This Table includes conditions of exposure, $D_0(H)$, $D_0(US)$, and $D_0(US+H)_{\exp}$, the corresponding mean lethal exposure time observed in experiments with hyperthermia alone, ultrasound treatment alone, and simultaneous heat and ultrasound application. Using these parameters, we calculated the mean lethal exposure time, $D_0(US+H)_{indep}$, expected under the independent action of both modalities in the following manner:

$$D_0(US+H)_{indep} = \frac{D_0(US)\cdot D_0(H)}{D_0(US)+D_0(H)}. \qquad (6.10)$$

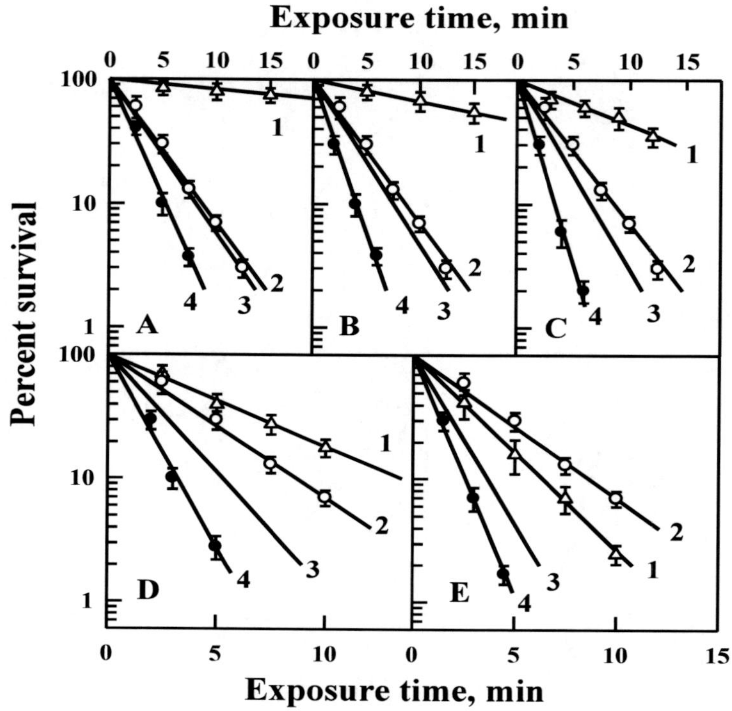

Figure 6.33. Survival curves of wild-type diploid yeast cells of *S. cerevisiae* (strain XS800) after exposure to hyperthermia (curves 1: A, 50.5 °C, B, 51.5 °C, C, 52.5 °C, D,– 53.5 °C, E, 54.5 °C), ultrasound treatment (0.195 W/cm^2) under room temperature (17 °C, curves 2), theoretical curves after combined action of these agents expected for additive interaction of hyperthermia and ultrasound (curves 3), experimental survival curves after simultaneous action of these agents (curves 4). Error bars represent the standard error of the mean.

These values are also included in Table 6.11. These data indicate that hyperthermia in conjunction with ultrasound exposure can increase the slope of experimental survival curve.

To quantify the synergistic effect, we apply the synergistic enhancement ratio k defined as the ratio of the calculated duration of treatment, assuming an independent effect of ultrasound and hyperthermia, to the duration of treatment in real experimental data after simultaneous action of ultrasound and hyperthermia at any fixed level of survival. As soon as all survival curves were exponential, this parameter doesn't depend on the survival level for which it can be calculated. So, in this study, the synergistic enhancement ratio was calculated as the ratio of the mean lethal exposure time assuming an independent sum of the effects produced by both modalities to that obtained in experiments with simultaneous action of ultrasound and hyperthermia:

$$k = \frac{D_0(US+H)_{indep}}{D_0(US+H)_{exp}}. \tag{6.11}$$

This parameter is also included in Table 6.11 and is depicted in Figure 6.34A as a function of the exposure temperature and ultrasound intensity. It is evident from the data given in this figure that, for the simultaneous action of heat and ultrasound, both temperature and ultrasound intensity affect the synergy. As can be seen, for a given intensity, there is a specific temperature that maximized the synergistic effect.

Table 6.11. Parameters of survival curves of yeast cells *S. cerevisiae*, strain XS800, exposed to heat and ultrasound alone or combined with each other; all values include the standard deviation of the estimates

Ultrasound intensity, W/cm^2	Tempe-rature, °C	$D_0(H)$, min	$D_0(US+H)_{exp}$ min	$D_0(US+H)_{indep}$ min	Synergistic enhancement ratio (k)
0.053	20	–	28.8 ± 3.0	28.8	1.00
0.053	44	2042 ± 186.0	19.0 ± 2.0	28.8	1.51 ± 0.10
0.053	45	490 ± 38.0	11.8 ± 1.3	27.8	2.37 ± 0.15
0.053	46	306 ± 31.0	9.6 ± 1.0	26.3	2.74 ± 0.15
0.053	47	122 ± 13.0	9.6 ± 0.9	23.3	2.43 ± 0.13
0.053	48	83 ± 7.4	9.4 ± 0.9	21.4	2.28 ± 0.15
0.053	49	70 ± 6.8	10.2 ± 1.0	20.4	2.00 ± 0.16
0.195	20	–	3.8 ± 0.4	3.8	1.00
0.195	50.5	54.3 ± 5.0	2.3 ± 0.3	3.5	1.53 ± 0.15
0.195	51.5	25.0 ± 3.0	1.8 ± 0.2	3.3	1.85 ± 0.13
0.195	52.5	11.9 ± 1.5	1.5 ± 0.1	2.9	1.93 ± 0.15
0.195	53.5	5.9 ± 0.7	1.4 ± 0.1	2.3	1.67 ± 0.14
0.195	54.5	2.8 ± 0.2	1.1 ± 0.1	1.6	1.47 ± 0.13

It is also clear from the data presented in Figure 6.34A that ultrasound intensity is also an important parameter and may serve as a determinant for synergy. The inserted panel (Figure 6.34B) shows the relationship between the ultrasound intensity and temperature during exposure provided the highest synergistic effect. From these findings, it can be inferred that the temperature at which ultrasound exposure is delivered should be elevated to obtain the maximum synergistic effect with increasing ultrasound intensity.

In previous Chapter, a simple mathematical model was proposed to describe the synergistic effect of simultaneous action of ionizing or UV radiation and high temperature. The model was shown to be capable of optimizing and predicting of condition under which the highest synergy can be achieved. Hence, it would be of interest to accommodate this model for description, optimization and prediction of synergistic interaction of ultrasound and hyperthermia.

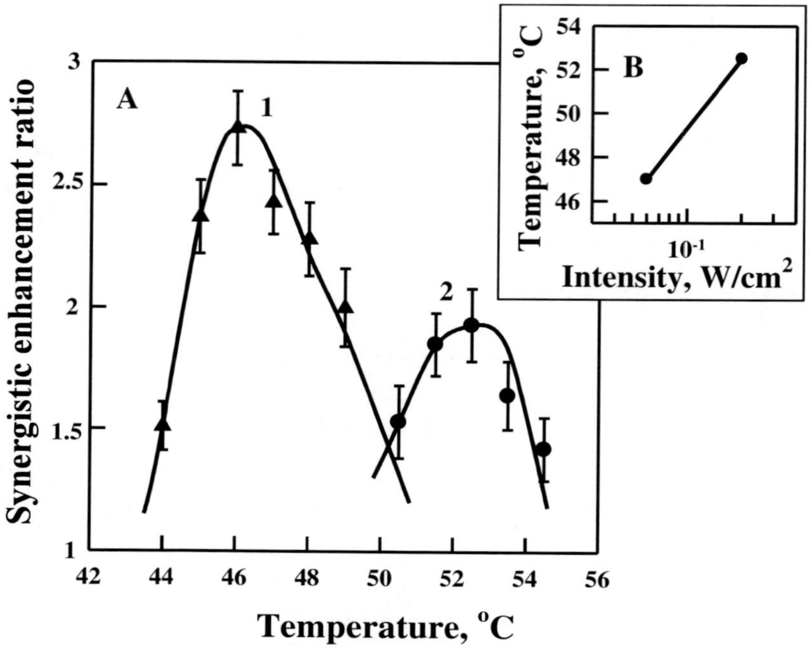

Figure 6.34. The dependence of synergistic enhancement ratio (k) upon exposure temperature and ultrasound intensity (curve 1, 0.053 W/cm^2, curve 2, 0.195 W/cm^2) for wild-type diploid yeast cells of *S. cerevisiae* (strain XS800). Error bars show inter-experimental standard errors.

Using the foregoing equations and the results described, we calculated the dependence of synergistic enhancement ratio k on the ratio N_2 / N_1 (Figure 6.35). The calculation of this parameter was described above. Here, the solid line is the prediction from our model (Eq. 5.5), and triangles denote the experimental value of k. Predicted values of k were estimated with p_1 = 1.73 and p_2 = 25.3. These values were evaluated by means of Eq. (5.14) and Eq. (5.15), and the value of k_1 and k_2 that were derived from real experiments performed at 48 and 45 °C, respectively. The maximal value of the synergistic enhancement ratio predicted by Eq. (5.11) (k_{max} = 2.62) is expected in accord with Eq. (5.9) at $N_2 / N_1 = 0.068$.

The synergistic effect appears to decline with a deviation of the ratio N_2 / N_1 from the optimal value. One can see that experimental results are comparable to the predicted values of the synergistic effect. This implies that the model considered is able to describe quantitatively the synergistic interaction of simultaneous action of ultrasound and hyperthermia on yeast cell inactivation. The degree of synergistic interaction was found to be dependent on the ratio of lethal damages (N_2 / N_1) induced by both agents applied.

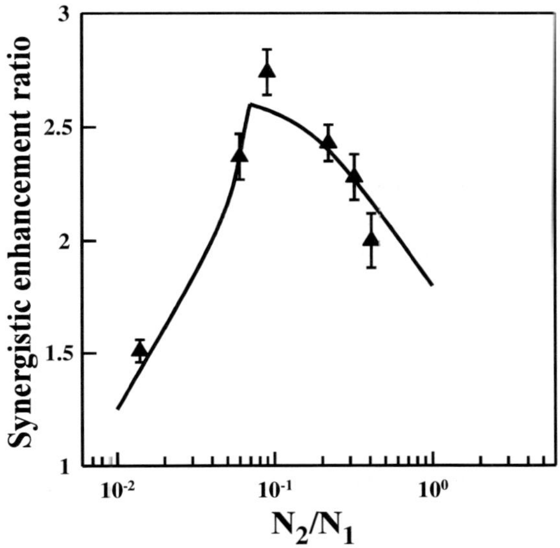

Figure 6.35. The dependence of the synergistic enchancement ratio (k) upon exposure temperature for *S. cerevisiae* diploid yeast cells (strain XS800), subjected to simultaneous action of ultrasonic and heat.

6.6. Prediction of the Thermal Enhancement Ratio after the Sequential Thermoradiation Action

The effect of sequence of interaction of hyperthermia and radiation has been the subject of a number of studies. A simple mathematical model has been described above for the synergistic interaction of two agents acting simultaneously. The model was shown above to provide a description of the observed dose rate and temperature dependences after the simultaneous action of hyperthermia with ionizing radiation, UV light radiation, ultrasound and some chemical compounds. However, the model has not been applied to predict the degree of interaction of heat and ionizing radiation after their sequential action. Extension of the model to the sequential treatment of heat and ionizing radiation seems to be of interest for theoretical and practical reasons. The influence of ionizing radiation dose rate to the effectiveness of heat and ionizing radiation interaction has been imperfectly understood [33; 34] for their sequential treatment. Thus, the purposes of the study can be formulated as follows: (i) to study experimentally the effectiveness of interaction of heat and ionizing radiation after their sequential treatment; (ii) to suggest the simplest mathematical model which would be able to account for the results obtained and the currently available experimental information.

In the experiments described here yeast cells were adopted as a model system of eukaryotic cells, not only to offer a practical advantage but also because it is possible to maintain them over a long period in a certain phase of cell cycle. Moreover, the simpler the biological system studied, the fewer complicated biological factors are involved. The wild-type diploid yeast *S. cerevisiae*, strain XS800, was used in this study. Yeast cells were incubated for 4 days at 30 °C on a complete nutrient agar layer before they were irradiated. Homogeneous cell populations were treated in the stationary phase of growth. Final suspensions contained 10^6 cells per ml. Two ^{60}Co γ-ray sources were used in this study: "Gammacell 220" (Atomic Energy of Canada LTD) and "Issledovatel" (Radon, Russia). The γ-ray dose rates, estimated by ferrous sulphate dosimetry and by a Siemens calibrated ionization chamber, were 2 and 80 Gy/min. Hyperthermia (50 and 58 °C) was administrated in a water bath in which a desired temperature ± 0.1 °C was maintained. At the end of the treatment, the samples were rapidly cooled to room temperature. The time interval between heating and irradiation was 0.5–1.0 min. All experimental series were repeated 3–5 times, and the mean survival values were calculated from at least three independent experiments. Error bars are standard deviations of the means. Exponential part of survival curves were constructed from data points using the method of least squares. We present here the averaged findings based on our previous publications [30; 49; 52].

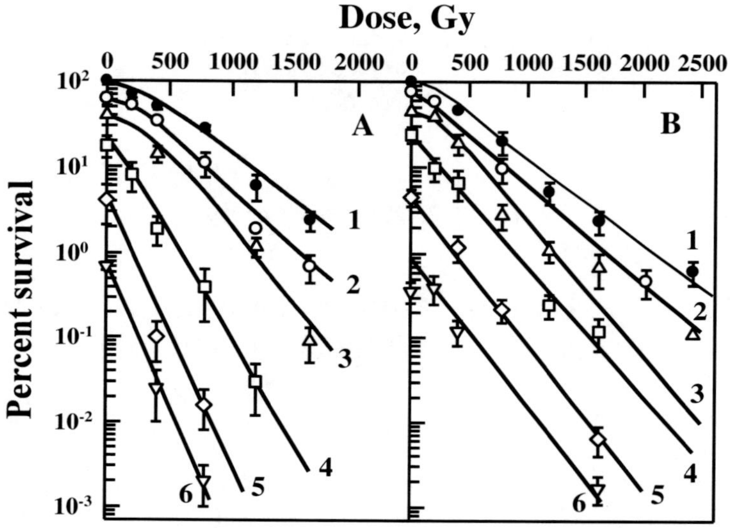

Figure 6.36. Survival curves for diploid yeast cells of *S. cerevisiae*, strain XS800. Cells were exposed to a sequential treatment with ionizing radiation (^{60}Co γ-rays, 2 Gy/min) and hyperthermia (50 °C) (A) and the reverse order of these agents (B). Durations of hyperthermia exposure were 0, 0.5, 1.5, 3.0, 6 and 9 hours, curves 1, 2, 3, 4, 5 and 6, respectively.

Figure 6.36 shows the survival curves for yeast cells irradiated with graded doses of γ-rays (2 Gy/min) applied alone or combined with heat exposure (50 °C). Cells were subjected to ionizing radiation before or after heat exposure. These data reveal some conclusions. Post heating and preheating cells markedly increased cell sensitivity to ionizing radiation. The amount of radiosensitization was quantitatively estimated by the thermal enhancement ratio (TER), calculated as the ratio of slopes. The degree of radiosensitization tends to increase with an increasing heat exposure. Figure 6.37 shows the dependence of the TER on the duration of heat exposure applied after (open circle) or before irradiation (closed circles). It is evident that in both cases there is a value of heat exposure duration (about 2–3 hours) after which no further increase in the cell radiosensitivity can be observed. Note that the sequence of the combination is very important. Ionizing radiation applied before heating was more effective than radiation after heating, i.e., a maximum of TER of 3.0 versus 1.7.

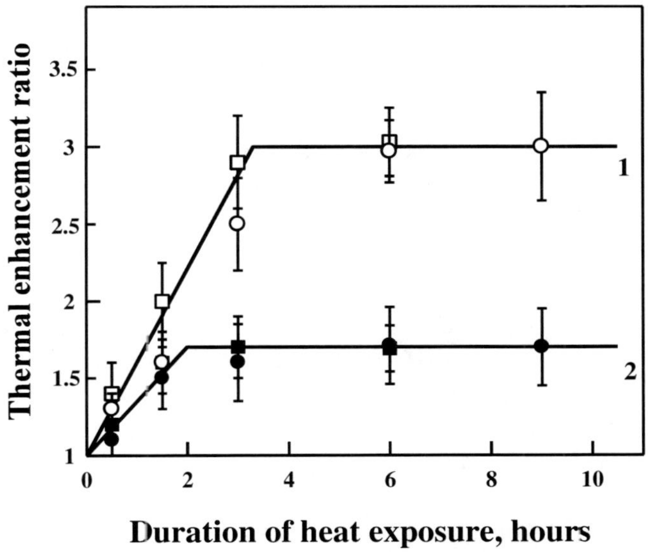

Figure 6.37. The dependence of the thermal enhancement ratio for diploid yeast cells of *S. cerevisiae*, strain XS800, on the duration of heat exposure (50 °C) after the sequential action of ionizing radiation (^{60}Co γ-rays) and hyperthermia. Curve 1, γ-rays at 2 or 80 Gy/min + 50 °C (open circles and squares, respectively). Curve 2, 50 °C + γ-rays at 2 or 80 Gy/min closed circles and squares, respectively).

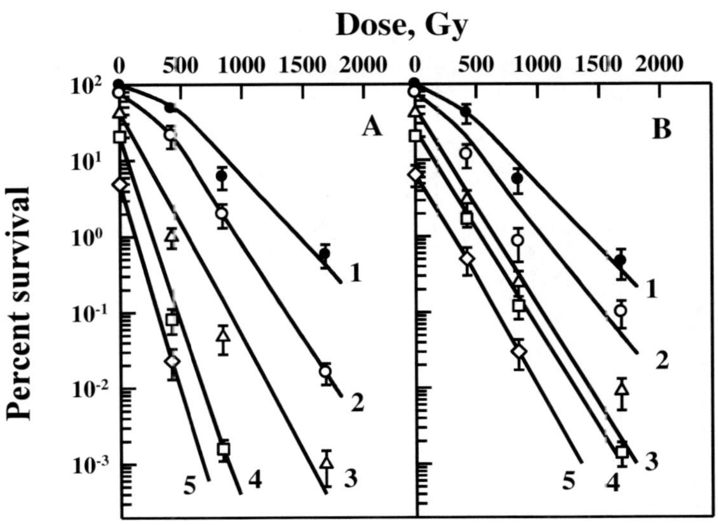

Figure 6.38. Survival curves for diploid yeast cells of *S. cerevisiae*, strain XS800, exposed to a sequential treatment with ionizing radiation (^{60}Co γ-rays, 80 Gy/min) and hyperthermia (50 °C) (A) and the reverse order of these agents (B). Durations of hyperthermia exposure were 0, 0.5, 1.5, 3.0, 6 and 9 hours, curves 1, 2, 3, 4 and 5, respectively.

Figure 6.38 shows the survival curves for the yeast cells irradiated with graded doses of γ-rays delivered at high dose rate (80 Gy/min) and applied alone or combined with the heat exposure (50 °C). Here again, hyperthermia after irradiation was much more effective than hyperthermia before irradiation. Figure 6.37 also includes these data showing the dependence of the TER on the duration of heat exposure applied after or before irradiation. It can be seen that the maximum TER was 3.0 for heating after irradiation versus 1.7 for heating before irradiation. Also, Figure 6.37 shows the TER was the same for 2 Gy/min and 80 Gy/min.

Figure 6.39 exhibits a set of survival curves for yeast cells irradiated with γ-rays (2 Gy/min) alone or combined with heat exposure (58 °C). It is evident that cells treated for 0.5 and 1.0 min after ionizing radiation induced a small amount of radioresistance. These results suggest that a certain level of cytotoxicity from hyperthermia alone is required to achieve a radioenhanced cell killing after combination of ionizing radiation at 2.0 Gy/min and acute hyperthermia (58 °C). The amount of the induced radiosensitization and radioresistance was estimated by TER.

Figure 6.40 shows the dependence of TER on the duration of heat exposure. The limit values of the TER were 1.5 independently of the sequence of combination. Figure 6.41 shows the survival curves for the yeast cells irradiated with γ-rays (80 Gy/min) applied alone or combined with heat exposure (58 °C). Figure 6.40 also includes these data. Here again, the limit values of TER was identical for treatment with γ-ray + hyperthermia and for the reverse order of treatment but their values were larger (TER = 1.9) than for low dose rate (TER = 1.5). The sequence of the combination was important for the rate of TER increase which was greater for the sequence 58 °C + γ-ray (80 Gy/min). Note that the combination of ionizing radiation either at a low or high dose rate with high temperature was never as effective as that for the greatest effect (TER= 3.0) obtained at either dose rate of radiation followed by a low temperature.

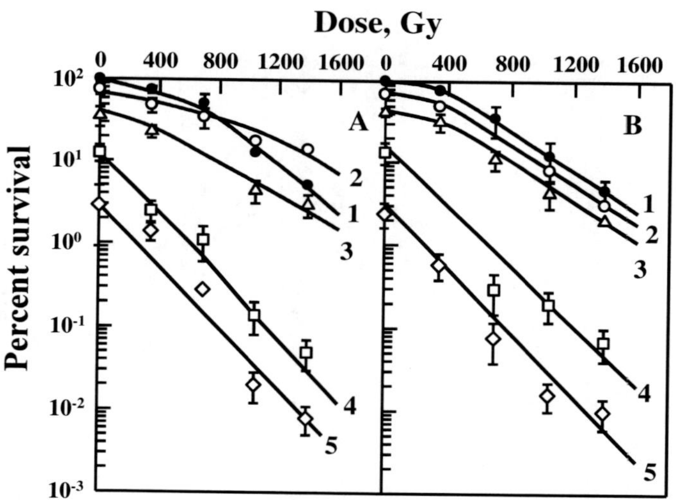

Figure 6.39. Survival curves for diploid yeast cells of *S. cerevisiae*, strain XS800, exposed to a sequential treatment with ionizing radiation (^{60}Co γ-rays, 2 Gy/min) and hyperthermia (58 °C) (A) and the reverse order of these agents (B). Durations of hyperthermia exposure were 0, 0.5, 1.0, 3.0 and 6 minutes, curves 1, 2, 3, 4 and 5, respectively.

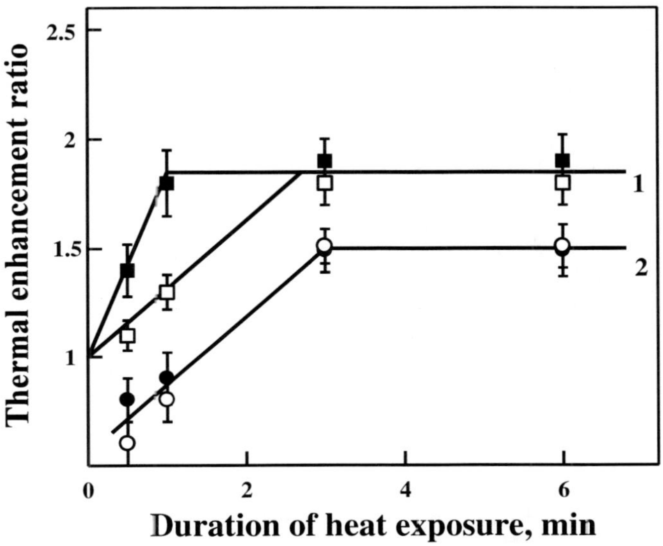

Figure 6.40. The dependence of the thermal enhancement ratio for diploid yeast cells of *S. cerevisiae*, strain XS800, on the duration of heat exposure (58 ºC) after the sequential action of ionizing radiation (^{60}Co γ-rays) and hyperthermia. Curve 1, γ-rays at 80 Gy/min + 58 °C (open squares), 58 °C + 80 Gy/min (closed squares). Curve 2, γ-rays at 2 Gy/min + 58 °C (open circles), 58 °C + γ-rays at 2 Gy/min (closed circles).

Mathematical model and its validation. We again suggest here that the thermal enhancement of ionizing radiation by hyperthermia might be expected to result from the additional lethal damage arising from the interaction of sublesions induced by both agents. Let N_1 and N_2 be the yield of some hypothetical lethal damage produced by ionizing radiation and hyperthermia, respectively. We hypothesize that the additional damage may be formed due to the interaction of sublesions induced by both agents. These sublesions are thought to be ineffective when each agent is applied separately. Let p_1 and p_2 be the mean numbers of the sublesions that arise for one effective damage induced by ionizing radiation and heat, respectively. Then the number of additional lethal damage (N_3) responsible for a thermal radiosensitization may be given by Eq. (5.3). The whole number of lethal damage (N_Σ) after combined action will be determined by

$$N_\Sigma = N_1 + N_2 + N_3 = N_1 + N_2 + \min\{p_1 N_1; p_2 N_2\}. \tag{6.12}$$

The effectiveness of cell radiosensitivity by hyperthermia was characterized in this study by the value of TER which was defined as the ratio of slope of the survival curves obtained after combined action to the slope after ionizing radiation applied alone. Inasmuch as the slope is determined by the amount of effective damage, we can write

$$TER = \frac{N_1 + N_3}{N_1}. \tag{6.13}$$

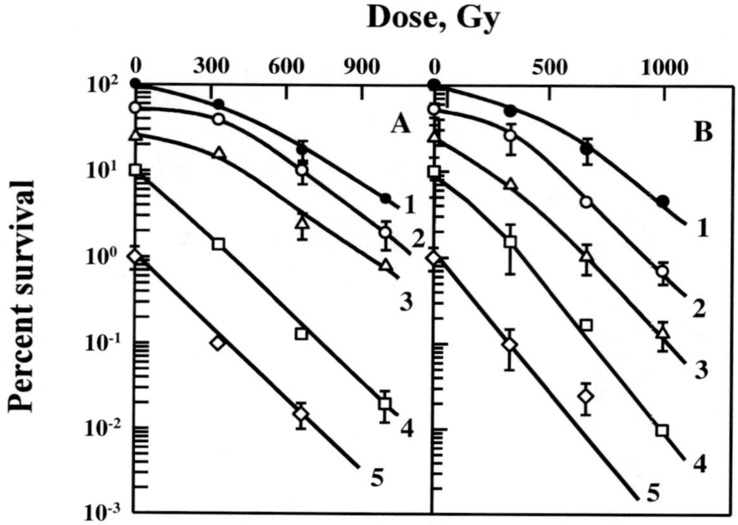

Figure 6.41. Survival curves for diploid yeast cells *S. cerevisiae*, strain XS800, exposed to a sequential treatment with ionizing radiation (^{60}Co γ-rays, 80 Gy/min) and hyperthermia (58 °C) (A) and the reverse order of these agents (B). Durations of hyperthermia exposure were 0, 0.5, 1.0, 3.0 and 6 minutes, curves 1, 2, 3, 4 and 5, respectively.

Taking into account Eq. (5.3), one can deduce

$$TER = 1 + \min\{p_1N_1; p_2N_2\} / N_1. \tag{6.14}$$

If follows from Eq. (6.14) that if $p_1N_1 > p_2N_2$ we have

$$TER = 1 + p_2N_2 / N_1. \tag{6.15}$$

Eq. 6.15 shows that the value of TER should increase with an increase in N_2, i.e., the duration of heat exposure, until the inequality $p_1N_1 > p_2N_2$ holds. This prediction is in accordance with the data obtained in this study (Figures 6.37 and 6.40). Eq. (6.15) also shows that the increase in TER growth should be greater for increased values of p_2 parameter. It follows from Eq. (6.14) that if $p_1N_1 < p_2N_2$, we have

$$TER = 1 + p_1, \tag{6.16}$$

i.e., the value of TER is a constant value from the moment when the inequality $p_1N_1 < p_2N_2$ occurs, as was observed in this study (Figures 6.37 and 6.40) at the point when the plateau in TER is reached. For this point Eq. (5.10) holds and shows that this moment corresponds to an equal number of sublesions produced by radiation and heat.

The basic parameters of the model (p_1 and p_2) can be estimated in the following way. If the inequality $p_1N_1 < p_2N_2$ is correct, we have from Eq. (6.12) that

$$p_1 = (N_\Sigma - N_1 - N_2)/N_1. \tag{6.17}$$

On the contrary, if the inequality $p_1N_1 > p_2N_2$ holds, we have from equation (6.12) that

$$p_2 = (N_\Sigma - N_1 - N_2)/N_2. \tag{6.18}$$

This mathematical formalism enables the determination of the basic model parameters p_1 and p_2 and allows for the prediction of the TER (Eq. 6.15) until the inequality $p_1N_1 > p_2N_2$ holds, i.e., the highest value of the TER (Eq. 6.16) and the condition (Eq. 5.10) at which it becomes apparent. Thus, this model would permit the interpretation of biological data in terms of the numbers of hypothetical sublesions responsible for the synergistic interaction of heat and ionizing radiation. This method was applied to the experimental data presented above. Table 6.12 includes the estimated basic model parameters and the greatest TER after different conditions of the sequential action of ionizing radiation and hyperthermia obtained in the experiments and predicted by the model described. A comparison between the theoretically predicted highest values of TER and the experimentally determined values shows a good correspondence in all cases.

Overall, we can conclude that in spite of the approximations used in the simplified model considered, the experimental results seem to be in reasonable agreement with the model predictions. The results of this study demonstrate the possibility of a reconcilable explanation and prediction of experimental data on the simultaneous and the sequential action of heat and ionizing radiation with yeast cells. One can expect the applicability of the model to predict the effectiveness of interaction of different agents applied sequentially to living cells of various origins.

Table 6.12. The dependence of the basic model parameters (p_1, p_2) and the greatest TER after different conditions of the sequential action of ionizing radiation (γ-rays) and hyperthermia (H)

Treatment conditions			Model parameters		The greatest value of TER	
Temperature, °C	Dose rate, Gy/min	Order of treatment	p_1	p_2	Experiment	Theory
50	2	γ-rays + H	1.9	4.4	3.0	2.9
50	2	H + γ-rays	0.8	2.9	1.7	1.8
50	80	γ-rays + H	1.9	4.4	3.0	2.9
50	80	H + γ-rays	0.8	2.9	1.7	1.8
58	2	γ-rays + H	0.5	0.9	1.5	1.5
58	2	H + γ-rays	0.5	0.9	1.5	1.5
58	80	γ-rays + H	0.8	0.9	1.9	1.8
58	80	H + γ-rays	0.8	1.9	1.9	1.8

CONCLUSION

In this Chapter, a simple semi-empirical mathematical model of synergistic interaction was applied to describe, predict and optimize the synergistic interaction of simultaneous action of ionizing radiation and heat to the simple microbiological objects: viruses, bacterial spores, streptococci, bacterial and yeast cells. In order to apply this model quantitatively to the experimental data, two basic parameters of the model were estimated from real experiments. They were calculated for such experimental condition, when one of the agents applied induces more sublesions than another. The model appears to be appropriate and the predictions are valid. Indeed, it was shown that the model describes the experimental data, predicts the greatest value of the synergistic effect and the condition under which it can be achieved. The remarkable aspect of these data is that there appears to be a sharp dependence of synergistic effect on the ratio of damages produced by both agents. The predicted bell-shaped dependence of synergy on the ratio of damages produced by acting modalities was confirmed for various biological objects simultaneous treated mainly with ionizing radiation and heat. It was shown that for a fixed intensity of physical or chemical agents there were specific temperatures that maximized the synergistic effect. Similarly, the dependencies of the synergistic effect on the intensity of physical factors have been obtained. It is evident that for a given temperatures, there is a specific dose rate that maximizes the synergistic interaction. Any deviation of temperature or dose rate from the optimal values results in a decrease in synergy. From the model standpoint, the highest synergy can be obtained for a specific condition when an equal number of sublesions are produced by each of the agents applied in combination. In other words, the two agents appear to be acting almost independently, with the effect of only one agent predominating. However, when the two agents are both causing considerable cytotoxicity, the interaction becomes synergistic.

An appreciable result of this Chapter is the dependence of synergy on the dose rate of applied agents which was confirmed for bacteriophage, bacterial spores and yeast cells simultaneous treated with ionizing radiation and heat. It was demonstrated that, with decreasing dose rate, to obtain maximum synergistic effect, the temperature at which radiation is delivered should be diminished.

A comparison between our different findings with yeast cells shows that the maximum synergistic enhancement ratio was almost identical (1.7 ± 0.06) for three dose rates (5, 10 and 25 Gy/min) but was less (1.46 ± 0.05) for the highest dose rate (250 Gy/ min). This difference is real and not solely due to a random variation (t = 3.07, P<0.015). A possible reason for this difference can be considered. Experimental data for the highest dose rate were obtained for temperatures above 52.5 °C. It is of interest that just at this temperature a distinct break of the Arrhenius plot for heat inactivation of this yeast strain has been demonstrated [48]. It is known that the interaction of ionizing radiation and hyperthermia at temperatures above and below the inflection point differs qualitatively. Therefore, it is possible that the nature and/or a number of the thermal sub-lesions induced by hyperthermia for temperatures above 52.5 °C may be changed, resulting in a decrease of the synergism.

It should be noted that the largest value of synergism attained in this study for yeast cells (1.82 ± 0.07) is significantly smaller in comparison with that (2.68 ± 0.15) observed when the cells were exposed to ^{60}Co γ-rays at various temperatures [47]. This difference is statistically significant [53] and also may be qualitatively interpreted by assuming that synergism is due to the interaction of sub-lesions induced by both heat and radiation. In particular, it may be associated with pulsed irradiation applied in this study. It would seem probable that some of the sub-lesions induced by either heat or radiation may be lost for the interaction during the time interval between the following pulses of radiation. As a result, a decrease of synergism should be observed.

The mechanism of cellular inactivation by heat and ionizing radiation has yet to be elucidated. There appear to be two possible explanations for increasing the overall biological effect under the combined action of these modalities. The first hypothesis is that the mechanism of the synergistic interaction of radiation and hyperthermia damage may be attributed to a reduced cellular capacity to repair radiation damage after combined action of both modalities. Another mechanism is based on the supposition that synergism may be related to additional lethal and/or potentially lethal damage arising from the interaction of sub-lesions induced by both agents, these sub-lesions being non-lethal when each agent is applied separately. Both points of view do not contradict one another. For example, if the additional lethal damage is non-repairable, the whole recovery after combined action should be reduced. This conclusion is supported by the evidence of the liquid-holding recovery for yeast cells exposed to heat combined with ionizing radiation (Figures 4.3 and 4.4) or UV light (Figure 4.11). The lack of synergism in radiosensitive mutants [48] is believed to reflect the absence of sub-lesions in such cells, perhaps because sub-lesions for wild-type cells may be lethal for mutant cells.

The molecular mode of the repair inhibition is still unclear. There is now abundant evidence pointing that hyperthermic inhibition of the rate and/or extent of repair of radiation-induced DNA damage which was observed with all the assays currently used for DNA damage assessment may reflect the production of irreversible damage rather than the impairment of recovery mechanisms themselves after the combined exposures. In fact, evidence has accumulated for cultured mammalian cells that hyperthermic radiosensitization might be due to the nuclear protein aggregation that damaged DNA becomes less accessible for the DNA repair machinery and this (rather than repair enzyme inactivation *per se*) is responsible for the observed thermal inhibition of DNA repair [26; 27; 58; 72]. Thermal repair inhibition is generally described as reduced repair rates leading to an increased number of residual strand breaks [70]. It was previously shown by Mills and Meyn [42] that for a combined treatment with X-irradiation and heat, the decrease in cell survival correlated with the increase in the number of non-rejoined strand breaks. The combined treatment appeared to affect the repair rates but also lead to the formation of additional strand breaks [12]. Residual additional strand breaks then can result in additional chromosome aberrations [11; 71] finally producing the enhanced cell killing. Considering all these data as a whole, we can conclude that they don't contradict the results obtained in this study. However, the analysis of the molecular mechanism of synergistic interaction and production of irreversible damage is beyond the main aim of this study. Moreover, it can be envisaged that they might be different for various biological objects, test effects and interactive agents. The major points to be inferred from the data presented here are that the mathematical model basing on the minimum initial postulates describes in general form the dependence of the synergistic effect on the ratio of the effective damages produced by agents used in combination, predicts the greatest synergistic interaction, condition under which it can be achieved and the portion of the irreversibly damage resulted from the combined exposures.

Our new data show the enhanced cell killing after simultaneous combined treatment with heat and ultrasound (20 kHz) in comparison with the additive effects produced by each agent applied alone. The following new peculiarities of synergistic interaction of these modalities have been revealed.

First, the synergism was not observed for any combination of ultrasound and heat and was obtained only within a definite range of the temperature at which the sonication was carried out. Secondly, for a given ultrasound intensity, there was a specific temperature at which the greatest synergistic effect was determined. Thirdly, any deviation in temperature from the optimal value resulted in a decrease in synergism. Finally, the correlation between ultrasound intensity and the temperature that ensured the maximum synergistic interaction was demonstrated. From these results, the conclusion can be drawn that the temperature at which ultrasound sonication is delivered should be elevated to obtain the maximum synergistic effect with increasing ultrasound intensity.

The crucial role of the dose rate of ionizing radiation, fluence rate of UV light and the intensity of ultrasound after their combined application with heat was demonstrated for yeast cells. The correlation between the optimal temperature that maximized the synergy and intensity of applied agents was elucidated: this temperature shifted towards lower temperature values with a decrease in the intensity. Just such responses should be expected according to the model considered. These findings are consistent with earlier extensive studies, which showed the importance of the dose rate of the ionizing radiation for the synergistic effect of simultaneous thermoradiation action for viruses [66], bacterial spores [4], and mammalian cells [3]. Taking these data and the results described in this Chapter as a whole, one can conclude the significance of the time factor under the simultaneous action of two harmful modalities. These results seem to be related to the fact that, if the ultrasound (or other agents) intensity increases, then the lethal dose is delivered in a short time, so the duration of heat treatment decreases and could explain the higher temperatures that should be applied in the cells. It may be concluded that, for a short duration of exposure, relatively higher temperatures and intensities of any agents mentioned should be used to provide the greatest synergy. On the contrary, for a long duration relatively smaller temperature and intensities must be used.

The present studies provide information for yeast cells concerning the dependence of the TER after the sequential action of heat and ionizing radiation on the dose rate of ionizing radiation as well as on the temperature and duration of heating. The results obtained were described and interpreted by means of a mathematical model in accordance with which the synergistic interaction of heat and ionizing radiation is expected to result from the additional lethal damage arising from the interaction of the sublesions induced by both agents. Since the magnitude of synergistic effect is directly dependent on the number of sublesions available, temporal effect such as a sublesion repair would contribute to a change in the effectiveness of interaction; therefore, a simultaneous exposure should be more effective than sequential one [74]. The decrease of synergy, when the effect of one agent predominates, simply reflects a small number of sublesions produced by other agent.

The major point to be inferred from the experimental data presented is that the extent of thermal radiosensitization was the greatest when the prolonged heat exposure (50 °C) was applied after irradiation than when applied before irradiation. The effectiveness of thermal radiosensitization at high dose rate was the same (TER = 3) as at a forty-fold lower dose rate. These observations suggest a longer preservation of sublesions produced by ionizing radiation than sublesions produced by hyperthermia. These observations suggest the elimination and/or repairing of thermal sublesions during the protracted heat exposure, and that similar elimination of thermal sublesions was not observed when the cells were exposed to ionizing radiation followed by a prolonged heat exposure, since the produced thermal sublesions could immediately interact with the long-lived sublesions produced by ionizing radiation.

The existence of the plateau in the curve describing the TER dependence on the duration of heat exposure may be brought about by an exhaustion of the ionizing radiation sublesions. It is worth noting that several studies have shown that there was an upper limit to the synergistic interaction of different agents acting sequentially [11; 38; 59; 64].

Acute heat exposure (58 °C) with ionizing radiation delivered at 2 Gy/min or 80 Gy/min resulted in a decrease in the highest values of TER (1.5–1.9) for various sequences of heat and ionizing radiation. The decrease of this value in comparison with the highest values (TER = 3.0) obtained for post-heating at 50 °C would be a reflection of a less number of heat sublesions produced for thermal damage at 58 °C. Hyperthermia at 58 °C could produce a greater number of severe thermal lesion, but a smaller number of sublesions. The p_1 parameter was greater when ionizing radiation at any dose rate was followed by a protracted action of heat (50 °C) in comparison with the heat at 58 °C (1.9 versus 0.5–0.8). The most likely explanation for these results is that the decrease in both the TER and p_1 after the acute thermal exposure (58 °C) might be attributed to a limitation of time needed for an interaction of the sublesions. A moderate decrease in the highest value TER for the acute heat action (58 °C) and the low dose rate compared with the high dose rate (1.9 versus 1.5) may be partly related with an antagonistic interaction of these agents clearly observed after small duration of heat. This effect, known as a thermal induction of radiation resistance, was described in literature [5; 43] and is beyond the main aim of this study.

It is worth noting that our data for yeast cells support studies with mammalian cells [11; 18; 21; 34 and many others]. Specifically, for high temperature where there is more killing from heat alone than for lower temperature where there is less killing from heat alone, cell killing for heat combined with ionizing radiation is independent of the order of treatment, i.e., heat before irradiation give the same killing as heat after irradiation. Our data make it clear that for the lower temperature the sequence γ-ray + heat give more cell killing than the inverse order of treatment, with the amount of cell killing for heat after exposure to ionizing radiation being greater for 50 °C than for 58 °C. Similar inferences were made for mammalian cells. It was concluded in just cited papers that when the lethal effect from hyperthermia alone is minimal, lethality from combined action of ionizing radiation and thermal treatment is greater when hyperthermia follows irradiation than when hyperthermia precedes irradiation. It indicates the general responses of yeast and mammalian cells to the combined action of heat and ionizing radiation.

Despite the fact that all data were received for unicellular organisms and exposure temperatures were far beyond the physiological range, the data may, in principle, serve as an indicator that low intensities of harmful agents existing in the biosphere may synergistically interact with each other or environmental heat even at physiological temperatures of homoiothermal animals and men and thereby to increase their inactivating and genetic consequences. From here, the biological meaning and significance of the results obtained in this study are readily apparent.

Since the basic model postulates do not concern the real biochemical mechanism, it is expected the model can be useful for description of synergistic effect of other physical and chemical agents combined with each other, providing the methodology of optimization and prediction of some non trivial results of synergistic interaction. It may be concluded on this basis that the general relationship for synergistic interaction of various agents may be exist for various biological systems. The finding that these patterns are so general and are seen for a set of completely different agents that kill cells *via* totally different mechanisms implies that a fundamental process of interaction must be involved.

Therefore, basic results of this Chapter may be compactly summed as follows.

1. Synergistic responses of viruses, *Bacillus subtilis* spore, *Streptococcus* and bacteria *Escherichia coli* were analyzed. In accordance with mathematical model, the synergistic effect should be observed within a definite ratio of damages produced by interacting agents. The model predictions were in an agreement with experimental results – the value of the greatest synergistic effect and condition under which it can be observed.
2. Peculiarities of the interaction of ionizing radiation and heat in yeast cells irradiated with various dose rates of ionizing radiation have been explored. It was demonstrated that, with decreasing dose rate to obtain maximum synergistic effect, the temperature at which radiation is delivered should be decreased. Optimum inactivation may be achieved under either a fixed dose rate with variation of exposure temperature, or *vice versa*.
3. Fluence rate significance in the synergistic interaction of UV light and heat was clearly demonstrated both for survival and recombination of yeast cells. Here again, with fluence rate decreasing the temperatures at which the synergy can be observed should be shifted to lower temperatures and *vice versa*.
4. The crucial role of ultrasound intensity in synergistic interaction of ultrasound and hyperthermia was demonstrated for yeast cells. The correlation between the optimal temperature that maximized the synergy and intensity of ultrasound was elucidated: this temperature shifted towards lower temperature values with a decrease in the intensity and *vice versa*.
5. The common pattern of the significance of the dose rate of ionizing radiation, the fluence rate of UV light, the intensity of ultrasound as well as the concentration of chemical agents in their synergistic interaction with heat indicates that for a long duration of interaction, which is important for problems of safety and health effects, one would expect a synergistic interaction of these factors with ambient temperatures. These data imply that the synergistic interaction may be important at low intensities of the acting agents really observed in biosphere.
6. The peculiarities of the sequential thermoradiation action on yeast cells after sharply differing dose rates and exposure temperatures have been ascertained. The effectiveness of treatment with heat and ionizing radiation was dependent on the duration of heat exposure. The greatest effect was observed for the prolonged heat exposure after ionizing radiation independently of dose rate. This observation testifies to a longer preservation of sublesions produced by ionizing radiation than sublesions produced by hyperthermia.

REFERENCES

[1] Ansari, AS; Ali, R. Synergistic action of ultraviolet radiation and hydrogen peroxide on citrulline. *J. Radiat. Res.*, 1984; 25, 283-289.

[2] Beam, CA; Mortimer, RK; Wolfe, RG et al. The relation of radioresistance to budding in *Saccharomyces cerevisiae. Archs. Biochem. Biophys.*, 1954; 49, 110-122.

[3] Ben-Hur, E. Mechanisms of the synergistic interaction between hyperthermia and radiation in cultured mammalian cells. *J. Radiat. Res.,* 1976; 17, 92-98.

[4] Brannen, JP. A temperature- and dose rate-dependent model for the kinetics of cellular response to ionizing radiation. *Radiat. Res.*, 1975; 62, 379-387.

[5] Bresler, SE; Beketova, AG; Noskin, LA et al. Temperature-induced radioresistance of *Escherichia coli* cells. *Radiobiology*, 1984; 24, 579-83.

[6] Burns, VW. Effects of sonic irradiation on yeast. *Radiat. Res.*, 1967; Suppl. 7, 231-238.

[7] Chang, PC; Li, HY; Tang, HJ et al. *In vitro* synergy of baicelein and gentamicin against vancomycin-resistant *Enterococcus. J. Microbiol. Immunol. Infect.*, 2007; 40, 56-61.

[8] Clarke, PR; Hill, CR. Physical and chemical aspects of ultrasonic disruption of cells. *J. Acoust. Soc. Amer.*, 1970; 47, 649-653.

[9] Conger, A.D. The cytogenetic effect of sonic energy applied simultaneously with X-rays. *Proc. Nat. Acad. Sci. U.S.*, 1948; 34, 470.

[10] Craig, AG; Tyler JM; Kay R. Combined ultrasonic and gamma-irradiation of *E. coli B. Int. J. Radiat. Biol.*, 1977; 22, 485-490.

[11] Dewey, WC; Sapareto, SA; Betten, DA. Hyperthermic radiosensitization of synchromous Chinese hamster cells: Relationship between lethality and chromosomal aberrations. *Radiat. Res.*, 1978; 76, 48-59.

[12] Dikomey, E; Jung, H. Correlation between thermal radiosensitization and slowly rejoined DNA strand breaks in CHO cells. *Int. J. Radiat. Biol.*, 1995; 8(3), 227-233.

[13] Dugan, VL. A kinetic analysis of spore inactivation in composite heat and gamma radiation environment. *Space Life Sci.*, 1971; 2, 498-505.

[14] Dugan, VL; Trujillo, R. Heat-accelerated radioinactivation of attenuated poliovirus. *Radiat. Environ. Biophys.,* 1975; 12, 187-195.

[15] Frankenberg, D. A ferrous sulphate dosimeter independent of photon energy in the range from 25 keV up to 50 MeV. *Phys. Med. Biol.*, 1969; 14, 597-605.

[16] Gerner, EW; Leith, JT; Boone, ML. Mammalian cell survival response following irradiation with 4 MeV X-rays or accelerated helium ions combined with hyperthermia. *Radiology*, 1976; 119, 715-720.

[17] ter Haar, G; Stratford, IJ. Evidence for non-thermal effect of ultrasound. *Br. J. Cancer*, 1982; 45 (Suppl. 5), 172-175.

[18] Hahn, GM. *Hyperthermia and Cancer*. New York: Plenum Press, 1982.

[19] Hand, JW; ter Haar, G. Heating techniques in hyperthermia. *Br. J. Radiol.*, 1981; 54, 443.

[20] Haynes, RH. Molecular localization of radiation damage relevant to bacterial inactivation. In: Angenstein L, Mason R, Rosenberg B editors. *Physical Processes in Radiation Biology*. New York: Academic Press, 1964; 51-72.

[21] Henle, KJ; Leeper, DB. The modification of radiation damage in CHO cells by hyperthermia at 40 and 45 °C. *Radiat. Res.*, 1977; 70, 415-424.

[22] Hoerter, J; Eisenstark, A. Synergistic killing of bacteria and phage by polystyrene and ultraviolet radiation. *Environ. Molec. Mutagen.*, 1988; 12, 261-264.
[23] Holmberg, M; Strausmains, R. The repair of chromosome aberrations in human lymphocytes after combined irradiation with UV-radiation (254 nm) and X-rays. *Mutat. Res.*, 1983; 120, 45-50.
[24] Jagger, J. *Introduction to Research in Ultraviolet Photobiology*. NJ: Prentice Hall, Englewood Cliffs, 1967.
[25] Joshi, DS; Haveman, I; Barendsen, GW. Influence of hyperthermia on the effectiveness of UV-radiation for induction of reproductive death of cultured mammalian cells. *Ind. J. Exp. Biol.*, 1984; 22, 248-250.
[26] Kampinga, HH. Cell biological effects of hyperthermia alone or combined with radiation or drugs: a short introduction to newcomers in the field. *Int. J. Hyperthermia*, 2006; 22, 191-196.
[27] Kampinga, HH; Dynlacht, JR; Dikomey, E. Mechanism of radiosensitization by hyperthermia (> or = 43 degrees C) as derived from studies with DNA repair defective mutant cell lines. *Int. J. Hyperthermia*, 2004; 20, 131-139.
[28] Kappos, A; Pohlit, W. A cybernetic model for radiation reactions in living cells. I. Sparsely ionizing radiation, stationary cells. *Int. J. Radiat. Biol.*, 1972; 22, 51-65.
[29] Kiefer, J; Kraft-Weyrather, W; Hlawica, MM. Cellular radiation effects and hyperthermia. Influence of exposure temperatures on survival of diploid yeast irradiated under oxygenated and hypoxic conditions. *Int. J. Radiat. Biol.*, 1976; 30, 293-300.
[30] Kritsky, RO; Belkina, SV; Zhurakovskaya, GP et al. Prediction of synergistic effects of sequentional thermoradiation action. *Radiation and Risk*, 2008; 17(4), 55–66.
[31] Kurpeshev, DC; Lebedeva, TV; Svetitsky, PV et al. *The Experimental Basis for the Use of Hyperthermia in Oncology*. Rostov-on-Don: "NOK", 2005.
[32] Kuzin, AM. *Synergism in Radiobiology*. Edition. Pushchino, 1990.
[33] Little, JB. Low-dose radiation effects: interaction and synergism. *Health Phys.*, 1990; 59, 49-55.
[34] Loshek, DD; Orr, JS; Solomonidis, E. Interaction of hyperthermia and radiation. *Br. J. Radiol.*, 1977; 50, 893-901.
[35] Loverock, P; ter Haar, G. Synergism between hyperthermia, ultrasound and γ irradiation. *Ultrasound Med. Biol.*, 1991; 17(6), 607-612.
[36] Maehara, H; Iwami, Y; Mayanagi, H et al. Synergistic inhibition by combination of fluoride and xylitol on glycolysis by mutans streptococci and its biochemical mechanism. *Caries Res.*, 2005; 39, 521-528.
[37] Mark, U; Tevini, M. Combination effects of UV-B radiation and temperature on sunflower (*Helianthus annuus L.*, cv. Polstar) and maize (*Zea mays* L., cv. Zenit 2000) seedlings. *J. Plant Physiol.*, 1996; 148, 49-56.
[38] Martignoni, KD; Smith, KC. The synergistic action of ultraviolet and X-radiation on mutants of *Escherichia coli* K-12. *Photochem. Photobiol.*, 1973; 18, 1-8.
[39] Martins, BI; Raju, MR; Hayes, TL et al. Survival of cultured mammalian cells exposed to ultrasound. *Radiat. Environ. Biophys.*, 1977; 14(3), 243-250.
[40] Matsumoto, M; Takagi, H; Yoshimura, N. Synergistic suppression of retinal pigment epithelial cell proliferation in culture by radiation and hyperthermia. *Invest. Opthalmol. Visual Sci.*, 1993; 34, 2068-2073.

[41] Melloni, E; Marchesini, R; Emanuelli, H et al. Hyperthermal effects in phototherapy with hematoporphyrin derivative sensitization. *Tumori*, 1984; 70, 321-325.

[42] Mills, MD; Meyn, RE. Effects of hyperthermia on repair of radiation-induced DNA strand breaks. *Radiat. Res.*, 1981; 87, 314-328.

[43] Mitchell, RE; Morrison, DP. Heat-shock induction of ionizing radiation resistance in *Saccharomyces cerevisiae*, and correlation with stationary growth phase. *Radiat. Res.*, 1982; 90, 284-291.

[44] Moustacchi, E; Waters, R; Heude, M et al. The present status of DNA repair mechanisms in UV irradiated yeast taken as a model eukaryotic system. In: Nygaard OF, Adler HI, Sinclai WK editors. *Radiation Research Biochemical, Chemical and Physical Perspectives.* New York: Academic Press, 1975; 632-649.

[45] Nakai, S; Matsumoto, S. Two types of radiation-sensitive mutants in yeast. *Mutat. Res.*, 1967; 4, 129-136.

[46] Pakhomova, ON; Tsyb, TS. Mutagenous effect and the mitotic crossing-over induction in yeasts under combined exposure to alpha-particles and gamma-rays. In: Chadwick KH, Cox R, Leenhouts HP, Thacker J editors. *Molecular Mechanisms in Radiation Mutagenesis and Carcinogenesis.* Brussels: European Commission, 1994; 203-206.

[47] Petin, VG; Berdnikova, IP. Effect of elevated temperatures on the radiation sensitivity of yeast cells of different species. *Radiat. Environ. Biophys.*, 1979; 16, 49-61.

[48] Petin, VG; Berdnikova, IP. Responses of yeast cells to heat applied alone or combined with gamma-rays. *Int. J. Radiat. Biol.*, 1981; 39, 281-290.

[49] Petin, VG; Kim, JK; Zhurakovskaya, GP et al. Some peculiarities of the sequential action of heat and ionizing radiation on yeast cells. *Int. J. Hyperthermia*, 2009; 25(1), 72-78.

[50] Petin, VG; Komarov, VP; Skvortzov, VG. Combined action of ultrasound and ionizing radiation on yeast cells. *Radiat. Environ. Biophys.*, 1980; 18, 45-55.

[51] Petin, VG; Pohlit, W. Influence of dose rate on survival and recovery of yeast cells. *Radiobiology*, 1969; 9, 492-499.

[52] Petin, VG; Zhurakovskaya, GP; Komarova, LN. *Radiobiological Basis of Synergistic Interaction in Biosphere*. Moscow: GEOS, 2012.

[53] Petin, VG; Zhurakovskaya, GP. The peculiarities of the interaction of radiation and hyperthermia in *Saccharomyces cerevisiae* irradiated with various dose rates. *Yeast*, 1995; 11(6), 549-554.

[54] Reddy, NM; Rao, BS; Murthy, MS. Liquid holding recovery in stationary and log phase culture of diploid yeast exposed to gamma and alpha radiations. *Rad. Environ. Biphys.* 1976; 13, 167-175.

[55] Reynolds, MC; Brannen, JP. Thermal enhancement of radiosterilization. *In: Radiation Preservation of Food.* Vienna: International Atomic Energy Agency, 1973; 165-176.

[56] Reynolds, MC; Garst, DM. Optimizing thermal and radiation effects for bacterial inactivation. *Space Life Sciences*, 1970; 2, 394-399.

[57] Rogers, AH; Bert, AG. Effects of xylitol and fluoride on the response to glucose pulses of *Streptococcus* mutans T8 growing in continuous culture. *Oral Microbiol. Immunol.*, 1992; 7, 124-126.

[58] Sakkers, RJ; Filon, AR; Brunsting, JF et al. Selective inhibition of repair of active genes by hyperthermia is due to inhibition of global and transcription coupled repair pathways. *Carcinogenesis,* 1995; 16, 743-748.

[59] Sapareto, SA; Hopwood, LE; Dewey, WC. Combined effects of X-irradiation and hyperthermia on CHO cells for various temperatures and orders of application. *Radiat. Res.*, 1978; 73, 221-33.

[60] Scheie, AA; Assev, S; Rölla, G. Combined effect of xylitol, NaF and $ZnCl_2$ on growth and metabolism of *Streptococcus sobrinus* OMZ 176. *APMIS*, 1988; 96, 761-767.

[61] Shore, RE. Overview of radiation-induced skin cancer in humans. *Int. J. Radiat. Biol.*, 1990; 57, 809-827.

[62] Soyfer, VN. Influence of physiological conditions on DNA repair and mutagenesis in higher plants. *Physiol. Plant.* 1983; 58, 373-380.

[63] Stapleton, AE. Ultraviolet radiation and plants: burning questions. *Plant Cell*, 1992; 4(11), 1353-1358.

[64] Streffer, C; Vaupel, P; Hahn, G. *Biological Basis of Oncologic Thermotherapy*. Berlin, Heidelberg, New York, London, Paris, Tokyo, Hong Kong: Springer, 1990.

[65] Todd, PW; Schroy, S. X-ray inactivation of cultured mammalian cells: enhancement by ultrasound. *Radiology,* 1974; 113, 445-447.

[66] Trujillo, R; Dugan, VL. Synergistic inactivation of viruses by heat and ionizing radiation. *Biophys. J.*, 1972; 12, 92-113.

[67] Tyrrell, RM. Synergistic lethal action of ultraviolet radiations and mild heat in *Escherichia coli*. *Photochem. Photobiol.*, 1976; 24, 345-351.

[68] Tyrrell, RM; Peak, MJ. Interaction between UV radiation of different energies in the inactivation of bacteria. *J. Bacteriol.*, 1978; 136, 437-440.

[69] Waldow, SM; Dougherty, TJ. Interaction of hyperthermia and photoradiation therapy. *Radiat. Res.*, 1984; 97, 380-385.

[70] Warters, RL. Persistence of radiation-induced double-strand breaks in the DNA of heated CHO cells. *Int. J. Radiat. Biol.*, 1993; 64, 669-676.

[71] Weissenborn, U; Obe, G. Modification of X-ray induced chromosome aberration frequency by pre- and postirradiation hyperthermia of human peripheral lymphocytes. *Int. J. Radiat. Biol.*, 1991; 59, 973-984.

[72] Wynstra, JH; Wright, WD; Roti, JL. Repair of radiation-induced DNA damage in thermotolerant and nonthermotolerant HeLa cells. *Radiat. Res.*, 1990; 85, 85-89.

[73] Zacharov, IA; Kovaltsova, SV; Marfin, SV. Strains of yeast *Saccharomyces cerevisiae* for studying induced mutagenesis and recombination. *Genetica,* 1979; 15, 41-48.

[74] Zaider, M; Rossi, HH. The synergistic effects of different radiations. *Radiat. Res.*, 1980; 83, 732-739.

Chapter 7

SYNERGISTIC EFFECTS DISPLAY FOR CULTURED MAMMALIAN CELLS, ANIMALS AND HUMANS

ABSTRACT

Synergistic interaction of hyperthermia with ionizing radiation or cytostatics on cultured mammalian cells is analyzed. In some temperature ranges, the synergistic enhancement ratio was shown to be increased or decreased with the increase in the exposure temperature. The dependence of the synergistic enhancement ratio of temperature shifted to lower temperatures with a decrease in dose rate of ionizing radiation. The relationship between the concentration of drugs, treatment temperatures and the synergistic enhancement ratio was established. Temperature range within which the synergistic interaction can be observed is shifted to lower temperatures with a decrease in the concentration of chemical agent. The increase in mammalian cell radiosensitivity by chemicals may be mainly related with the production of irreversible damage, which cannot be repaired, with partially impairment of the recovery capacity itself, and sometimes with the higher probability to produce primary radiation damage. To demonstrate the admissibility of the synergy model for animal, the interaction of damage produced by simultaneous action of ambient temperature and microwave was studied for temperature increment in rabbit body. For this case, the model was adjusted to describe the synergistic interaction, to determine its greatest value and the condition under which it can be achieved. The prediction of the model was shown to be consistent with experimental data on rabbit heating. Some examples of synergistic effect observed for man is presented and discussed.

INTRODUCTION

In the preceding Chapters some new common patterns of synergistic effects display have been described. They were universal and did not depend on the biological object and test being studied as well as on the agents used in combined actions. These data strongly invoked the need to elaborate a new theoretical conception of the synergy which, being useful for environmental radiation protection, took into account the regularities revealed. Such conception was advanced. In accordance with this conception, the mechanism of synergy is not related with the impairment of the recovery capacity *per se* and may be attributed to the enhanced yield of irreversible damages. A great number of experimental data obtained in this area stress the need for a mathematical approach in efforts to optimize and predict the interaction of environmental agents. It was assumed that some additional effective damage produced due to the interaction of sublesions during combined action, may be responsible for the synergistic interaction. Mathematical model based on this conception was shown to prognosticate the value of the highest synergy, condition under which it can be achieved and the dependence of synergy on the intensity of the agents applied. The model plausibility was demonstrated for a number of microorganisms. It is of obvious interest to demonstrate the universality of this model and the possibility of its application to describe and optimize the synergistic effects observed for cultured mammalian cells, animals and humans. For example, hyperthermia and radiation are known to interact synergistically in cultured mammalian cells. Data pertaining to this problem have been published by many authors [5–7; 16; 24; 25; 50]. Temperature dependence in the cytotoxicity of the chemical agents as well as the synergistic interaction of the combined action of hyperthermia with a number of chemical environmental pollutions or therapeutic agents has been reported for cultured mammalian cells [18; 29; 69 and many others]. But the authors of these papers did not describe their results in terms of a quantitative description and prediction of synergistic effects, including the prognosis of the maximum synergy, the condition of its achievements and the dependence of synergy on the intensity of the applied factors. In particular, it would be of interest to analyze the dependence of synergism on the intensity of some chemical agents. The treatment duration of chemical agents may be considered as an analog of the dose while the concentration of compound may serve as a measure of its intensity.

A lot of chemicals are known to enhance the inactivation effect of ionizing radiation on cultured mammalian cells. It is assumed that drug radiosensitization may be brought about by the inhibition of repair on cellular level including the recovery from potentially lethal radiation damage. The inhibition of cell recovery [36; 37; 42; 70] and DNA double strand breaks repair [9; 68; 73] by chemicals is expressed both as a retarded recovery rate and a lesser extent of recovery. As was mentioned above for yeast cells, these data may be interpreted by the damage or impairment of the recovery process itself or by the increase in the portion of irreversible damage. As was shown above, the inhibition of the recovery from potentially lethal damage in yeast cells exposed to hyperthermia and ionizing radiation or hyperthermia and UV light was realized only through the enhanced yield of irreversible damage whereas the recovery capacity itself was not damaged and impaired. There are no reports in the literature on a separate estimation of chemical influence on the process of recovery itself and the production of irreversible damage. Therefore, it would be of interest to carry out such study and to investigate the role of recovery inhibition in cancer research.

Synergistic interaction of different environmental factors is known at the level of animals and humans. For instance, the ambient temperature had a profound effect on the thermogenic effect of microwave radiation of various animals [21; 49] and men [2]. In these experiments, agents could interact with each other in a synergistic manner. However, nobody was capable to demonstrate the dependence of the synergistic effect on the intensity of microwaves and to show the existence of the optimal power flux density, which maximizes the synergy. These studies are important from theoretical and practical perspectives. First of all, it concerns a potentially increased health risk to the general public and occupationally exposed population. Secondly, there is a great difference in safety standard of microwave irradiation in various countries. Thirdly, these agents are explored in cancer treatment.

Synergistic interactions of various harmful agents, which could reveal for people, are also known. The ambient temperature had a profound effect on the thermogenic effect of microwave radiation of men [2]. It is known from epidemiological and toxicological studies that the interaction between tobacco smoking and radon may be synergistic [3; 12; 55]. A survey of data on the combined effect of cigarette smoke and alcohol on the formation of cancer in men was published [35]. The synergistic interaction of smoking and alcohol on the formation of esophageal cancer has been also studied [15]. The authors come to the conclusion that the combined use of alcohol and smoking may increase cancer incidence. Therefore, it is of interest to give an overview of the known synergistic effects in humans and to analyze the possibility of their interpretation with mathematical model of synergy.

Thus, the main aims of this Chapter may be formulated as follows.

1. To study the synergistic interaction of hyperthermia with ionizing radiation or cytostatics on cultured mammalian cells.
2. To elucidate the role of recovery inhibition in cancer research.
3. To estimate experimentally rabbit heating by microwave exposure at various ambient temperatures.
4. To test the possibility of the synergy model to describe and predict the results of simultaneous action of microwave and ambient temperature on rabbit heating.
5. To present an overview of the known synergistic effects to humans and to analyze the possibility of their interpretation with our model of synergy.

7.1. Synergistic Interaction of Hyperthermia with Ionizing Radiation or Cytostatics on Cultured Mammalian Cells

Heat is known to be one of the most potent modifiers of radiation effect in mammalian cells. A variety of biological objects are killed more readily by ionizing radiation under hyperthermia compared with normal conditions. Hyperthermia and radiation are known to interact synergistically in cultured mammalian cells. Data pertaining to this problem have been published by many authors [5–7; 16; 24; 25; 50]. Using the averaged results of these publications for Chinese hamster cells X-irradiated at various temperatures with different dose rates of X-ray (250 kV) and the equation presented herein, we calculated the synergistic enhancement ratio. Table 7.1 reports the results of these calculations. The synergistic enhancement ratio k was predicted by Eq. (5.5) with p_1 = 4.6 and p_2 = 30. The first value was obtained from Eq. (5.14) for cell inactivation at 42 °C and a dose rate of 0.033 Gy/min and the second one from Eq. (5.15) and experimental data for cell inactivation at 40 °C and a dose rate of 0.033 Gy/min. It is evident from the data given in Table 7.1 that the experimental values are in agreement with those predicted from the model discussed. The reason for the somewhat large deviation of theoretical results from experimental data may be found in the fact that initial data were taken from different papers and for various cell lines. One can see from the data presented in Table 7.1 that, in accordance with the model, the temperature range, synergistically increasing the effect of ionizing radiation, was shifted toward higher temperatures when dose rate was increased. More clearly it can be seen from Figure 7.1. Just such a dependence was predicted by the mathematical model of synergy presented above.

Table 7.1. Parameters defining Chinese hamster inactivation by simultaneous action of high temperature and ionizing radiation

Dose rate, Gy/min	Parameters	Temperature, °C					
		39.0	40.0	41.0	42.0	42.5	43
0.033	N_2/N_1	0.002	0.015	0.097	0.650	–	–
	k_{exp}	1.28	1.45	2.54	3.75	–	–
	k_{theory}	1.07	1.45	3.50	3.75	–	–
0.12	N_2/N_1	–	0.003	0.020	0.138	0.36	0.92
	k_{exp}	–	1.40	2.10	3.0	3.0	3.60
	k_{theory}	–	1.09	1.60	4.6	4.4	3.40
3.6	N_2/N_1	–	–	–	0.003	–	0.018
	k_{exp}	–	–	–	1.45	–	2.10
	k_{theory}	–	–	–	1.10	–	1.50

Temperature range studied was not sufficient for a clear manifestation of the maximum synergy. It is evident from Figure 7.1 that the synergistic enhancement ratio increased with the increase in the exposure temperature. The second important pattern emerging from these data is that the effectiveness of synergistic interaction depends on dose rate. It is clear from Figure 7.1A that the curves demonstrating the dependence of the synergistic enhancement ratio of temperature shifted to lower temperatures with a decrease in dose rate of ionizing radiation. To display the same synergistic effect, for example for k = 2.2, one should decrease the exposure temperature ($T_1<T_2<T_3$) with a decrease in dose rate of ionizing radiation. Figure 7.1B shows the correlation between acting temperature and dose rate providing the same synergistic effect (k = 2.2). Similar response was described above for viruses, bacterial spores and yeast cells. The finding that this pattern is general for different biological objects implies that a fundamental process of interaction is involved. Just such dependence was predicted by the mathematical model of synergy presented above.

Since chemical reactions in general are accelerated by increasing temperature, one would expect that the rate of cell injury by chemical agents would also be accelerated at elevated temperatures. Temperature dependence in the cytotoxicity of the chemical agents as well as the synergistic interaction of the combined action of hyperthermia with a number of chemical environmental pollutions or therapeutic agents has been reported [18; 29; 69 and many others]. Some data on the synergistic interactions of chemical agents with each other or with ionizing radiation and hyperthermia have already been described above. It would be of interest to analyze the dependence of synergism on the intensity of some chemical agents. The treatment duration of chemical agents may be considered as an analog of the chemical dose while the concentration of compound may serve as a measure of its intensity.

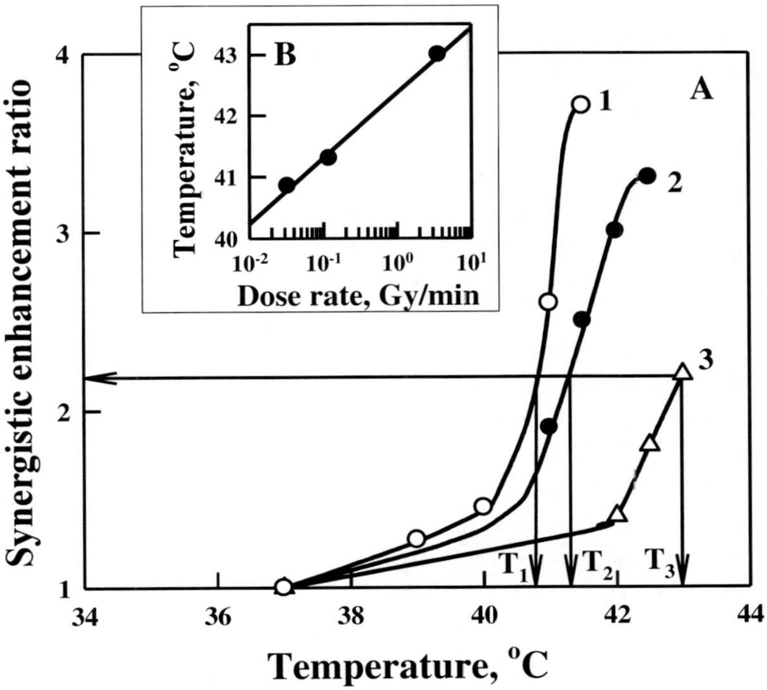

Figure 7.1. The results of the simultaneous thermoradiation action on Chinese hamster cell (CHO) inactivation. A, the dependence of the synergistic enhancement ratio on action temperature and dose rate of ionizing radiation (curve 1 – 0.0033, curve 2 – 0.12 and curve 3 – 3.6 Gy/min); B, the relationship between dose rate and temperature providing the equal synergistic effect (k = 2.2).

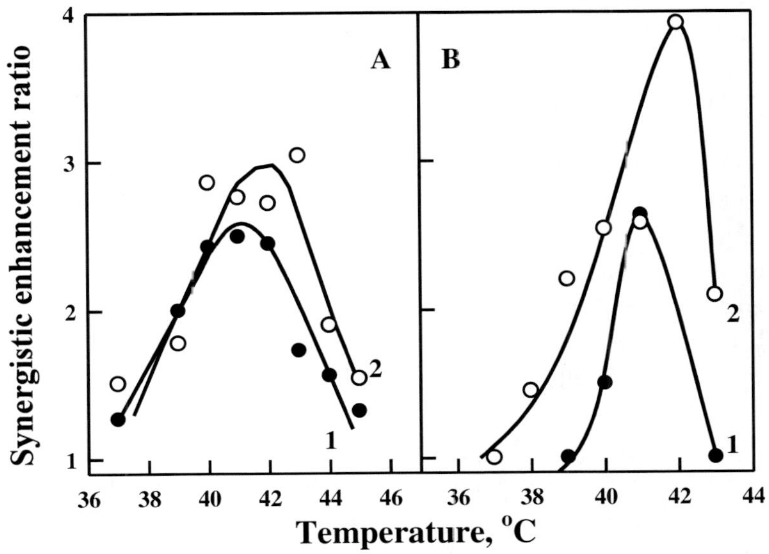

Figure 7.2. The dependence of the synergistic enhancement ratio (k) upon exposure temperature and concentration of thio-TEPA (A: curve 1, 5 μg/ml, curve 2, 10 μg/ml) and cis-DDP (B: curve 1, 3 μM, curve 2, 6 μM) for Chinese hamster cells.

Johnson and Pavelec [29] obtained the relationship between temperature and the cytotoxicity of tris(1-aziridinyl)-phosphine sulfide (thio-TEPA) *in vitro* with the inactivation of clone-forming hamster cells. Chinese hamster cells were exposed to thio-TEPA in 2 concentrations, 5 and 10 μg/ml at various temperatures. The effect of various concentrations of cis-diamminedichroloplatinum (II) (cis-DDP) and hyperthermia on cultured Chinese hamster ovary (CHO) cells was investigated by Urano et al. [69]. In both these papers, a precise and useful relationship between temperature and cytotoxic effect was defined, and in addition, the thermodynamic characteristics of the main chemical reactions involved in the cytotoxicity of the agents have been published. However, the authors of the cited papers did not estimate the dependence of synergism on the concentration (intensity) of the drugs. Using their data, including extrapolation which was made for higher temperatures, we have calculated the relationship between the concentration of drugs, treatment temperatures and the synergistic enhancement ratio. The results of the calculations are given in Figure 7.2A for thio-TEPA and in Figure 7.2B for cis-DDP.

It is apparent that all curves are bell-shaped. It can also be seen that with a decrease in concentration of both drugs the exposure temperature should also be decreased to display the highest synergistic interaction. Again, just such a dependence was predicted by the mathematical model of synergy presented above. To emphasize the general importance of these rules it would be of interest to analyze the dependence of synergism on the concentration (intensity) of another chemical compound.

In particular, cisplatin is a synthetic substance widely used in chemotherapy both individually and in combination with local hyperthermia. The results of a through study of the simultaneous combined action of cisplatin at different concentrations (3, 6 and 12 μM) and elevated temperatures (36–43 °C) on the inactivation of Chinese hamster cells (CHO) have been published [69]. However, the authors of the cited work did not use their data to estimate synergistic effect quantitatively and its dependence on the exposure temperature and cisplatin concentration. Using their data and some extrapolation, we carried out these calculations.

Figure 7.3 exhibits the dependence of the synergistic enhancement ratio on cisplatin concentration and temperature at which the simultaneous exposure of these agents on Chinese hamster cells (CHO) was carried out. We suggested that the action of cisplatin at 36 °C corresponds to the individual action of this drug. As can be seen from these data, the synergistic interaction increased up to a maximum value with increase in the exposure temperature and then the effectiveness of interaction decreases. Just such a dependence could be expected from the model of synergy presented above. This effect was observed for cisplatin concentrations of 3 and 6 μM. For concentration of 12 μM the highest synergistic effect has not been achieved.

The data presented in Figure 7.3A shows that the temperature range within which the synergistic interaction can be observed is shifted to lower temperatures with a decrease in the concentration of chemical agent. In particular, the synergistic enhancement ratio $k = 3$ (Figure 7.3A, ascending parts of the curves) is observed at 38.3, 39.8 and 40.8 °C ($T_1<T_2<T_3$) for cisplatin concentration of 3, 6 and 12 μM, respectively. This correlation between cisplatin concentration and the acting temperature providing this synergistic effect is reflected in Figure 7.3B. Obviously, the less concentration of cisplatin is applied, the less temperature will provide the specified value of synergistic enhancement ratio. Again, this dependence was

predicted by our model and has already been described above for the simultaneous application of hyperthermia with ionizing radiation, UV light and ultrasound.

Eichholtz-Wirth and Hietel [18] have obtained findings on combined action of hyperthermia (37–43 °C for 1 h) and cisplatin (0–6 μg/ml) on Chinese hamster fibroblasts (line B14F28). Having used their results we estimated the synergistic enhancement ratio (Eq. 5.5) in the dependence of the treatment conditions. The data obtained for cisplatin concentration of 1, 2, 4 and 6 μg/ml are depicted in Figure 7.4. It is clear that in all cases there is an optimal temperature at which the greatest synergy can be achieved. Any deviation of acting temperature from this optimum results in a decline of the synergy. These data also indicate that the temperature range within which the synergistic interaction can be observed is shifted to lower temperatures with a decrease in cisplatin concentration. This relationship is more evident from data represented in Figure 7.5. Just such patterns should be expected from the model of synergy presented above.

Thus, the previously identified new patterns of synergistic effect display for the simultaneous action of hyperthermia with ionizing radiation, ultraviolet light or ultrasound on cells of different origin may be extended to the simultaneous action of elevated temperature with ionizing radiation and various chemical compounds. They include the following facts. Firstly, for any constant concentration of chemical agents, the synergy can be observed only within a certain temperature range. Secondly, inside this range, there is a specific temperature that maximizes the synergistic effect. And thirdly, a decrease in the chemical concentration necessitates reducing an exposure temperature to provide the highest or any required level of synergy.

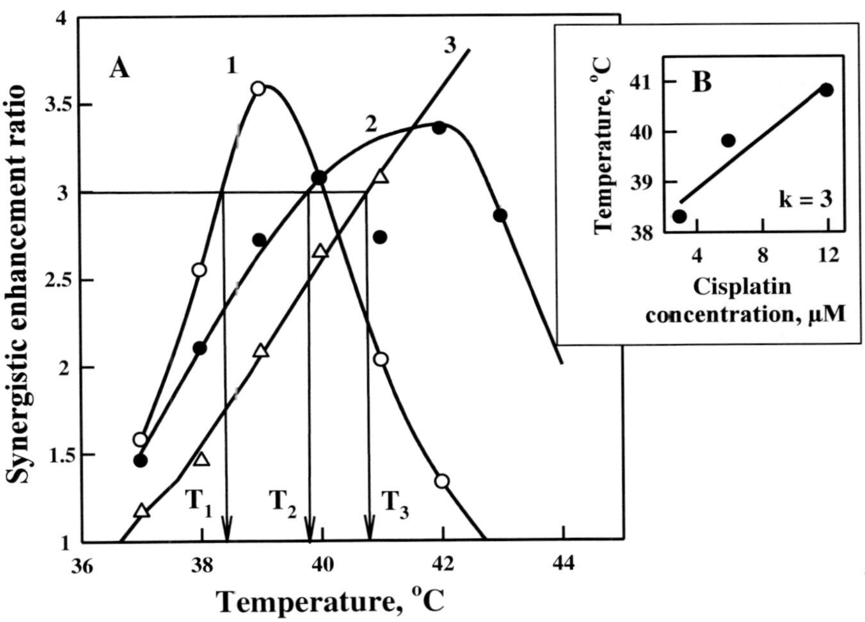

Figure 7.3. The dependence of the synergistic enhancement ratio of cisplatin (concentration 3, 6 and 12 μM, curves 1, 2 and 3, respectively) and temperature at which the simultaneous exposure of these agents on Chinese hamster cells (CHO) was carried out (panel A); B, the relationship between cisplatin concentration and temperature providing specific synergistic enhancement ratio ($k = 3$).

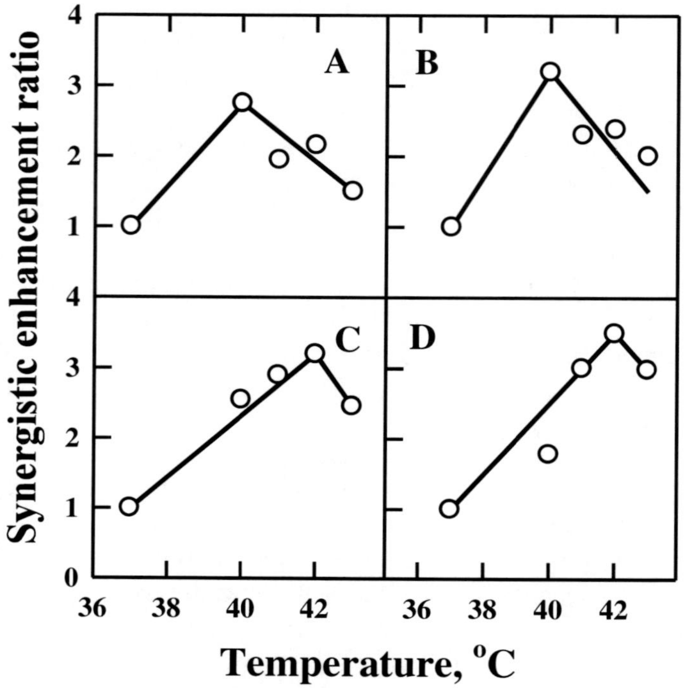

Figure 7.4. The dependence of the synergistic enhancement ratio on the acting temperature for Chinese hamster fibroblasts (line B14F28) simultaneously exposed to hyperthermia and various cisplatin concentration: 1, 2, 4 and 6 μg/ml, panels A, B, C and D, respectively.

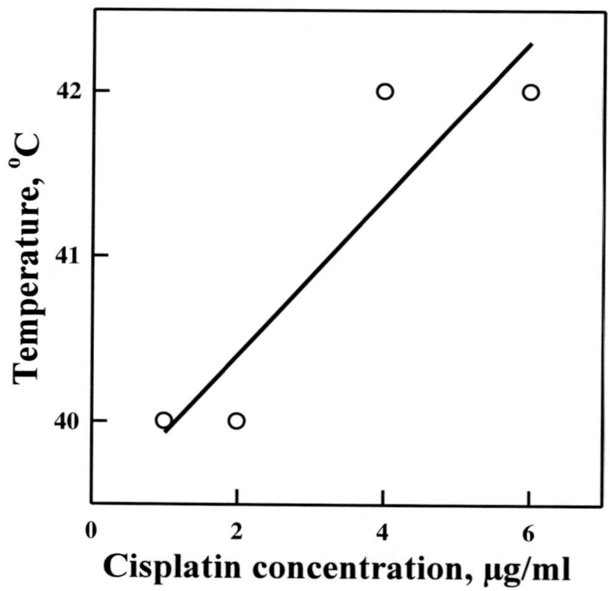

Figure 7.5. The relationship between cisplatin concentration and exposure temperature providing the greatest synergistic interaction for Chinese hamster fibroblasts (line B14F28).

7.2. THE ROLE OF RECOVERY INHIBITION IN CANCER RESEARCH

It is believed that DNA repair is one of the most important factors in determining cellular sensitivity to ionizing radiation and some other cytotoxic agents. Many tumors are known to become resistant to ionizing radiation due to the increased efficiency of DNA repair. Therefore, the impairment of cell ability to recover from radiation damage would be of great relevance in cancer treatment. A lot of chemicals are known to enhance the inactivation effect of ionizing radiation on various cellular systems. It seems generally accepted now that the enhancing effects may be due to both direct drug toxicity and to the enhancement of the cellular radiosensitivity. Analogously with the description of thermoradiation action, we assumed that drug radiosensitization may be brought about by the inhibition of repair on cellular level including the recovery from potentially lethal radiation damage. This type of recovery was demonstrated in cells of various origins and may play a role in the treatment of tumors with ionizing radiation [72]. The inhibition of cell recovery [36; 37; 42; 70] and DNA double strand breaks repair [9; 68; 73] by chemicals is expressed both as a retarded recovery rate and a lesser extent of recovery. It is obvious that these observations, as we supposed above, may be caused by the following reasons: (i) the damage or impairment of the recovery process itself, (ii) the increase in the portion of irreversible damage, (iii) both of these reasons. Although there are considerable interests in combining ionizing radiation with chemicals in order to improve tumor treatment, there were no reports in the literature on a separate estimation of chemicals influence on the process of recovery itself and the production of irreversible damage. Therefore, it would be of particular interest biophysically to implement such a study. Here, a quantitative approach describing the recovery of cell from potentially lethal damage as a decrease in the effective dose (Eq. 1.10) will be used to estimate separately the probability of recovery per unit time (Eq. 1.15) and the fraction of irreversible damage (Eq. 1.12). This approach will be applied to experimental data published by others [27; 28; 36; 37; 42; 71] and related with the inhibition of recovery from potentially lethal damage by various chemicals in cultured mammalian cells.

Figure 7.6 shows survival and recovery curves of stationary phase cells of Chinese hamster V79 cells irradiated (300 kV X-rays, dose rate being 1.25 Gy/min) and recovered without chemicals. Arrows indicate an example of the effective dose estimation. Both these curves were obtained by the averaging of the six dose-effect and four time-effect curves published by other authors [36; 37].

To determine the time course of inhibition of recovery from potentially lethal damage, immediately after irradiation with ionizing radiation stationary phase Chinese hamster cells were incubated with 0, 10, 20 mM of pyruvate, lactate or 3-aminobenzamide, 0, 5, 10, 20 μM of novobiocin or nalidixic acid during 6, 12 and 24 h before they were plated without chemicals to determine their survival by colony-forming ability. Besides, cells were treated with 20 mM of pyruvate and lactate and 20 μM of novobiocin and nalidixic acid starting 1 h after irradiation for 6, 12 and 24 h and then replated colony formation without chemicals. Chemicals were added following irradiation in order to limit their effect to the recovery period only.

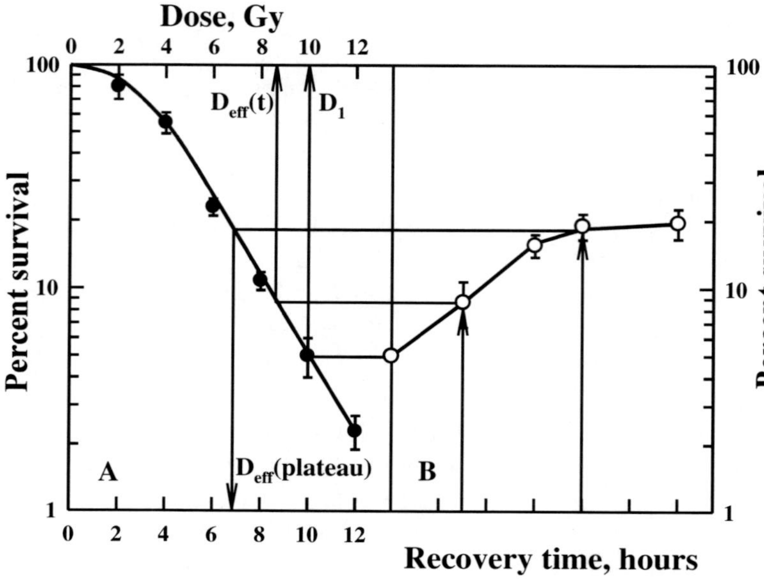

Figure 7.6. Survival of Chinese hamster V79 cells in the dependence of X-ray dose (A) and the duration of recovery from potentially lethal damage (B). Cells were X-irradiated and recovered without chemicals. Arrows indicate examples of the initial dose D_1 as well as the effective doses $D_{eff}(t)$ and $D_{eff}(plateau)$ determination. *Points*, mean; *bars*, SE.

Kinetics of the recovery from potentially lethal radiation damage showed that the survival increase due to recovery observed in the controls was gradually reduced as the chemical concentration increased, i.e., the inhibition of recovery was drug concentration dependent. Using these results and the data presented in Figure 7.6, we calculated the dependence of the relative fraction of irreversible damage $K(t)$ (Eq. 1.11) on the duration of recovery time of the Chinese hamster V79 cells recovering after irradiation without chemicals and in the presence of various chemical inhibitors of cell recovery. The results are shown in Figure 7.7.

It can be noted that the untreated cells subjected to post-irradiation recovery showed an appreciable decrease in $K(t)$ whereas this effect became gradually worse as the chemical concentration increased. It appears that the inhibition of recovery depends on drug concentration and is almost complete with 20 mM of pyruvate and lactate and 20 μM of novobiocin and nalidixic acid. For instance, the limited values of $K(t)$, i.e., the values of irreversible component are equal to 0.60, 0.75 and 0.92 for cells recovering from radiation damage without drug and in the presence of 10 and 20 mM pyruvate, respectively. Qualitatively similar results were obtained for other chemicals. Figure 7.8 shows the calculated irreversible component of radiation damage, $K(plateau)$ (Eq. 1.12), in Chinese hamster V79 cells recovering from potentially lethal damage with and without various inhibitors of recovery. The obvious increase in the irreversible component with drug concentration should certainly lead to a decrease in the recovery rate because of the decrease in a number of cells capable of recovery.

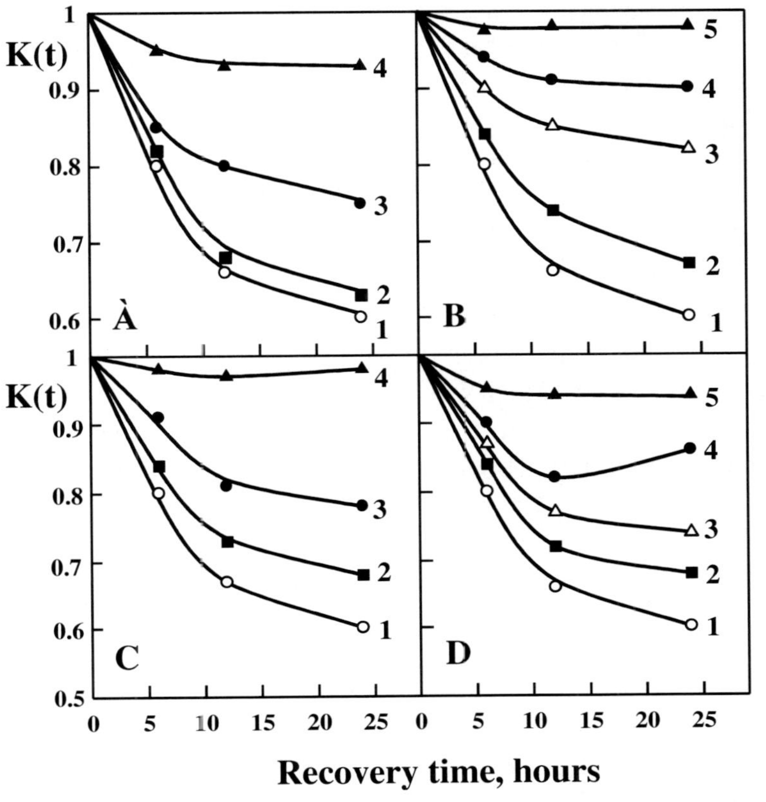

Figure 7.7. The dependence of the relative fraction of irreversible damage $K(t)$ on the duration of recovery time of Chinese hamster V79 cells recovering after irradiation without chemicals (curves 1) and in the presence of chemical inhibitors of cell recovery. A, pyruvate: 20 mM 1 h after irradiation (curve 2), 10 and 20 mM immediately after irradiation (curves 3 and 4). B, novobiocin: 20 μM 1 h after irradiation (curve 2), 5, 10 and 20 μM immediately after irradiation (curves 3, 4 and 5). C, lactate: 20 mM 1 h after irradiation (curve 2), 10 and 20 mM immediately after irradiation (curves 3 and 4). D, nalidixic acid: 20 μM 1 h after irradiation (curve 2), 5, 10 and 20 μM immediately after irradiation (curves 3, 4 and 5).

To clarify whether or not this decrease could completely explain the observable deceleration of the recovery rate, we estimated the probability of recovery for various conditions of recovery. The experimental data make it possible to calculate the function $A(t)$ (Eq. 1.14) in the dependence of the recovery time of Chinese hamster cells recovering after irradiation without chemicals and in the presence of various chemical inhibitors. The results are shown in Figure 7.9.

One can see that this function declines exponentially with the recovery time independently of whether the recovery took place without chemicals or with the increasing concentration of various drugs. By taking Eq. (1.15) and the results shown in Figure 7.9, we calculated the recovery constant β for all recovery conditions being tested. The total set of parameters describing the recovery of Chinese hamster cells under different postirradiation conditions are summarized in Table 7.2. This Table also includes the parameters describing the recovery of Chinese hamster cells in the presence of different concentrations of 3-

aminobenzamide estimated on the basis of experimental data published by others [37] but not presented here in Figures. It can be seen that in all cases the recovery constant was independent of recovery conditions (β = 0.15 ± 0.01 h^{-1}) whereas the irreversible component was gradually enhanced as the chemical concentration increased.

To illustrate the correctness of the approach used here, Figure 7.10 exhibits the survival of Chinese hamster V79 cells in dependence of the duration of recovery from potentially lethal damage recovering after irradiation without chemicals and in the presence of some chemical inhibitors of cell recovery. Experimental points were taken from [36]. Solid lines were calculated in the following way. The effective dose $D_{eff}(t)$ was computed in accordance with Eq. (1.10) and parameters of the recovery obtained in this study (Table 7.2). Then the corresponding survival $S(t)$ was estimated following the example shown in Figure 7.6 by arrows. One can see a close fit of the predicted curves to the experimental data. Thereby the applicability of the mathematical approach to describe the recovery from potentially lethal damage was demonstrated for cultured mammalian cells.

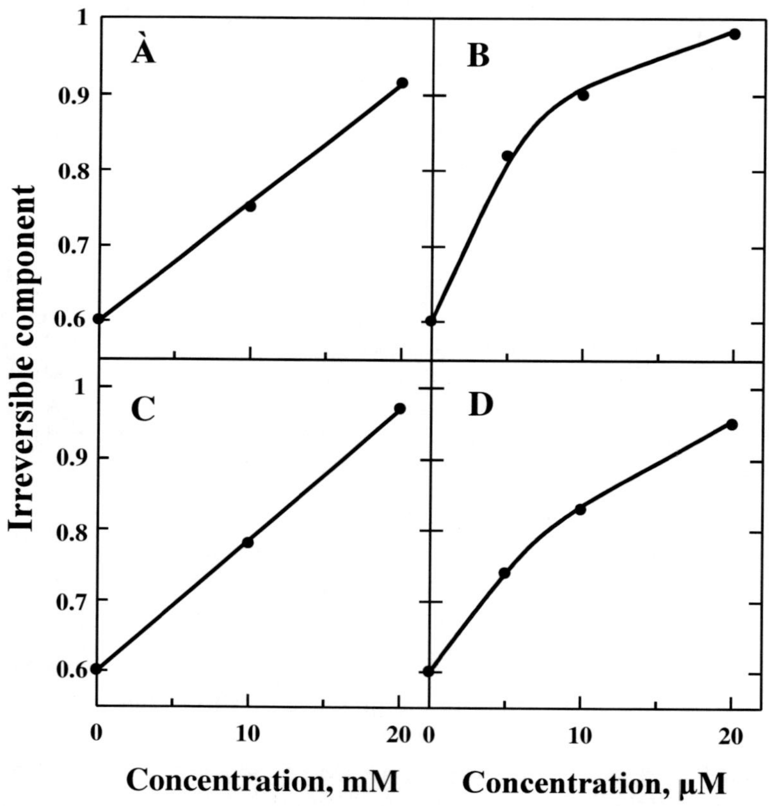

Figure 7.8. Irreversible component, $K(plateau)$, of radiation damage in Chinese hamster V79 cells recovering from potentially lethal damage in the presence of various concentrations of chemicals: pyruvate (A), novobiocin (B), lactate (C) and nalidixic acid (D).

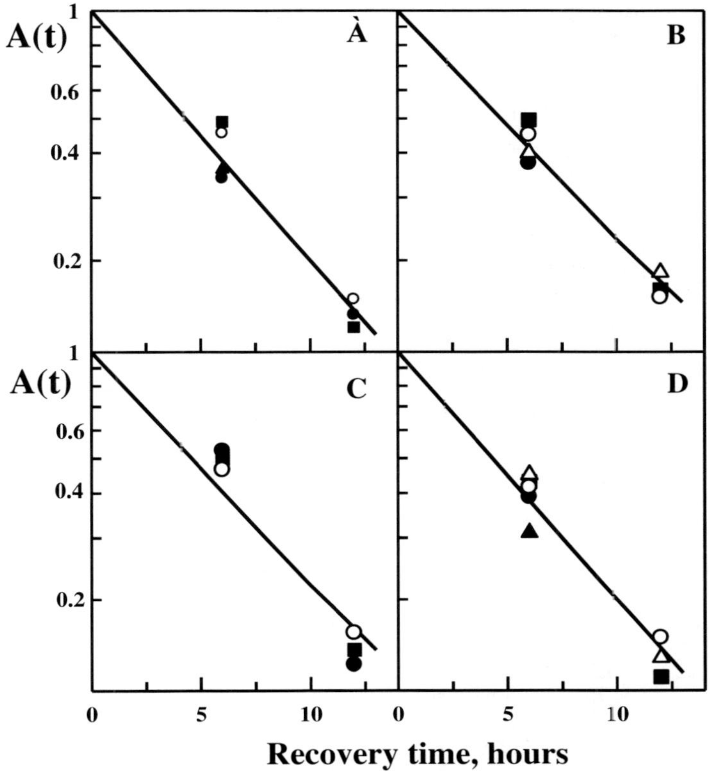

Figure 7.9. The dependence of the relative fraction of repairable damage $A(t)$ on the duration of recovery time of Chinese hamster V79 cells recovering after irradiation without chemicals (open circles) and in the presence of chemical inhibitors of cell recovery. A, pyruvate: 20 mM 1 h after irradiation (closed squares), 10 and 20 mM immediately after irradiation (closed circles and triangles, respectively). B, novobiocin: 20 μM 1 h after irradiation (closed squares), 5, 10 and 20 μM immediately after irradiation (open triangles, closed circles and triangles, respectively). C, lactate: 20 mM 1 h after irradiation (closed squares), 10 and 20 mM immediately after irradiation (closed circles and triangles, respectively). D, nalidixic acid: 20 μM 1 h after irradiation (closed squares), 5, 10 and 20 μM immediately after irradiation (open triangles, closed circles and triangles, respectively).

It would be of interest to find out whether or not this conclusion can be justified for other cells and chemicals used in combination with ionizing radiation. The inhibition of the repair of potentially lethal damage by novobiocin at nontoxic concentrations was obtained with V79 Chinese hamster cells, clone V79-B310H [70]. After the overnight growth of initially single cells to yield groups of cells, cells were X-irradiated when they were in active growth. Following irradiation, growth medium was returned to the dishes with or without novobiocin. If the inhibitor had been added, the medium was removed at the appropriate time, the dishes were rinsed, and fresh medium was returned to them. After incubation for 7–10 days at 37 °C for colony formation, colonies were stained with methylene blue, dried, and counted for the determination of surviving fraction.

Table 7.2. Radiobiological parameters of Chinese hamster cells recovery

Chemicals	Conditions of recovery	Irreversible component K	Recovery constant β, h^{-1}
Without chemicals	Without chemicals	0.60	0.15
Pyruvate	20 mM 1h after irradiation	0.63	0.16
	10 mM immediately after irradiation	0.75	0.16
	20 mM immediately after irradiation	0.92	0.16
Novobiocin	20 μM 1h after irradiation	0.68	0.14
	5 μM immediately after irradiation	0.82	0.14
	10 μM immediately after irradiation	0.90	0.14
	20 μM immediately after irradiation	0.98	–
Lactate	20 mM 1h after irradiation	0.67	0.14
	10 mM immediately after irradiation	0.78	0.14
	20 mM immediately after irradiation	0.99	–
Nalidixic acid	20 μM 1h after irradiation	0.68	0.15
	5 μM immediately after irradiation	0.74	0.15
	10 μM immediately after irradiation	0.82	0.15
	20 μM immediately after irradiation	0.94	–
3-amino-benzamide	20 mM 3h after irradiation	0.68	0.15
	10 mM 1 h before + after irradiation	0.71	0.15
	20 mM 1 h before + after irradiation	0.95	–

Differential response of human and rodent cell lines to chemical inhibition of the repair of potentially lethal damage was investigated by Little et al. [42]. They have examined the effects of several classes of metabolic inhibitors on the repair of potentially lethal damage. The following cell lines were used in this investigation: human squamous cell carcinoma SCC-61, human diploid fibroblast cell strain AG1522, Chinese hamster ovary (CHO) cells, and mouse C3H 10T1/2 cells. Cells were incubated with various inhibitors (3-aminobenzamide, hydroxyurea, 9-β-D-arabinofuranosyladenine – ara-A, 1-β-D-arabinofuranosylcytosine – ara-C) and their concentrations beginning 18–24 h prior to X-irradiation. Survival was determined by a standard colony forming ability.

Effects of 5'-iododeoxyuridine (IdU) on the repair of radiation induced potentially lethal damage were considered by Wang and Iliakis [71]. Experiments were carried out with CHO cells, strain 10B. Before X-irradiation cells were allowed to grow to a plateau-phase for 5 days. IdU was supplied to the culture medium at the time of culture preparation, in the appropriate amounts. Cultures were resupplied with IdU in two days. The rationale for this refeeding protocol has been outlined [28]. Repair of potentially lethal damage was measured by delayed plating of plateau-phase cells. To assay for colony forming ability, cells were enzymatically detached from the culture dishes, counted and plated at the appropriate numbers to generate 25–400 colonies per dish.

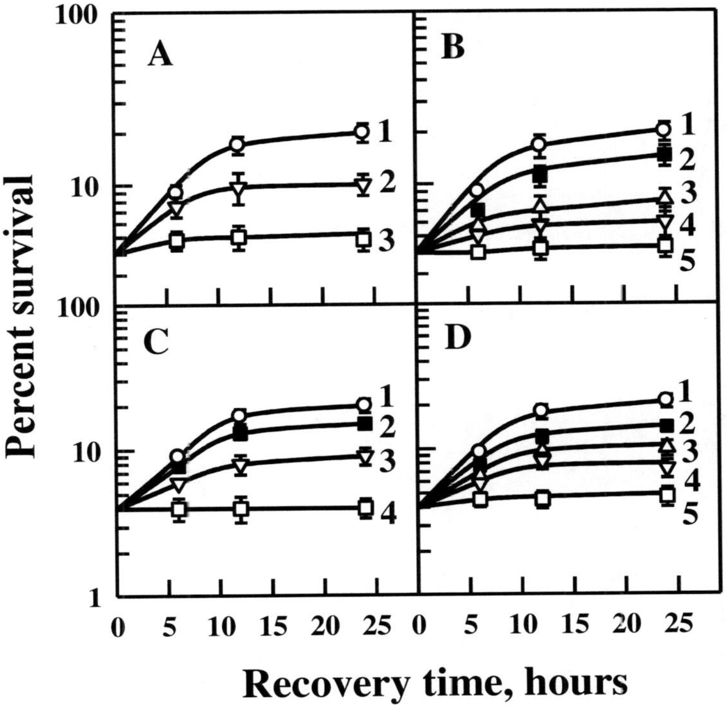

Figure 7.10. Survival of Chinese hamster V79 cells in dependence of the duration of recovery from potentially lethal damage recovering after irradiation without chemicals (curves 1) and in the presence of chemical inhibitors of cell recovery. A, pyruvate: 10 and 20 mM immediately after irradiation (curves 2 and 3). B, novobiocin: 20 μM 1 h after irradiation (curve 2), 5, 10 and 20 μM immediately after irradiation (curves 3, 4 and 5). C, lactate: 20 mM 1 h after irradiation (curve 2), 10 and 20 mM immediately after irradiation (curves 3 and 4). D, nalidixic acid: 20 μM 1 h after irradiation (curve 2), 5, 10 and 20 μM immediately after irradiation (curves 3, 4 and 5).

Experimental results describing the influence of chemical compounds on the recovery of V79 Chinese hamster cells, clone V79-B310H, Chinese hamster ovary cells (CHO), and SCC-61 human tumor cells have been published by others. The inhibition of the increase in the survival of X-irradiated (1250 cGy) V79 cells by the topoisomerase II inhibitor novobiocin was published by Utsumi et al. [70]. Novobiocin was used at concentrations that do not interfere with cell proliferation. The effect of incubation of X-irradiated (900 cGy) CHO cells with 5 mM 3-aminobenzamide on potentially lethal damage was published by Little et al. [42]. These authors also published data for SCC-61 human tumor cells recovering after irradiation without chemicals and in the presence of chemical inhibitors of cell recovery: 5 mM 3-aminobenzamide, 1 mM ara-C, 5 mM hydroxyurea, and 10 mM 3-aminobenzamide. Using the equations developed above, the relative fraction of unrepaired damage $K(t)$ (Eq. 1.11) and the relative part of the repairable damage $A(t)$ (Eq. 1.14) were assessed in the dependence of the duration of recovery time (Figure 7.11).

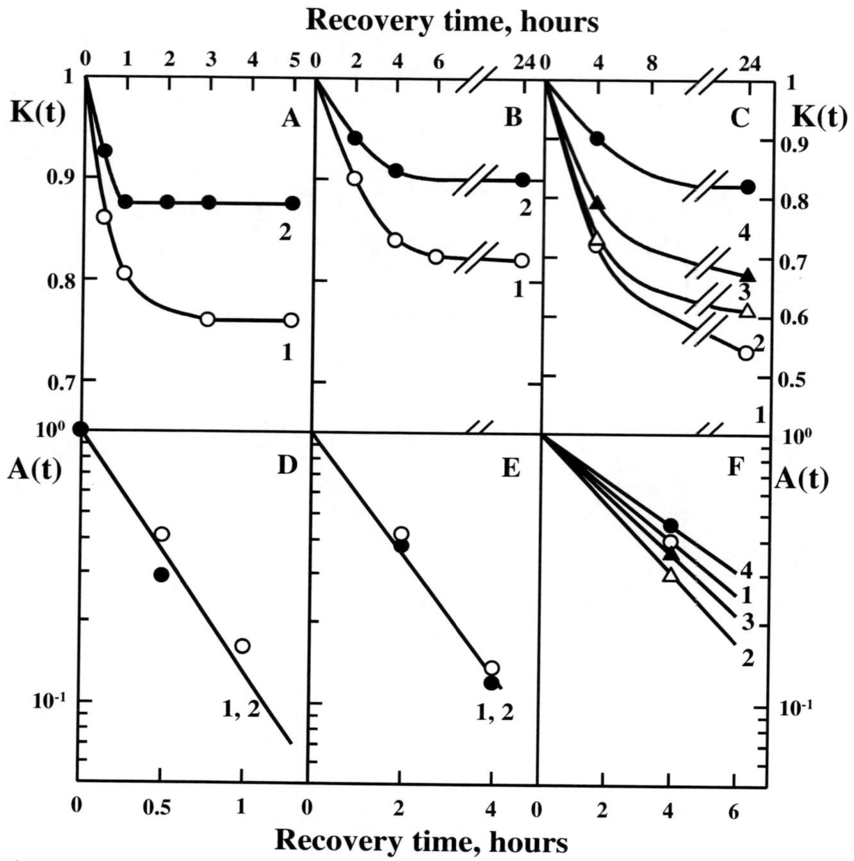

Figure 7.11. The dependence of the relative fraction of unrepaired damage $K(t)$ (A, B, C) and the relative fraction if repairable damage $A(t)$ (D, E, F) on the duration of recovery time of Chinese hamster V79-B310H cells (A, D), CHO cells (B, E) and SCC-61 human tumor cells (C, F) recovering after irradiation without chemicals (curves 1, open circles) and in the presence of various chemical inhibitors of cell recovery. A, D – novobiocin in nontoxic concentration (curve 2, closed circles). B, E – 5 mM 3-aminobenzamide (curve 2, closed circles). C, F – 1 mM ara-C (curves 2), 5 mM hydroxyurea (curves 3), and 10 mM 3-aminobenzamide (curves 4).

It is evident from these data that the irreversible component K is slightly increased from 0.76 to 0.88 for V79 cells and from 0.83 to 0.9 for CHO cells after irradiation without and with repair inhibitors – novobiocin and 3-aminibenzamid, respectively. The similar results were obtained for SCC-61 human tumor cells. The irreversible component K (Eq. 1.15) was estimated to be of 0.54, 0.61, 0.68 and 0.82 for ionizing radiation applied alone and combined with ara-C, hydroxyurea, and 3-aminobenzamide respectively. It means that all these chemicals resulted to an increased yield of irreversible damage in comparison with cell exposure and recovery without chemical inhibitors of recovery. Having used the data depicted in Figure 7.11D, E and F, we estimated the recovery constant β (Eq. 1.15). The values of this parameter were calculated to be independent on the conditions of recovery and equal to 0.03 and 0.01 min^{-1} for V79 and CHO cells. For SCC-61 human tumor cells recovering after irradiation without chemicals β = 0.004 min^{-1}. The corresponding values of this parameter were estimated to be of 0.005, 0.004, and 0.003 min^{-1} in the presence of chemical ara-C, hydroxyurea, and 10 mM 3-aminobenzamide, respectively. It is worth to note that the authors [42] analyzed recovery kinetics for SCC-61 human tumor cells rather rarely so we estimated the recovery constant only due to the single experimental point (Figure 7.11D).

So it is not excluded that the recovery constant variations obtained in our calculations are a simple statistical spread (β = 0.004 ± 0.0005 min^{-1}). Hence, the observations depicted in Figures 7.9 and 7.11 support the conclusion that all chemicals investigated have no influence to the repair process itself and chemical inhibition of the recovery of potentially lethal damage can be completely explained by the increase in the formation of irreversible damage The major point to be inferred from the data presented is that the basic effect of the chemicals tested at a concentration sufficient to inhibit recovery from potentially lethal damage appears to be the reduction in the number of cells capable to recover owing to the increase in the irreversibly damaged cells. It seems plausible that the enhanced yield of irreversible damage may be formed due to the synergistic interaction of damages produced by ionizing radiation and chemicals.

It would be of interest to search for chemical agents increasing cell radiosensitivity by inhibition of cell recovery both by the increase in the irreversible component and due to the decrease in the recovery constant. An example is human diploid fibroblast cells (strain AG1522) exposed to ionizing radiation alone and in the presence of 5 mM 3-aminobenzamide, 0.12 mM ara-A, and 5 mM hydroxyurea. The results were published by Little et al. [42]. These data also enable assessment the relative fraction of unrepaired damage $K(t)$ (Eq. 1.11) and the relative part of the repairable damage $A(t)$ (Eq. 1.14) in the dependence of recovery time (Figure 7.12A, C). On this basis we calculated the dependence of irreversible component K , i.e., the lowest value of $K(t)$, and the recovery constant β (Eq. 1.15) on the condition of irradiation and recovery. Here again we observed an increase of K from 0.5 for exposure with ionizing radiation alone to 0.59–0.61 for all chemical agents tested. The recovery constant β was equal to 0.005 min^{-1} after ionizing radiation applied alone or combined with 3-aminobenzamide and ara-A, while hydroxyurea resulted in a decrease of this parameter to 0.0036 min^{-1}. It means that hydroxyurea had a slight influence to the process of recovery itself.

The results presented in Figure 7.12B, D provide an additional confirmation of this conclusion. Original data for our calculation have been published by others [42]. Taking these data into account, we estimated (as was already mentioned) the dependence of the relative fraction of unrepaired damage $K(t)$ and the relative fraction of repairable damage $A(t)$ on the duration of recovery time of mouse cells, line C3H10T1/2 recovering after irradiation without chemicals and in the presence of 5 mM 3-aminobenzamide. It is evident that both the irreversible component K and the recovery constant β for mouse cells depend on the condition of irradiation and recovery. Here again the irreversible component K ranges from 0.5 for exposure with ionizing radiation alone to 0.81 for postirradiation recovery with 5 mM 3-aminobenzamide. The recovery constant β was calculated to be of 0.005 min^{-1} after ionizing radiation applied alone while the combined action of ionizing radiation and 3-

aminobenzamide resulted in a decrease in this value – 0.003 min^{-1}. Hence, the results presented in Figure 7.12 reveal that some inhibitors can simultaneously act both through the production of a greater part of irreversible damage and through a certain disturbance of cell repair.

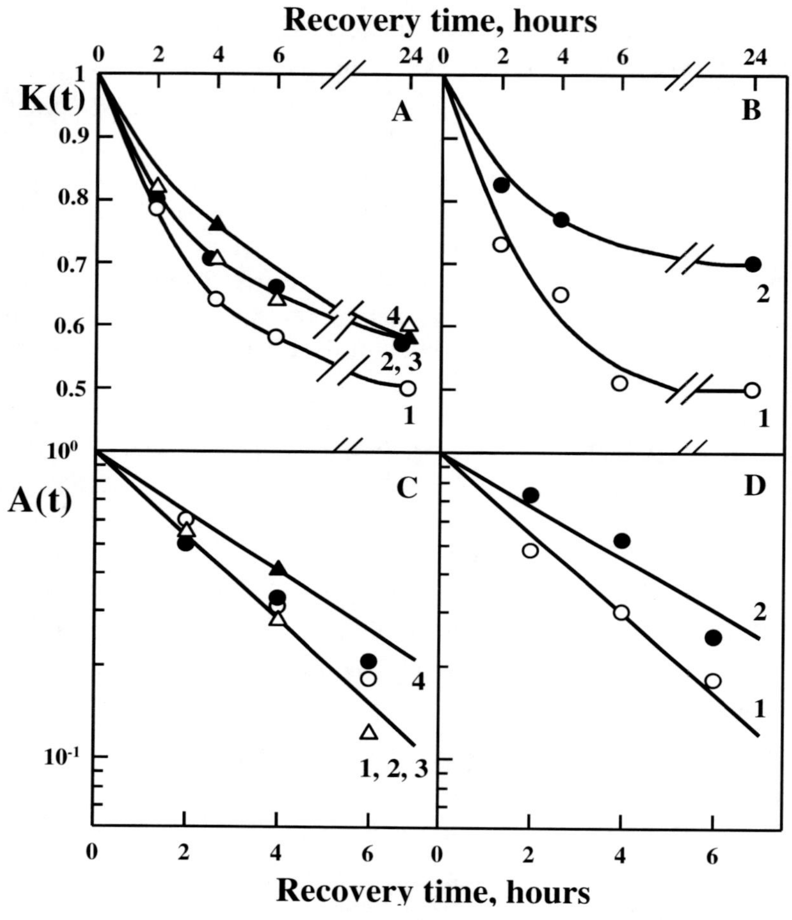

Figure 7.12. The dependence of the relative fraction of unrepaired damage $K(t)$ (A, B) and the relative fraction of repairable damage $A(t)$ (C, D) on the duration of recovery time of human diploid fibroblast cell strain AG1522 (A, C) and mouse cells line C3H10T1/2 (B, D), recovering after irradiation without chemicals (curves 1, open circles) and in the presence of various chemical inhibitors of cell recovery. A, C – 5 mM 3-aminobenzamide (curves 2, closed circles), 0.12 mM ara-A (curves 3, open triangles), 5 mM hydroxyurea (curves 4, closed triangles). B, D – 5 mM 3-aminobenzamide (curves 2).

It would be of interest to consider an example of chemical compounds acting through the enhanced probability to produce the primary radiation damage without influence both on the recovery constant and irreversible component of radiation damage. Halogen pyrimidines are used as radiosensitizers in the treatment of poorly radioresponsive tumors [27]. The effect of 5'-iododeoxyuridne (IdU) incorporated into DNA on radiation sensitivity and cellular repair capability was published [71] for plateau-phase Chinese hamster ovary cells (strain 10B) exposed to X-rays. It is of interest to apply our methodology discussed here to estimate differentially the influence of this compound on the irreversible component and the recovery constant. Using the survival and recovery curves reconstructed and averaged on the basis of the results published by others [27; 28; 71], we calculated the relative fractions of unrepaired $K(t)$ and repairable $A(t)$ radiation damage in the dependence of the recovery time. The results are presented in Figure 7.13. It is obvious that the irreversible component $K = 0.68 \pm 0.02$ independently on the condition of irradiation, i.e., whether it was occurred without or with various concentration of IdU. Having used the data depicted in Figure 7.13B, we estimated the recovery constant β (Eq. 1.15): for cells irradiated and recovered without IdU $\beta = 0.007$ min^{-1}. This parameter was slightly changed for cells grown for 5 days in the presence of 2 μM and 8 μM IdU. The corresponding values were estimated to be of 0.006 and 0.01 min^{-1}. It indicates that in comparison with control cells the recovery constant was only slightly decreased for 2 μM IdU and even increased for 8 μM IdU. It is not excluded that these changes may reflect the possible variability of experimental data around the control value especially as the results were published from one experiment [71]. At the same time the incorporation of IdU into DNA resulted in the considerable increase in cell radiosensitivity. The dose modifying factors estimated by us as the ratio of the doses at 0.2 per cent survival in the absence and presence of the drug on the basis of the published results [71] were 1.4 and 2.4 for 2 and 8 μM IdU, respectively.

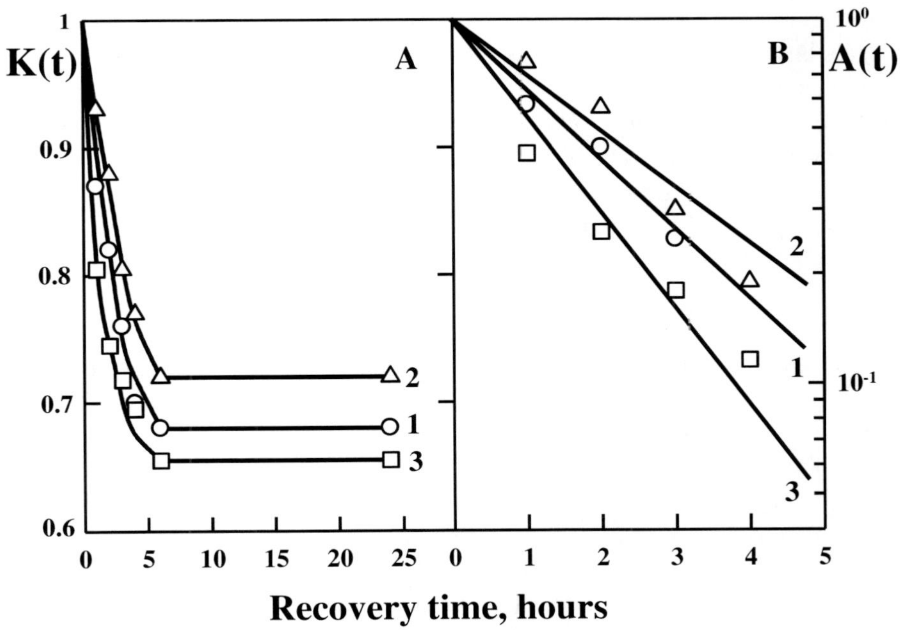

Figure 7.13. The dependence of the relative fraction of unrepaired damage $K(t)$ (A) and the relative fraction of repairable damage $A(t)$ (B) on the duration of recovery time of Chinese hamster cells, strain CHO 10B irradiated and recovered without IdU (curves 1) and for cells grown for 5 days before irradiation in the presence of 2 and 8 μM IdU (curves 2 and 3, respectively).

Thus, it can be concluded that the considerable effect of 5'-iododeoxyuridne incorporated into DNA on radiation sensitivity of CHO cells might realized not through the increase in the irreversible component of radiation damage or through the decrease in the recovery constant. It means that the increase in cell radiosensitivity may happen through the higher probability to produce the primary radiation damage.

7.3. Rabbit Heating by Microwave Exposure at Various Ambient Temperatures

There is a wealth of data on the combined action of various physical and chemical agents on animals. Smoke exposures given before the exposures to radon did not influence radon-induced tumor incidence, but smoke exposures given after radon exposure increased the tumor incidence by a factor of 2-3 over rats receiving radon alone [12]. Shellabarger et al. [64] investigated the influence of the interval between neutron irradiation and diethylstilbestrol on mammary carcinogenesis in female rats. Both radiation and drug are carcinogens for the mammary gland of rats and act in a synergistic manner, particularly with regard to the number of mammary adenocarcinomas per rat when the drug was given at about the same time as radiation. The possibility of synergistic effect in carcingenesis due to changes in cellular kinetics brought about by combined treatment with radiation and N-ethyl-N-nitrosourea was studied by Seyama et al. [61]. Synergistic interaction between low-LET ionizing radiation and the carcinogen 1,2-dimethylhydrazine in the induction of colonic tumors in rats has been described by Sharp and Grouse [63]. When the protocol of radiation plus this chemical was repeated three times at monthly intervals, a 15-fold increase in tumor incidence (from 5% to 74%) was observed at six months post-treatment. As compared with hamsters exposed only to ionizing radiation (lung cancers incident in about 3% of the animals) or only to benzo(a)pyrene (no incident cases in over 280 treated animals), animals receiving benzo(a)pyrene instillations after exposure to ionizing radiation were at a much higher risk (about 50%) of developing a lung tumor [41]. The effect of X-irradiation and hyperthermia at 43 ºC both individually and in combination on mouse embryos were investigated by Nakashima et al. [52]. Combined treatments produced higher frequencies (22-100%) of malformations that would have been expected from considering the sum of the separate treatments (0-42%). Incidence of lung tumors in female rats exposed to plutonium dioxide in combination with varying levels of cigarette smoke was published in [19]. Synergistic enhancement ration was about of 20 at low smoke and of about of 10 for high smoke.

There are many other similar reports on the manifestation of synergy of various environmental factors on animals. Even more papers have been published regarding to the absence of synergy. However, in all these publications the range of the ratio of damage induced by both agents was small and did not allow us to apply the proposed model for their quantitative description. Therefore, we present here our own data on rabbit heating by microwave exposure at various ambient temperatures.

One of the main biological effects of microwave radiation on living matter is an increase in the temperature of the irradiated object, which is directly proportional to the amount of energy absorbed. This is caused by the conversion of electromagnetic power to thermal energy. Hence, both the intensity of the incident field and exposure duration is important parameters that determine the quantity of energy absorbed by tissue. There is an increasing interest to the problem of synergistic effects observed after combined action of various agents. As we indicated above, many studies have shown that numerous physical and chemical agents combined with heat can interact in a synergistic manner when the effect produced by both agents used in combination exceeds the expected results from a simple summation of the individual effects produced by heat and particular agent. Some general rules have been revealed above for the synergistic interactions of a simultaneous action of hyperthermia with ionizing radiation, ultraviolet light, ultrasound, and some chemicals. First, for any constant rate of exposure, the synergy can be observed within a certain temperature range inside which there is a specific temperature that maximizes the resultant synergistic effect. Second, as the rate of exposure is reduced, this temperature range and the temperature providing the greatest synergistic effect decrease and *vice versa*. Third, for a constant temperature at which the irradiation occurs, the synergy can be observed within a certain range of exposure rate inside which an optimal intensity may be revealed that maximizes the synergistic effect. Fourth, as the exposure temperature is reduced, this range and the optimal intensity decrease and *vice versa*.

The widespread and increasing use of radio-frequency electromagnetic radiation (RFR) has greatly increased the possibility of exposure of both occupational and general population to RFR. With the rapid expansion of technologies causing increasing exposure to RFR, especially in mobile telecommunications, there are now challenges to the limit values of safety standards. Analysis of the literature on biologic effects of microwaves reveals several areas of established effects and mechanisms on the one hand and speculative effects and mechanisms on the other [48]. This has created a dilemma in trying to assess the real or imaginary potential hazard to man from exposure to RFR. With the accumulation of hygienic, experimental and clinical data it became apparent that more strict assessment of occupational conditions is necessary for those working with RFR sources. Under natural conditions, microwave irradiation is often accompanied by other factors existing in the environment, such as a high ambient temperature, noise and toxic compounds. Studies of the biological effects resulting from such combined influences are important from theoretical and practical perspectives. First of all, it concerns a potentially increased health risk to the general public and occupationally exposed population. Secondly, there is a great difference in safety standard of microwave irradiation in various countries. Thirdly, these agents are explored in cancer treatment. It was found that ambient temperature had a profound effect on the thermogenic effect of microwave radiation of various animals [21; 49] and men [2] exposed to microwaves. These agents were shown could interact with each other in a synergistic manner. However, none of these papers enable to demonstrate the dependence of the synergistic effect on the intensity of microwaves and to show the existence of the optimal power flux density, which maximizes the synergy. Thus, this study was designed to implement the next purposes: (a) to reveal the synergistic interaction of microwave and environment temperature; (b) to demonstrate the dependence of synergistic interaction on the microwave power flux density; (c) to show the existence of the optimal microwave intensity, which provide the greatest synergistic effect; (d) to apply our mathematical model of synergy to describe the experimental results.

The RFR exposure system has been described [33; 74]. Exposure to RFR were performed in a temperature-controlled anechoic chamber with dimensions of 6.5 (length) x 5.0 (width) x

3.5 (height) m, lined with commercial microwave absorbing material. The air temperatures in the chamber were initially adjusted to 22 ± 0.2 °C, 30 ± 0.5 °C, 38 ± 0.5 °C. Animals used in this study were Shinella rabbits with body masses of 2.5 to 3.2 kg both males and females. The rabbits were purchased from Russian Academy of Medical Sciences nursery at an age of 60 days. They were housed individually at an ambient temperature of 22 °C, with a relative humidity of 50% under natural lighting. Rabbit chow and water were provided *ad libitum*. Animals were pre-handled and gentled for 10 days prior to exposure. The body temperature was allowed to equilibrate in the chamber for at least one hour before irradiation. Rabbits were individually exposed to continuous-wave microwaves at a frequency of 7 GHz. The irradiation was carried out in the far zone and E-orientation of body at (55 ± 5)% humidity. All experiments were performed with the animals without general anesthesia.

In this study, rectal temperature was measured in rabbits during and immediately following exposure to 7 GHz microwave radiation at an ambient temperature of 22, 30 or 38 °C. Microwave irradiation at 38 °C was performed only for one power flux density (100 mW/cm^2). Individually each rabbit in sitting position was confined in a Plexiglas cage. The walls of the cage were performed with holes (15 mm in diameter spaced 3 cm) and the top of the cage was covered with cellular net made of cotton fabric to enhance thermal equilibrium with the anechoic chamber. The rabbit was not able to move in the plexiglas cage. Experiments were performed on 71 rabbits.

Rabbits were individually irradiated to the following power flux density: 0 (sham irradiation), 10, 15, 20, 30 and 100 mW/cm^2. Power flux density measurements were made using P3-9 power meter (Russia). The corresponding specific absorption rates (SAR) were 0, 0.75, 1.12, 1.50, 2.25, and 7.5 W/kg. To calculate SAR, the amount of energy absorbed was determined by measuring the amount of heat generated in a fresh animal carcass irradiated with microwaves by use of a differential or a quasi-adiabatic calorimeter. The heating pattern should be measured before it is altered by blood flow and heat diffusion. Therefore dead animals have been used as a phantom material, eliminating the problems of physiological heat production and loss. Calorimetric techniques are usually used to measure the mean whole-body SAR. We have used a twin-well calorimeter similar to that used by McRee [47] but with some modifications. Twin-well calorimeters are instruments which measure mean whole-body SAR very accurately. The accuracy of these techniques has been estimated to be about of 5%. Because one of the purposes of this work was to compare the dependence of animal lethality on the average SAR and on the power flux density, the last value was obtained by P3-9 apparatus, the accuracy of measurement of the power flux density being only ± 30%.

A modified medical thermometry unit (model TPE-1M, Russia), with a thermistor probe, which was shielded with plexiglas and fixed in the rectum, was used to monitor colonic temperature during exposure. In this unit, the measured temperature range was enlarged up to 50 °C, the accuracy was improved within to 0.05 °C, and the diameter of temperature sensor was decreased to 1 mm. Special experiments have demonstrated that the presence of the thermistor probe during the irradiation did not exert influence upon the registered rectal temperature. During exposure, the colonic temperature was read within every 10 minutes or more frequent. Animals were not re-exposed to RFR. During irradiation animals were watched by TV. We used ambient temperature of 30 ± 0.5 °C because of two thoughts: first of all, this temperature may exist at a working place of microwave equipments; secondly, in accordance with literature [33], it is an upper limit of relatively comfort temperature both for rabbit and slightly-dressed man.

It worth to note that here we present the ultimate averaged findings based on our previous experimental and theoretical data on this subject, which has been published previously [33; 34; 56; 74]. Figure 7.14 shows the results of simultaneous action of microwave and ambient temperature on rabbit heating. Microwave irradiation was performed at 22 and 30 °C at power flux density of 0 (sham irradiation), 10 and 15 mW/cm^2 (Figure 7.14A) as well as at 20, 30 and 100 mW/cm^2 (Figure 7.14B). In both parts of this Figure the increment of rectal temperature exhibited by animals are depicted. One can see that at an ambient temperature of 22 °C, the rectal temperature increment was negligible after 180 min exposure at power flux densities of 10 and 15 mW/cm^2. Exposure intensity of 20 mW/cm^2 led to an increment of rectal temperature approximately 0.3 ± 0.1 °C while further increase in the incident power density resulted in a noticeable elevation in deep body temperature after exposure time used here. It can also be seen, that the raising of ambient temperature from 20 to 30 °C in the most cases would greatly enhance rabbit heating because of a reduced rabbits ability to dissipate heat accrued from RFR exposure. In the absence of microwave exposure, rectal temperature exhibited by animals maintained at ambient temperature of 30 °C increased slightly (0.2 ± 0.1 °C) over a 3-hour period.

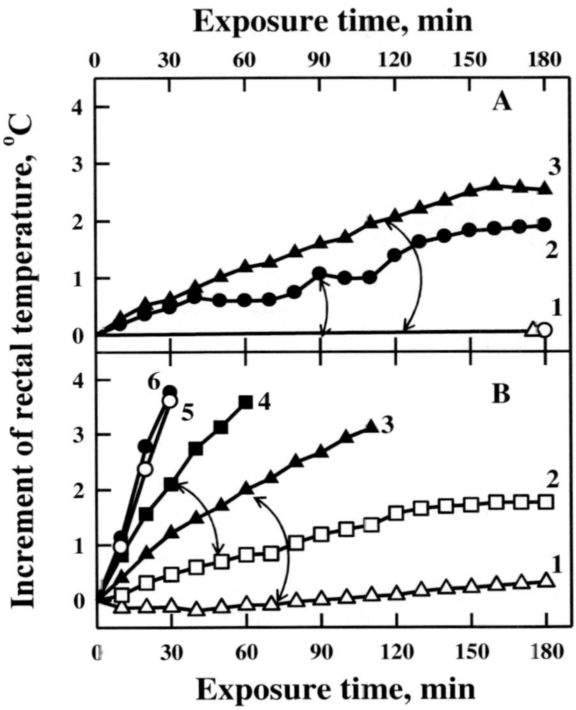

Figure 7.14. The change in rectal temperature response exhibited by rabbits exposed to 7 GHz microwaves (constant waves) at various environmental temperatures and power flux density. A: 10 (circle) and 15 (triangle) mW/cm^2 at 22 °C; curve 2, 10 mW/cm^2 at 30 °C; curve 3, 15 mW/cm^2 at 30 °C. B: curve 1, 20 mW/cm^2 at 22 °C; curve 2, 30 mW/cm^2 at 22 °C; curve 3, 20 mW/cm^2 at 30 °C; curve 4, 30 mW/cm^2 at 30 °C; curve 5, 100 mW/cm^2 at 22 °C; curve 6, 100 mW/cm^2 at 30 °C.

Basing on the data presented it can be concluded that the observed temperature increment after simultaneous action of high ambient temperature and RFR exposure was in most cases greater than that expected for additive effects of these factors. It means that the synergistic interaction between microwave energy and thermal load induced by high ambient temperature is observed.

To estimate the effect of synergy quantitatively we used the synergistic enhancement ratio k which can be determined in these experiments by the ratio of temperature increments observed in experiments after simultaneous action of both factors $\Delta T(RFR, 30°C)$ to that expected for additive summation of thermal effects from microwaves alone $\Delta T(RFR, 22°C)$ and high ambient temperature $\Delta T(30\,°C)$ estimated for any identical duration of exposure:

$$k = \frac{\Delta T(RFR, 30°C)}{\Delta T(RFR, 22°C) + \Delta T(30°C)}. \qquad (7.1)$$

It worth to note that the increment of rectal temperature exhibited by animals exposed to microwaves alone was estimated after irradiation at a comfort temperature (22°C). The synergistic enhancement ratio, determined by Eq. (7.1), shows how many times the temperature increment after simultaneous application of both modalities exceeds the sum of increments expected from individual action of each agent. In fact, the synergistic enhancement ratio is directly proportional to the angle between heating curves obtained at normal (22 °C) and high (30 °C) ambient temperatures (Figure 7.14).

To calculate the synergistic enhancement ratio we approximated experimentally obtained curves by straight lines. In other words, we did not take into consideration the reduction of heating effectiveness observed in some cases after a long irradiation and which might be related with operation of some additional mechanisms of temperature regulation. On the contrary, the adaptation period of rabbit to the experimental conditions was taken into consideration. It can be seen (Figure 7.14B) that colonic temperature was not increased immediately after the beginning of low intensity exposure but approximately after 1 hr. Therefore in our approximation we believe that the beginning of low intensity exposure was conditionally delayed for 30 min. That is why the values the synergistic enhancement ratio obtained here and published before were somewhat different.

Figure 7.15 shows the dependence of the synergistic enhancement ratio on the power flux density calculated on the basis of Eq. (7.1) and the results presented in Figure 7.14. It is clearly seen that the synergy takes place for some power densities while for others the interaction is small or only additive. It means that the synergistic interaction of microwave energy with high environmental temperature was observed for rabbits only within a definite ratio of incident power density of electromagnetic radiation. One can see that there is an optimal intensity at which the highest synergy is achieved.

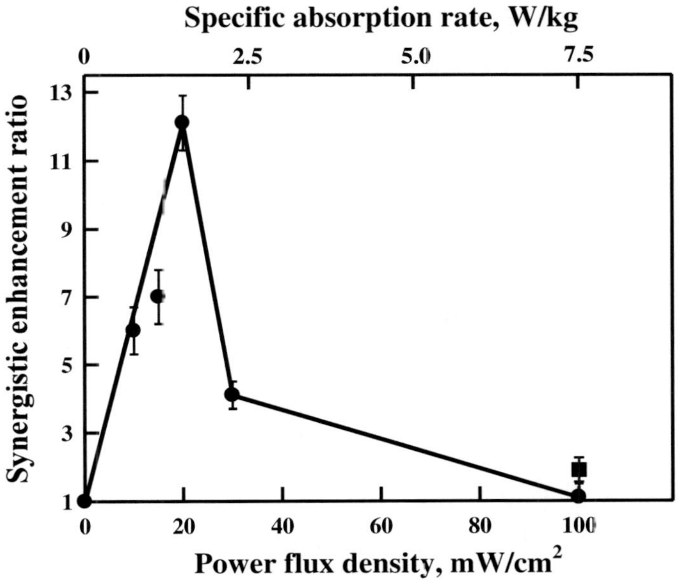

Figure 7.15. Relationship between the synergistic enhancement ratio and power flux density after simultaneous action of microwave (7 GHz) exposure and high environment temperature (30 °C) on rabbit heating.

In accordance with our previously described results with unicellular organisms, for a constant dose rate of ionizing radiation, the synergistic effect should depend on the exposure temperature. Therefore it would be of interest to test this conclusion for rabbit heating. We measured the change in rectal temperature exhibited by rabbits exposed to RFR at 100 mW/cm^2 and various ambient temperatures. The results are shown in Figure 7.16. As it can be seen, for this power flux density there is almost no difference in the efficiency of rabbit heating both at 22 and 30 °C. Inversely, microwave exposure of rabbit at a higher ambient temperature (38 °C) was more effective, the synergistic enhancement ratio, estimated by Eq. (7.1), was equal to 1.9. This result is depicted in Figure 7.15 by a square sign. It turned out that further investigation of the synergy for lower intensities of microwaves delivered at 38 °C was embarrassed because animals did not sustain this high environmental temperature for a long time. Nevertheless, the expected increase in a synergy with an increase of exposure temperature was fulfilled.

In this study, basic features of synergistic effect between radio-frequency electromagnetic radiation and ambient temperature on rabbit heating have been revealed. The results show that for a constant ambient temperature the effectiveness of synergistic interaction is observed only inside a certain range of the power flux density. Moreover, within this range an optimal value of the power flux density can be indicated, at which the greatest synergistic effect can be achieved. Both increase and decrease in the power flux density from this optimal value resulted in a decrease of the synergy. It is of interest that just such a dependence of synergy on the intensity of various physical agents was predicted by a mathematical model describing the synergistic effects of hyperthermia applied together with ionizing or ultraviolet radiations as well as with ultrasound and some chemicals (Chapter 6). The results obtained here are in qualitative agreement with the model prognosis and would be useful to develop optimal regimens of thermoradiation action in cancer treatment as well as for further exploiting unified radiofrequency radiation safety standards.

It is known for a long time that different animals after combined action of RFR and high ambient temperature may be overheated more rapidly comparing with the RFR exposure at comfort conditions because the heat dissipation at a high ambient temperature is awkward [22; 33]. Human beings and animals have similar physiological mechanisms of body heat homeostasis [1]. Therefore, one can expect the synergistic responses of man to the combined action of RFR and heat. The professional work at heavy heat exchange conditions is in a particular danger. Physical working load in the condition of microwave irradiation and high environmental temperature can also result to an additional rise in body temperature [74]. Thus, these data indicate that safety limits of microwave radiation should take into account a possible synergistic interaction of ambient temperature and RFR described here.

As was mentioned above, an ambient temperature had a profound effect on the thermoregulatory responses of various animals and men exposed to microwaves. In these experiments, the ambient temperature either exerts a microwave action or both these agents can interact with each other in a synergistic manner [1; 2; 21; 22; 74]. However, none of these papers enables us to optimize the synergistic interaction and demonstrate the existence of the greatest synergistic effect. The availability of a number of common features of synergistic effect display indicates the existence of a certain general idea regarding the mechanism of synergistic interaction, which would not be dependent of the object, biological test and agents applied. Since our model of synergy is not based on a particular nature of effective damage or sublesions responsible for the synergy and the mechanism of their interaction was not elucidated, it would be of interest to adopt this model for a description, optimization and prognosis of the synergistic effects between microwave and an ambient temperature on rabbit heating.

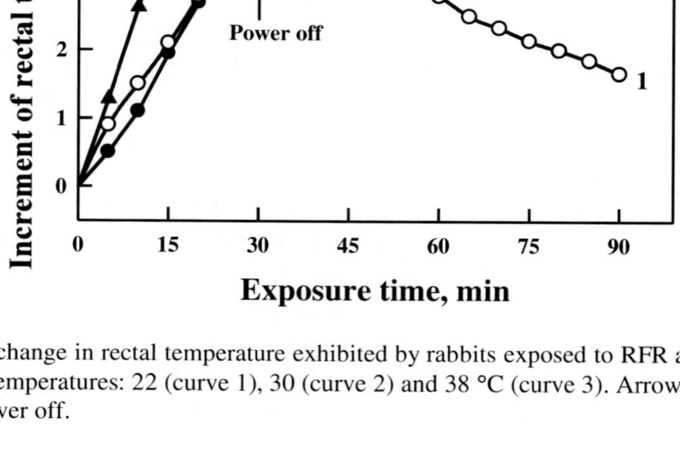

Figure 7.16. The change in rectal temperature exhibited by rabbits exposed to RFR at 100 mW/cm^2 and various ambient temperatures: 22 (curve 1), 30 (curve 2) and 38 °C (curve 3). Arrows show the moment of microwave power off.

Let N_1 and N_2 be the yield of some hypothetical effective damage produced by microwaves and ambient temperature, respectively, which resulted in an increase of the rectal temperature. The synergistic interaction takes place when the combined action of two agents exceeds the sum of their individual effects. With the use of this definition, it might be reasonable to assume that some additional effective damage has arisen during the simultaneous action of these modalities. Our mathematical model of synergistic interaction suggests that the additional damage may be formed due to the interaction of some sublesions induced by both agents. These sublesions are supposed to be ineffective after each agent is applied separately. We believe here that the effective damage resulting in the rabbit heating and the sublesions responsible for the synergy can be presented by real molecular damage as well as by a disturbance of a regulator mechanism which can be estimated quantitatively. Let p_1 and p_2 be the mean numbers of the sublesions that arise for one effective damage induced by microwaves and an ambient temperature, respectively.

Eq. (5.5) shows that the synergistic enhancement ratio should increase first with N_2 / N_1, and then reach a maximum, which should be followed by a decrease. It means that for a constant ambient temperature there an optimal power flux density for maximizing the synergy should exist. Besides, Eq. (5.5) shows that for any fixed values of p_1 and p_2, the synergism depends only on the ratio N_2 / N_1, which is the ratio of the effective damage produced by each of the acting agents applied alone. We can believe that within a certain exposure time, where the temperature increment is linearly related to the duration of an exposure, N_1 should be in direct proportion to an increment of the rectal temperature $\Delta T(RFR + 22^o C)$ induced by microwaves only and N_2 should be in direct proportion to the corresponding value of $\Delta T(30^o C)$ induced by an ambient temperature only. If this is the case, we can write

$$\frac{N_2}{N_1} = \frac{\Delta T(30^o C)}{\Delta T(RFR + 22^o C)} = \frac{dT / dt(30^o C)}{dT / dt(RFR + 22^o C)}. \tag{7.2}$$

It means that the ratio N_2 / N_1 can be estimated by the ratio of a rabbit heating rate induced by an ambient temperature to that produced by microwaves.

If the number of sublesions from microwaves is less or equal to the number of sublesions induced by an ambient temperature, i.e., $p_1 N_1 \le p_2 N_2$, then k (Eq. 5.5) may be expressed as

$$k_1 = 1 + \frac{p_1}{1 + \dfrac{dT / dt(30^o C)}{dT / dt(RFR + 22^o C)}}. \tag{7.3}$$

From here, parameter p_1 may be represented by

$$p_1 = (k_1 - 1)\cdot\left(1 + \frac{dT/dt(30^o C)}{dT/dt(RFR + 22^o C)}\right). \tag{7.4}$$

In contrast, if $p_2 N_2 \le p_1 N_1$, then k (Eq. 5.5) may be expressed as

$$k_2 = 1 + \frac{p_2 \dfrac{dT/dt(30^o C)}{dT/dt(RFR + 22^o C)}}{1 + \dfrac{dT/dt(RFR + 22^o C)}{dT/dt(30^o C)}}. \tag{7.5}$$

From here, parameter p_2 may be represented by

$$p_2 = (k_2 - 1)\cdot\left(1 + \frac{dT/dt(RFR + 22^o C)}{dT/dt(30^o C)}\right). \tag{7.6}$$

A curious feature of Eqs. (5.5, 7.2–7.6) is that they do not include an absolute amount of the damage produced by microwaves (N_1) and an ambient temperature (N_2) which would not be evaluated directly but only their ratio N_2 / N_1 that can be, under the suggestion made here, found using Eq. (7.2). By knowing experimental values of k_1 and k_2, obtained from two experiments which satisfy the requirements of Eq. (7.3) and Eq. (7.5), one can estimate the basic model parameters p_1 and p_2 using Eq. (7.4) and Eq. (7.6), respectively. Then using Eq. (5.5) one can predict the k values for any N_2 / N_1 ratio, the highest synergy (Eq. 5.11) and condition (Eq. 5.10) at which it can be achieved. The last equation means that the highest synergy should be expected when an identical number of sublesions are produced by both agents used in combination. It is of interest that the dependency of the synergy on the power flux density follows from Eq. (5.10). Indeed, let Eq. (5.10) holds for simultaneous action of some flux power density and ambient temperature. Then for any higher ambient temperature a smaller time is needed to obtain N_2 which results in a smaller value of N_1. Then to preserve Eq. (5.10) one must increase N_1 which can be done only by an increase in the power flux density. Analogously, the whole relationship between the synergy and the power flux density should be shifted to higher intensities with an increase in the ambient temperature.

Thus, the considered model predicts the following non-trivial results. First, for every constant ambient temperature, the synergy should be observed only within a certain range of the power flux density that could be different for various biological objects and test effects.

Second, within this range, there is a specific power flux density that maximizes the synergistic effect. Any deviation of microwave intensity from the optimal one would result in a reduction in the synergy. Third, the relationship between the synergy and the power flux density should be shifted to higher intensities with an increase in the ambient temperature. It is curiously that quite the same peculiarities have been described above. Therefore our next purpose is to apply this model to experimental results describing rabbit heating during simultaneous action of microwave and ambient temperature.

Taking into consideration the results presented in Figures 7.14 and 7.16, we estimated rabbit heating rates for different conditions of exposure to microwaves and ambient temperatures (22 and 30 °C). Under conditions of experiment, the first temperature (22 °C) was the room temperature. We used ambient temperature of 30 °C because of two thoughts: first of all, this temperature may exist at a working place of microwave equipments; secondly, in accordance with literature [22; 49], it is an upper limit of relatively comfort temperature both for rabbit and slightly-dressed man. Table 7.3 summarizes the following parameters: power flux density and the corresponding specific absorption rate, heating rates (*dT/dt*) at 22, 30 and 38 °C, N_2 / N_1 ratio calculated by Eq. (7.2) and the experimental values of the synergistic enhancement ratio Eq. (7.1). It is evident that the increase in the ambient temperature from 20 °C either to 30 °C or 38 °C in most cases would greatly enhance the rabbit heating because of rabbit's reduced ability to dissipate the heat accrued from RFR exposure.

Theoretically expected values of the synergistic enhancement ratio are inserted in Table 7.3 and presented by solid line in Figure 7.17, where the synergistic enhancement ratio k of a simultaneous action of microwaves and a high ambient temperature (30 °C) is shown as a function of the ratio N_2 / N_1 for rabbit heating. One can see that the theoretically predicted interaction (solid line), calculated by means of Eq. (5.5), corresponds fairly well with the experimental data (circles). The solid lines were calculated with the following values for the basic model parameters: p_1 = 15.5, p_2 = 37. This set of parameters was obtained from a real experimental value of the synergistic effect for the power flux densities of 10 and 30 mW/cm^2 and Eqs. (7.4) and (7.6), respectively.

Table 7.3. Heating rate (*dT/dt*) and synergistic enhancement ratio (k) after different conditions of rabbit exposure

Power flux density, mW/cm^2	Specific absorption rate, W/kg	*dT/dt* RFR at 22 °C, °C/hr	*dT/dt* RFR at 30 °C, °C/hr	*dT/dt* RFR at 38C °C, °C/hr	N_2 / N_1	Synergistic enhancement ratio, k	
						Experiment	Theory
0	0	0	0.067	0.7	–	–	–
10	0.75	0.033	0.6	–	2.0	6.0	6.0
15	1.12	0.067	1.1	–	1.0	8.2	8.8
20	1.5	0.23	1.7	–	0.61	12.1	10.8
30	2.25	0.74	3.3	–	0.09	4.1	4.1
100	7.5	6.9	7.6	–	0.01	1.1	1.2
100	7.5	–	–	14.4	0.1	1.9	–

It can be inferred that the synergy occurred only within a definite ratio of the effects (N_2 / N_1) produced by each agent. It is also apparent that there is an optimal ratio of N_2 / N_1 which results in the greatest synergistic interaction. It follows from the model that the highest synergistic effect, k_{max} = 12, should be expected (Eq. 5.11) under N_2 / N_1 = 0.41 (Eq. 5.9). Figure 7.17 demonstrates that a good agreement exists between these predicted values and the experimental data. It means that the model presented here can be used to predict and optimize the synergistic effect between radio-frequency electromagnetic radiation and an environmental temperature for rabbit heating.

To test the conclusion indicating that the relationship between the synergy and the power flux density should be shifted to higher intensities with an increase in the ambient temperature, a change in the rectal temperature exhibited by rabbits exposed to RFR at 100 mW/cm^2 and 38 °C was measured. As can be seen (Table 7.3), for this power flux density there was only a negligible difference in the efficiency of rabbit heating both at the comfort (22 °C) and higher temperature (30 °C): the synergistic enhancement ratio k = 1.1. Microwave exposure of rabbits at an ambient temperature of 38 °C was more effective. As was estimated, the heating rate was 0.7 and 14.4 °C/hr when animals were exposed to the high temperature alone (38 °C) and combined with microwaves at an exposure intensity of 100 mW/cm^2. For this particular case, it might also be thought that the first value is proportional to N_2, while the second one – to the sum of $(N_1 + N_2 + N_3)$. It gives N_2 / N_1 = 0.1 (Eq. 7.2) and k = 1.9 (Eq. 5.5).

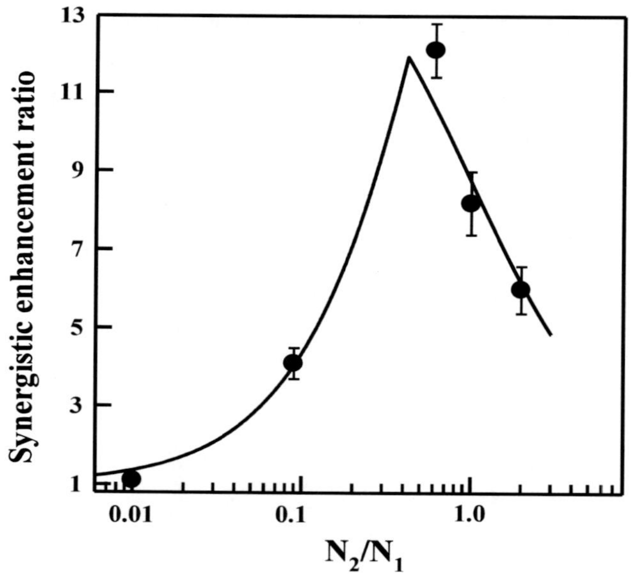

Figure 7.17. The experimental (solid circles) and theoretical (solid lines) dependencies of the synergistic enhancement ratio of a simultaneous action of microwaves (7 GHz) and high ambient temperature (30 °C) on the ratio of N_2 / N_1 for rabbit heating.

Figure 7.18 demonstrates the relationships between the synergistic enhancement ratio and the power flux density after a simultaneous action of microwave exposure and an ambient temperatures (30 °C, curve 1), and 38 °C (curve 2) for rabbit heating. The only experimental point obtained for rabbits exposed to an exposure intensity of 100 mW/cm^2 and an ambient temperature of 38 °C, shows that the synergistic enhancement ratio was increased for these conditions of irradiation. It was shown that a further investigation of the synergy for lower intensities of the microwaves delivered at 38 °C was difficult because the animals did not sustain this high environmental temperature for a long time. The increase in the ambient temperature from 30 to 38 °C led to a 10.5-fold increase in the heating rate caused by these temperatures (0.067 and 0.7 °C/hr, respectively). At an ambient temperature of 30 °C the greatest synergy was observed at 20 mW/cm^2, when the heating rate due to microwaves only (exposure at 22 °C) was 0.23 °C/hr. Assuming the basic parameters of the model might be unchanged and taking into account Eq. (5.10), one can conclude that to keep the same greatest synergistic effect (k = 12.1) observed in the experiments we must provide a 10.5-fold increase in the amount of damage produced by RFR alone, i.e., heating rate should be 2.4 °C/hr. The data presented in Table 7.3 shows that this heating rate due to microwaves alone could be observed at about of 40 mW/cm^2. Curve 2 in Figure 7.18 was constructed by using this predicted point (open square) together with the experimentally obtained result (closed square).

It is evident that the results obtained are in accordance to the consequence from the model indicating that the relationship between the synergy and the power flux density should shifted to higher intensities with an increase in the ambient temperature.

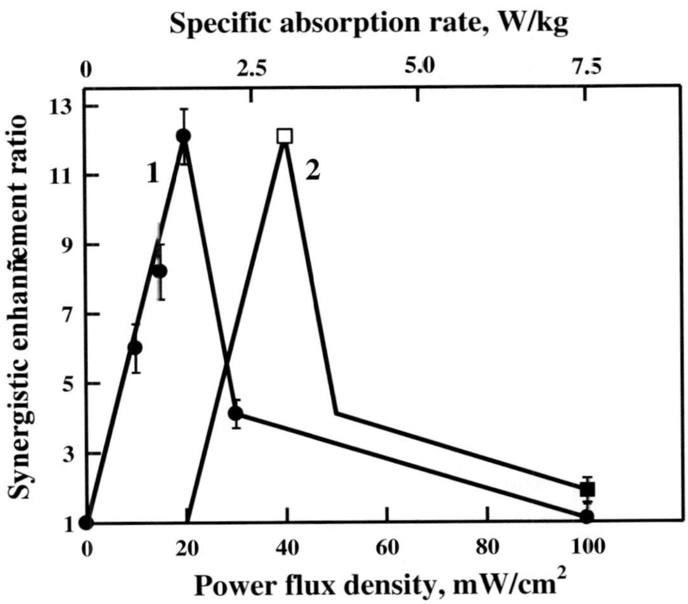

Figure 7.18. The relationships between the synergistic enhancement ratio and the power flux density after a simultaneous action of microwave exposure and ambient temperatures (30 °C, curve 1), and 38 °C (curve 2) for rabbit heating.

In conclusion, the widespread and increasing use of RFR has greatly increased the possibility of an exposure of both an occupational and general population to RFR combined with various environmental factors including a higher ambient temperature. The presented experimental results and mathematical model, predicting the effectiveness of the interaction of microwaves and an ambient temperature, are of important from theoretical and practical perspectives. First is the concept of the interaction to describe and predict the synergistic effect. Second is the concern for an evaluation of a potentially increased health risk to the general public and occupationally exposed population. And third, as these agents are being explored for a cancer treatment, there is a possibility to optimize the acting agents to provide the greatest synergistic effect.

7.4. Significance of Synergistic Effects for Humans

A certain level of background exposure to ionizing radiation and natural or man-made chemicals is always presented in the environment. Thus, all assessments of toxicity, carcinogenicity or mutagenicity of chemical, physical and biological agents should be performed as studies of their combined effects. The assessment of possible synergistic effects at the exposure levels relevant for risk estimation remains very difficult because of the high exposures used in experimental systems and the possible threshold levels. There are several examples regarding the manifestation on synergistic interaction of harmful physical and chemical agents in humans. Some well-known examples are worth to mention.

UV radiation is recognized as a most important initiator and co-factor for human skin carcinogenesis. It is mainly the skin that is exposed to UV radiation. A study of combined exposures to gamma radiation and UV radiation was presented by Shore [65], who analyzed 12 studies on the incidence of skin cancer in populations irradiated with known skin doses. In the absence of a proper control, it was concluded that the interaction is unclear.

Ultrasound has achieved widespread use in medical diagnostic and therapeutic procedures. Studies have shown that at the intensities used for diagnostic purposes, ultrasound does not interact with ionizing radiation to cause cytogenetic damage in treated cells, although the yield of sister chromatid exchanges was observed to be slightly increased in one study [40].

The combination of radiation exposure and exposure to dusts and fibres is quite common in important industrial environments such as mining, metallurgical industries, and power plants. Some dusts and fibres are pathogenic or carcinogenic by themselves. Both experimental results from mammals and epidemiological evidence are available indicating the synergy [4; 10; 14; 32; 54; 58]. Silicosis was shown to be a risk factor for human lung cancer in metal miners and is implicated as a modifier of lung cancer risk in radon exposed underground workers [38]. Combined exposure to phosphate ore dust, gamma radiation, and radon daughter products also resulted in elevated lung cancer risks in earlier practices [8]. The synergistic enhancement ratio was about of 3 for the effect of cytosine arabinozide on the frequency of X-ray-induced chromosome aberrations in normal human leukocytes [59].

Especially in underground mining, possible effects from epidemiologically proven lung carcinogens arsenic, chromium, nickel, and antimony [45] have to be assessed together with high-LET radiation from radon. Arsenic in particular is a major risk factor in combined exposures to mineral dust, radon, metals, and diesel fumes [39; 57]. The risk enhancing effects of iron dust seem to be limited to very high dust concentrations, leading to changes in lung function [8].

People are exposed to an indoor radon concentration combined at most time with other agents including tobacco smoking. It is well known from epidemiological and toxicological studies that the kind of interaction which occurs between tobacco smoking and radon may be synergistic, i.e., can lead to the effects, the sum of which is greater than that expected from the addition of the effects from separate exposures to the individual agents [3; 12; 55]. This problem is especially important for high natural background areas [44; 51].

The most comprehensive and complete analysis of radon-induced health risks was published by Lubin et al. [43]. The review contains a joint analysis of original data from 11 studies of male underground miners. For all studies taken together, the combined influence of smoking and radon progeny exposures on lung cancer was clearly more than purely additive [11]. A best estimate from miner data indicates that the lung cancer risk for smokers expressed in absolute terms is higher by a factor of about 3. Smoking is also of great importance for non-occupational radon exposures in the indoor environment. Only one large study [55] was indicative of an indoor radon risk and its modification by tobacco that is comparable to what is predicted from miner studies.

Numerous published data on combined action of radon and smoking allow applying the synergistic model presented above for qualitatively explanation the observed effects. There are two classes of epidemiological data, related with this problem: affect both on the miners and on the people most of time conducting indoors with high level of radon. The second class connects with the situation where people are exposed to relatively low concentrations of radon. The first case reflects the situation when radon is mainly responsible for lung cancer induction while the tobacco smoke inhalation is chiefly produces lung cancer. Seemingly, one can use these data to identify the basic parameters of the model discussed above. It was found [55] that Sweden's population is exposed to high concentrations of radon than the population of many other countries. The average concentration of radon in Swedish dwellings is 100 Bq/m^3, while in the UK this figure is only 20 Bq/m^3. In the U.S., it is recommended to resettle people from areas where the radon concentration exceeds 148 Bq/m^3.

Data base of the study published by Pershagen et al. [55] includes 1500 patients with lung cancer who were selected from the Swedish Cancer Registry. The original data presented in this paper demonstrate the dependence of the tumor frequency on the radon concentration. Using these data and Eq. (3.5) we calculated experimental values of the synergistic enhancement ratio for situations when people were exposed to radon concentrations in the range of 50 to 400 Bq/m^3 and were smokers (smoking more than 10 cigarettes per day). The greatest synergistic effect (k = 3.5) was shown to observe at the highest radon concentration. Smaller radon concentrations showed smaller values of the synergy that was expected from our model. This makes it possible to assess one of the basic parameters of our model, when the effect of tobacco smoke is prevailing. Another parameter of the model need to be calculated for the opposite situation when lung cancer is mainly produced by radon and its short-lived decay products. We used a survey consisting of periodic sputum cytology evaluation and which was performed among 249 underground uranium miners and the corresponding controls [3]. Cytological sputum samples yielding moderate atypia, marked atypia, or the presence of cancer cells were classified as being abnormal. They presented data for the combined action of agents and the effects of tobacco smoke inhalation on lung cancer induction in miners for a range of radon concentrations (from less than 2.6 kBq/m^3 to more than 15.7 kBq/m^3). Potentiating action of cigarette smoking and radiation was evidenced from a sputum cytology survey among uranium miners and controls. In accordance with published results, the frequencies of abnormal cytology were 5% after radon exposure alone, 7% after cigarette smoking alone and 28% after combined action of these modalities. Hence, the synergistic enhancement ratio $k = 2.3$. Shortage of data on the effect of radon applied alone allowed us to evaluate only a single experimental value of the synergistic enhancement ratio ($k = 2.3$) that was not sufficient to use our mathematical model to describe completely the synergistic interaction of radon and tobacco smoking. Nevertheless, we can conclude that these data demonstrate clearly the synergistic interaction of radon and smoking in the induction of lung cancer in humans.

Other harmful agents are also known, the synergistic interaction of which can revealed for people. In particular, a survey of US data on the combined effect of cigarette smoke and alcohol on the formation of laryngeal cancer in men was published [35]. The author comes to the conclusion that the combined use of alcohol and smoking may increase cancer incidence by a factor of 20. Let us try to apply the discussed model of synergy to these results. Assume that parameters of the model describing the effects of alcohol have an index of "1", and the data related with tobacco smoke have an index "2". Table 7.4 shows the values of the relative risk of cancer developing after exposure to tobacco smoke alone, and after the combined action of cigarette smoking at a constant frequency of alcohol intake (15–29 times a week). It is evident that the relative risk of cancer increases with the number of cigarettes smoked up to 7.4, and the relative risk after the combined action of these factors is 20.2. Accept that the number of induced damage responsible for cancer induction is proportional to the relative risk. We suppose that the increment of the relative risk over unity is determined by some effective damage and sublesions responsible for the synergy. The corresponding values of N_1 and N_2 as well as the ratio of N_2 / N_1 are summarized in Table 7.4 in the dependence of the frequency of smoking under constant frequency of alcohol intake.

Table 7.4. The combined effect of alcohol and tobacco smoke on the induction of oral cavity and pharynx cancer of men

Smoking, cigarettes/day	The relative risk		N_1	N_2	N_2 / N_1	N_{real}	N_{indep}	k
	Smoking alone	Smoking plus alcohol						
0	1.0	1.4	0.4	0	0	0.4	0.4	1.0
1–19	1.7	5.4	0.4	0.7	1.75	4.4	1.1	4.0
20–39	1.9	7.2	0.4	0.9	2.25	6.2	1.3	4.8
>40	7.4	20.2	0.4	6.4	16.0	19.2	6.8	2.8

This Table also includes the number of damage induced by the combined action of agents observed really (N_{real}) and expected after independent action of the agents ($N_{indep} = N_1 + N_2$). The values of the synergistic enhancement ratio (k) were estimated by the ratio N_{real} / N_{indep}. It is evident that this ratio increases with increase in the ratio of

N_2 / N_1, reaches a maximum and then decreases. Therefore, it is of interest to apply our mathematical approach to these results.

We believed that for smoking more than 40 cigarettes a day and taking alcohol 15–29 times a week $p_1 N_1 < p_2 N_2$. For such a situation in accordance with Eq. (5.14) parameter $p_1 = 31$. When about 25 cigarettes were smoked a day and at the same frequency of alcohol intake carcinogenesis is caused in a great extent by the influence of alcohol, i.e., $p_1 N_1 > p_2 N_2$. For this case parameter p_2, i.e., the number of sublesions induced simultaneously with one effective damage from tobacco smoke, can be estimated. The value of this parameter (p_2 = 5.1) was chosen for optimal description of both points obtained from epidemiological data. Taking into account the basic parameters of the model and Eq. (5.5), we calculated the theoretically expected dependence of the synergistic enhancement ratio on the ratio of damages produced by smoking and alcohol intake N_2 / N_1. The outcome is presented by solid curve in Figure 7.19 together with the results of epidemiological data (circles). Having used the basic parameters of the model (p_1 and p_2) and Eq. (5.11), it can be predicted that the highest synergistic effect (k_{max} = 5.4) should be attained at N_2 / N_1 = 6.1 (Eq. 5.9) and any deviation of this ratio from the optimal value should result in a decrease in the effectiveness of synergistic interaction. A good agreement appears to exist between theoretically predicted values and epidemiological data.

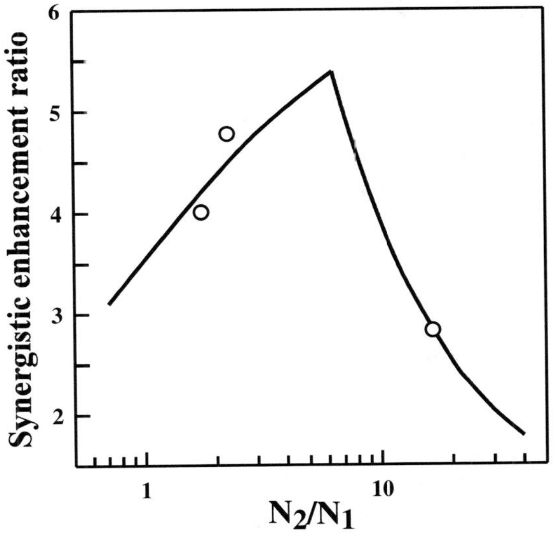

Figure 7.19. The dependence of the synergistic enhancement ratio on the ratio of damage produced by smoking (N_2) and alcohol intake (N_1) for combined action of these harmful agents on the induction of oral cavity and pharynx cancer of men; the theoretically predicted data are presented by solid curves and circles denote the results of epidemiological data.

The synergistic interaction of smoking and alcohol on the formation of esophageal cancer has been studied by other authors [15]. This epidemiological study included a survey of 200 persons with malignant disease and 775 control subjects. Alcohol intake ranged from 0 to more than 121 g of alcohol per day, the consumption of tobacco has been fixed. Applying these data and methodology, which has been just described, we assessed the dependence of the synergistic enhancement ratio of the ratio of damage induced by tobacco smoke and alcohol (N_2 / N_1).

The results are summarized in Table 7.5 and depicted in Figure 7.20. Parameter p_2, the mean number of sublesion that arise for one effective damage from tobacco smoke, was calculated for the situation where the effect is due mainly to the influence of alcohol. When tobacco consumption constitutes more than 30 g per day and alcohol intake is over 121 g per day, k = 2.8 and N_2 / N_1 = 0.14. Substituting these values into the Eq. (5.15), we obtain p_2 = 14.7. For the same tobacco consumption and alcohol intake of 41–120 g per day, the esophageal cancer may be caused largely influence of tobacco smoke. According to the model, for this situation one can estimate parameter p_1, the mean number of sublesions that arise for one effective damage induced by alcohol (Eq. 5.14). The value of this parameter (p_1 = 5.7) was chosen for optimal description of both points obtained from epidemiological data. Having used the basic parameters of the model (p_1 and p_2) and Eq. (5.11), it can be predicted that the highest synergistic effect (k_{max} = 5.1) should be attained at N_2 / N_1 = 0.39 (Eq. 5.9) and any deviation of this ratio from the optimal value should result in a decrease in the effectiveness of synergistic interaction. Here again, a fairly well agreement appears to exist between theoretically predicted values and epidemiological data.

Table 7.5. The combined effect of alcohol and tobacco smoke on the induction of esophageal cancer of men

Alcohol intake, g/day	The relative risk		N_1	N_2	N_2 / N_1	N_{real}	N_{indep}	k
	Alcohol alone	Smoking plus alcohol						
0–40	1	7.8	0	6.8	0	6.8	6.8	1.0
41–80	7.3	35	6.3	6.8	1.08	34	13.1	2.6
81–120	11.1	83	10.1	6.8	0.67	82	16.9	4.9
>121	49.6	155.6	48.6	6.8	0.14	154.6	55.4	2.8

It is of interest to analyze the importance of the synergistic interactions of environmental risk factors for the possible increase in the consequences of radiation accidents. The majority of synergistic data presented in this monograph was obtained for combined action of hyperthermia with various physical and chemical deleterious agents, the used temperature being far beyond the normal physiological temperature. A great complex of experimental data demonstrated the potential role of synergistic interaction at low intensities of physical factors or small concentration of chemical agents existing in biosphere. Therefore, it can be expected the existence of optimal intensities of ionizing radiation, UV light, ultrasonic, non-ionizing electromagnetic radiation as well as the concentration of chemical pollutants which provide the highest synergistic interaction with body temperatures of homoiothermic animals and humans.

Previously we attempted to extrapolate the dependence of the greatest synergistic effect of simultaneous thermoradiation exposure of cultured mammalian cells to low dose rates [57]. These extrapolations indicated the existence of a certain range of dose rates (10^{-5}–10^{-3} Gy/min), within which the maximum synergistic effect may be detected at physiological temperatures typical for animals and humans. The predicted range was too wide because of the large uncertainties. It is curious that such dose rates encountered in some areas contaminated by the Chernobyl accident. This means that the effect of ionizing radiation on organisms at the optimum dose rate of ionizing radiation may be higher by a factor of 3–5 compared with the forecast of the biological effects induced by an ionizing radiation without taking into consideration its synergistic interaction with metabolic heat of irradiated object.

It is of interest that direct experimental studies [17] of mutation frequency in plants at real dose rates of ionizing radiation taking place after the accident within 30-km zone around the Chernobyl reactor demonstrated a synergistic effect of chronic exposure to ionizing radiation and a subsequent exposure to lead nitrate on the mutagenic damage in *Arabidopsis thaliana* seeds. Seeds were picked out from wild populations growing for 5 years in places with different levels of radioactive contamination at various dose rates inside the 30-km zone around the Chernobyl nuclear power station. After that the seed were exposed to lead nitrate at a constant concentration (3.39 g/l). The frequency of mutant embryous and the proportion of lethal embryos were estimated. The effect was dependent on the mean dose rate (Figure 7.21). This dependence was calculated by us basing on the published results [17].

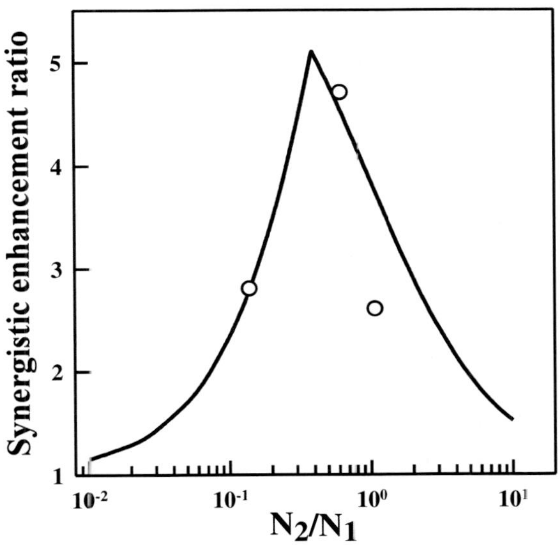

Figure 7.20. The dependence of the synergistic enhancement ratio on the ratio of damage produced by smoking (N_2) and alcohol intake (N_1) for combined action of these harmful agents on the induction of esophageal cancer of men; the theoretically predicted data are presented by solid curves and circles denote the results of epidemiological data.

In this case, the synergistic enhancement ratio was defined as the ratio between the increment of the mutant and lethal embryos after the combined action (compared to controls) to the sum of these increments after the action of each agent separately. As it can be seen, there is an optimal dose rate (about of 5 μGr/h) providing the highest synergy. Any increase or decrease in this dose rate results in a reduction of the efficiency of the interaction of damage induced by these agents, just as it was predicted by the mathematical model of synergy discussed here. Thus, the predicted regularities of synergistic interaction have been demonstrated in real objects and under real environmental conditions. Therefore these results support the general significance of the revealed regularity: the lower is the intensity of one of the agent applied in combination, the lower should be the intensity of another agent in order to provide their maximal synergistic interaction. This implies a fundamental possibility of mutual enhancement of the effects induced by harmful environmental factors at the intensities really occurring in biosphere. Such an enhancement may be a result of synergistic interaction of different physical and chemical agents with each other or with the temperature of homoiothermal animals and human beings.

It is not excluded that the strengthening effects of the Chernobyl and other radiation accidents may be partly related with the synergistic interaction of ionizing radiation with other physical or chemical agents. For example, it is known that iodine deficiency may be an important modifier of radiation-induced thyroid cancer. Some social factors, including economic situation or radiophobia, – sense of fear towards ionizing radiation, – may also synergistically interact with ionizing radiation.

Experimental studies in rats have shown that iodine deficiency may be an important modifier of radiation-induced thyroid cancer [31; 53]. These authors noted that iodine deficiency by itself is a weak carcinogen, but may be an effective promoter of cancer induced by ionizing radiation. According to our mathematical model, the interaction promoter and radiation is a particular case of synergy when the number of effective damage produced by promoter (N_2) is small but the mean number of sublesion from the promoter (p_2N_2) may be sufficient to exhibit a synergistic interaction due to a large value of p_2. Indeed, it was shown on the basis of epidemiological studies that synergistic interaction of ionizing radiation and iodine deficiency may be of important in the development of thyroid cancer in the areas affected by the Chernobyl accident [62]. This study was carried out in the Bryansk region (Russia) and included districts with various contaminations of ^{137}Cs up to 40, 5 and less than 1 Ci/km^2. 119,785 children and adolescents born in the years 1968–1986 have been examined. The greatest synergistic effect was obtained for moderate iodine deficiency. In this case, the maximum ratio of the observed number of cancers (per 100,000) to the expected one was equal to 7.3. With a decrease in the degree of iodine deficiency this ratio progressively decreased to 1.7. This ratio was somewhat reduced (6.9) also for the increased level of iodine deficiency. These data indicate a possible existence of optimal balance between radiation dose and the degree of iodine deficiency that is consistent with the set of data presented above.

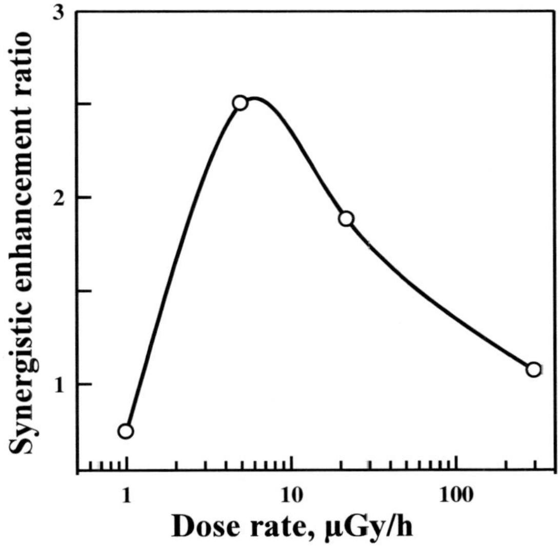

Figure 7.21. The dependence of the synergistic enhancement ratio on dose rate of ionizing radiation for combined action of lead nitrate and chronic irradiation of *Arabidopsis thaliana*.

The presented mathematical model of synergism predicts that an unusual situation in radiobiology may be observed when the biological effects produced by ionizing radiation at the smaller contaminations may be worse than that obtained at higher ones. It may be realized if there is a synergistic interaction with another harmful factor. We can demonstrate the validity of this statement by simple calculations. Assume that at non-optimal dose rate the synergistic enhancement ratio for the interaction of ionizing radiation with another agent k = 1.5. This means that the mean number additional effective damage (N_3) produced due to synergistic interaction

$$N_3 = 0.5 \cdot (N_1 + N_2), \tag{7.7}$$

where N_1 and N_2 are the mean number of effective damage produced by ionizing radiation nad another agent, respectively. Then the common number of damage after combined action will be presented as

$$N_{comb} = (N_1 + N_2 + N_3) = 1.5 \cdot (N_1 + N_2). \tag{7.8}$$

Suppose that for the reduced dose rate, for example by a factor of 2, the greatest synergistic interaction took place. Suppose that for this case the synergistic enhancement ratio k =6. Then the number of effective damage induced by ionizing radiation should be reduced by a factor of 2 and be equal of $0.5N_1$ because of the reduced dose rate. For this case, an additional number of effective damage formed due to the synergistic interaction of both agents will be

$$N_3 = 5 \cdot (0.5N_1 + N_2) \tag{7.9}$$

and the common number of damage after combined action may be presented as

$$N_{comb} = \left(N_1 + N_2 + N_3\right) = 6 \cdot \left(0.5N_1 + N_2\right) = 3N_1 + 6N_2. \quad (7.10)$$

Comparing equations (7.8) and (7.10), it can be easily seen that the decrease in the dose rate of ionizing radiation resulted to an increase in the total number of the effective damage due to the greatest synergistic interaction. This paradox could not be explained from the view of classical radiobiology without utilization of synergistic interaction ideas.

CONCLUSION

In this Chapter, the synergistic interaction of hyperthermia with ionizing radiation or cytostatics in cultured mammalian cells has been studied. It was demonstrated that the universal rules of synergistic effect display described for the simultaneous action of hyperthermia with ionizing radiation, ultraviolet light or ultrasound on simplest cells of different origin may be extended to the simultaneous action of elevated temperature with ionizing radiation and various chemical compounds on cultured mammalian cells. First of all, for any constant dose rate of ionizing radiation or constant concentration of drug the synergy can be observed only within a certain temperature range. Second, inside this range, there is a specific temperature that maximizes the synergistic effect. And at last, a decrease in the dose rate of ionizing radiation or in chemical concentration necessitates a decrease in exposure temperature to provide the highest or any required level of synergy.

The main points which emerge from the findings pertinent to the role of recovery inhibition in cancer treatment can be summarized as follows. The increase in cell radiosensitivity by chemicals may be causally related to the following reasons: (i) the impairment of the recovery capacity itself, (ii) the production of irreversible damage, which cannot be repaired at all, (iii) the higher probability to produce primary radiation damage, (iv) all or some of these issues may occur simultaneously. The next aim of this study here was to determine which of these points are involved in the inhibition of cell recovery. To distinguish these possibilities, a mathematical model describing the process of recovery as a decrease in the effective radiation dose was used. The results of this study reveal that the basic effect of the chemicals tested at a concentration sufficient to inhibit recovery from potentially lethal damage appears to be the reduction in the number of cells capable to recover owing to the increase in the irreversibly damaged cells. This effect may be interpreted as being due to the conversion by drugs of radiation induced repairable damage so that enzymes could then not deal with the lesions [26; 66]. It would seem probable also that chemicals could interfere with the synthesis of requisite enzymes [9; 68]. The independence of the recovery constant on the presence of chemical inhibitors during irradiation and/or postirradiation recovery period, observed in the most cases, would imply that the same portion of repairable damage is eliminated for unit time independently of the recovery conditions investigated. This result strongly suggests that the most analysed chemicals don't damage repair enzymes responsible for recovery. Similar results have been presented above for diploid yeast cells exposed to hyperthermia and ionizing radiation or hyperthermia and UV light. It follows that some general mechanism of radiosensitization or synergistic effects may underlie the interaction of heat and some chemicals with ionizing radiation. The mechanism should imply the failure of direct interference with the repair process itself and favor a role of hyperthermia and chemicals analyzed in the decreasing rate and extent of repair by facilitating either the production of irreversible damage or early radiation damage fixation before recovery processes would be over or occur.

Some examples presented here indicate that the recovery constant can be slightly decreased due to the influence of chemical inhibitors to the recovery process itself. It was expressed in the decreasing of the probability of cell recovery per unit time. Hence, this observation supports the conclusion that chemical inhibition may cause the enhanced cell killing due to both interference with repair of potentially lethal damage and enhancement of expression of irreversible radiation damage. A suitable suggestion concerning the effect of 5'-iododeoxyuridne incorporated into DNA on radiation sensitivity and cellular repair capability of CHO cells could be put forward. As it turned out, such a DNA modification resulted in a great increase of cell radiosensitivity without influence both on the recovery constant and irreversible component of radiation damage. It means that the effect of 5'-iododeoxyuridne might be expressed through the enhanced yield of primary radiation damage rather than due to the formation of irreversible damage and/or its influence on recovery process itself. Detailed analysis of molecular mechanisms involved is beyond the main aim of this study. However, it can be noted that many mechanisms have been discussed by different authors [13; 36; 37; 46; 60; 70]. Only as examples, the following possibilities can be mentioned.

The enhancing effect seems to be accountable by direct chemical toxicity, cell recovery inhibition or the higher probability to produce primary radiation damage without changing the recovery process [23; 36; 37; 66]. Repair of DNA, as reflected in rejoining of strand breaks, may be of relevance to the cell recovery from sublethal and potentially lethal damage [20; 23]. Preferential impairment of DNA repair in malignant cells would be of great relevance in improving of microorganism sterilization and cancer treatment. The inhibition of cell recovery [36; 37; 42; 70] and DNA single and double strand breaks repair [9; 30; 68; 73] by chemicals is expressed both as a deceleration of recovery rate and a lesser extent of recovery.

Because of critical role of the DNA topoisomerases in the synthesis and conformation of DNA, and well-known information that ionizing radiation inhibits replicative DNA synthesis, there is a possibility that inhibitors of these enzymes might influence the radiation lethality [70]. It can be admitted that a decrease in a quantity of these enzymes wouldn't interfere with the probability of recovery but result in a greater portion of irreversible damage. A similar situation may be realized with insufficient energy metabolism. For instance, novobiocin and nalidixic acid have been shown to inhibit DNA, RNA, and protein synthesis in several mammalian cell lines [36; 37; 46; 70], their activity may be also expressed through the interfere with the function of topoisomerase II in an early stage of DNA repair [13; 36; 37]. It was presumed that the inhibition of recovery from potentially lethal damage by lactate and pyruvate may be due to severe metabolic changes such as a decrease in the intracellular ATP concentration [36]. These authors postulated that the recovery inhibition may be occurred due to the raising of lactate and pyruvate levels complicating the repair of DNA damage. This view is strengthened by the data showing that specific inhibitors of poly(ADP-R) synthesis enhance cell killing and inhibit the DNA strand break rejoining induced by ionizing radiation. 3-Aminobenzamide has been shown to be a putative specific inhibitor of poly(ADP-R) synthetase [60]. The results obtained by Kumar et al. [37] favor a possible role of the chemical in preventing repair by facilitating an early damage fixation before repair can occur, simultaneously reducing G_2-arrest.

The results of this study indicate to the opportunity to search agents, selectively or simultaneously acting on the probability of recovery and the yield of irreversible radiation damage. The recognition that specific inhibitors of recovery may exist, such as an inhibitor of recovery process itself and that resulting in the increased yield of irreversible damage, would provide both a possibility to analyze the mechanism of a drug and ionizing radiation interaction from this point of view and an expectation that useful regimens in cancer research may be devised to make use of these inhibitors.

To demonstrate the admissibility of the synergy model for animal, the interaction of damage produced by simultaneous action of ambient temperature and microwave was studied for temperature increment in rabbit body. The data presented here revealed that for a constant ambient temperature the effectiveness of the synergistic interaction of microwaves and a high ambient temperature is observed only inside a certain range of the power flux density. Moreover, within this range an optimal value exists, at which the synergy is the highest one. Both an increase and decrease in the power flux density from this optimal value resulted in a decrease of the synergistic effect. Qualitatively similar results have been observed for the synergy expression in various cellular systems. It was demonstrated that slightly modified mathematical model of synergy describes the experimental data for rabbit heating, predicts the greatest value of the synergistic effect, the condition under which it can be achieved as well as the dependency of the synergy on microwave power flux density and ambient temperature. From the model standpoint, the highest synergy can be obtained for a specific power flux density at which an equal number of sublesions are produced by each of the agents applied in combination.

A survey of the known synergistic effects in humans has been presented and the possibility of their interpretation with our model of synergy has been discussed. Despite the limited number of published data on the synergistic interaction of environmental factors to humans, the data presented in this Chapter are in a qualitative agreement with the model prognosis and would be, in principle, useful for further solution of the problems related with the synergistic interaction of environmental agents. Moreover, the presented mathematical model of synergism was used to explain an unusual situation in radiobiology when the biological effects produced by ionizing radiation at smaller contaminations may be worse than that obtained at higher ones.

Thereby several basic points presented here may be mentioned in short as the next resumes.

1. The universal rules of synergistic effect display described for the simplest unicellular organisms have been extended to cultured mammalian cells. For any constant dose rate of ionizing radiation or concentration of drug the synergy is observed only within a certain temperature range, inside which there is a specific temperature that maximizes the synergy.
2. It was shown for cultured mammalian cells that the basic effect of the chemicals tested at a concentration sufficient to inhibit recovery from potentially lethal damage appeared to be the reduction in the number of cells capable of recovery owing to the increase in the irreversibly damaged cells. The findings revealed the probability of

recovery was independent of whether the process of recovery happened with or without chemicals sensitizing the radiation effect.

3. To demonstrate the acceptability of the synergy model for animal, the interaction of simultaneous action of ambient temperature and microwave was studied for temperature increment in rabbit body. The data revealed that for a constant ambient temperature of the synergistic interaction of microwaves and a high ambient temperature is observed only within a certain range of the power flux density, inside of which an optimal value exists, at which the synergy is the highest.
4. It was shown that slightly modified mathematical model of synergy describes the experimental data for rabbit heating, predicts the greatest value of the synergistic effect, the condition under which it can be achieved as well as the dependency of the synergy on microwave power flux density and ambient temperature. From the model standpoint, the highest synergy can be obtained for a specific power flux density at which an equal number of sublesions are produced by each of the agents applied in combination.
5. A survey of the known synergistic effects in humans has been presented and the possibility of their interpretation with our model of synergy has been discussed. Despite the limited number of published data on the synergistic interaction of environmental factors to humans, the data presented are in a qualitative agreement with the model prognosis and would be, in principle, useful for further solution of the problems related with the synergistic interaction of environmental agents.

REFERENCES

[1] Adair, ER. *Microwaves and Thermoregulation*. Edition. New York: Academic Press; 1983.

[2] Adair, ER; Cobb, BL; Mylacraine, KS et al. Human exposure at two radio frequencies (450 and 2450 MHz): similarities and differences in physiological response. *Bioelectromagnetics*, 1999; 20 (Suppl. 4), 12-20.

[3] Band, P; Feldstein, M; Saccomanno, G et al. Potentiation of cigarette smoking and radiation: evidence from a sputum cytology survey among uranium miners and controls. *Cancer*, 1980; 45(6), 1273-1277.

[4] Barrett, JC; Lamb, PW; Wiseman, RW. Multiple mechanisms for the carcinogenic effects of asbestos and other mineral fibers. *Environ. Health Perspect.*, 1989; 81, 81-89.

[5] Ben-Hur, E. Mechanisms of the synergistic interaction between hyperthermia and radiation in cultured mammalian cells. *J. Radiat. Res.*, 1976; 17, 92-98.

[6] Ben-Hur, E; Bronk, BV; Elkind, MM. Thermally enhanced radiosensitivity of cultured Chinese hamster cells. *Nature New Biol.*, 1972; 238, 209-211.

[7] Ben-Hur, E; Elkind, MM; Bronk, BV. Thermally enhanced radio-response of cultured Chinese hamster cells – Inhibition of repair of sublethal damage and enhancement of lethal damage. *Radiat. Res.*, 1974; 58, 38-51.

[8] Bolck, G; Matanoski, GM; Seltser, R et al. Cancer morbidity and mortality in phosphate workers. *Cancer Res.*, 1988; 48, 7298-7303.

[9] Boothman, DA; Trask, DK; Pardee, AB. Inhibition of potentially lethal DNA damage repair in human tumor cells by β-lapachone, an activator of topoispomerase I. *Cancer Res.*, 1989; 49, 605-612.

[10] Brown, CC; Chu, KC. Additive and multiphcative models and multistage carcinogenesis theory. *Risk Anal.*, 1989; 9, 99-105.

[11] Burkart, W; Jung, T. Health risk from combined exposures: mechanistic considerations on deviations from additivity. *Mutat. Res.*, 1998; 411, 119-128.

[12] Chameaud, J; Perraud, R; Chretien, J et al. Lung carcinogenesis during in vivo cigarette smoking and radon daughter exposure in rats. *Cancer Res.,* 1982; 82, 11-20.

[13] Cleaver, JE. Specificity and completeness of inhibition of DNA repair by novobiocin and aphidicolin. *Carcinogenesis*, 1982; 3, 1171-1174.

[14] Cross, FT; Palmer, RF; Filipy, RE et al. Carcinogenic effects of radon daughters, uranium ore dust and cigarette smoke in beagle dogs. *Health Phys.*, 1982; 42(1), 33-52.

[15] Day, NE; Miňoz N. Pseophagus. In: Schottenfeld D, Fraumeni JF editors. *Cancer Epidemiology and Preventation*. Philadelphia: WD Saunders Co; 1982, 596-623.

[16] Dewey, WC; Hopwood, LE; Sapareto, SA et al. Cellular responses to combinations of hyperthermia and radiation. *Radiology*, 1977; 123(2), 463-474.

[17] Dineva, SB; Abramov, VI; Shevchenko, VA. The genetic effects of treatment of *Arabidopsis thaliana* seeds by the sodium lead of chronic irradiatied population. *Genetics*, 1993; 29, 1914-1920.

[18] Eichholtz-Wirth, H; Hietel, B. Heat sensitization to cisplatin in two cell lines with different drug sensitivities. *Int. J. Hyperthermia*, 1990; 6(1), 47-55.

[19] Finch, GL; Lundgren, DL; Barr, EB et al. Chronic cigarette smoke exposure increases the pulmonary retention and radiation dose of ^{239}Pu inhaled as $^{239}PuO_2$ by F344 rats. *Health Phys.*, 1998; 75(6), 597-609.

[20] Friedberg, EC; Walker, GC; Siede, W et al. *DNA repair and mutagenesis, 2nd edition.* Washington DC: ASM Press, 2006.

[21] Gordon, CJ; Ali, JS. Comparative thermoregulatory response to passive heat loading by exposure to radiofrequency radiation. *Comp. Biochem. Physiol.*, 1987; 88(1), 107-112.

[22] Gordon, CJ; Long, MD; Fehlner, KS et al. Temperature regulation in the mouse and hamster exposed to microwaves in hot environments. *Health Phys.*, 1986; 50(6), 781-787.

[23] Hall, EJ. *Radiobiology for the Radiologist*. Lippincott: Williams and Wilkins, 2000.

[24] Henle, KJ; Dethlefsen, LA. Time-temperature relationship for the induced killing of mammalian cells. *Ann. N. Y. Acad. Sci.*, 1980; 335, 234-253.

[25] Henle, KJ; Leeper, DB. The modification of radiation damage in CHO cells by hyperthermia at 40 and 45 °C. *Radiat. Res.*, 1977; 70, 415-424.

[26] Hill, BT; Bellamy, AS. An overview of experimental investigations of interaction between certain antitumor drugs and X-irradiation *in vitro*. *Adv. Radiat. Biol.*, 1984; 11, 211-267.

[27] Iliakis, G; Kurtzman, S. Application of non-hypoxic cell sensitizers in radiobiology and radiotherapy: rationale and future prospects. *Int. J. Radiat. Oncol. Biol., Phys.*, 1988; 15, 1235-1241.

[28] Iliakis, G; Kurtzman, S; Pantelias, G et al. Mechanism of radiosensitization by halogenated pyrimidines: effect of BrdU on radiation induction of DNA and

chromosome damage and its correlation with cell killing. *Radiat. Res.*, 1989; 119, 286-304.

[29] Johnson, HA; Pavelec, M. Thermal enhancement of thio-TEPA cytotoxicity. *J. Natl. Cancer Inst.*, 1973; 50, 903-908.

[30] Jorritsma, JB, Konings, AW. Inhibition of repair of radiation-induced strand breaks by hyperthermia, and its relationship to cell survival after hyperthermia alone. *Int. J. Radiat. Biol.*, 1983; 43(5), 505-516.

[31] Kanno, J; Onodera, H; Furuta, K et al. Tumor-promoting effects of both iodine deficiency and iodine excess in the rat thyroid. *Toxicol. Pathol.*, 1992; 20(2), 226-235.

[32] Knizhnikov, VA; Grozovskaya, VA; Litvinov, NN et al. Appearance of malignant tumors in the lungs of mice after inhalation of shistous ash, 3,4-benzopyzene and ^{210}Po. In: Iljin LA, Knizhnikov VA editors. *Hygienic Problems of Radiation and Chemical Carcinogenesis*. Moscow: Institute of Biophysics, 1979.

[33] Kolganova, OI; Zhavronkov, LP; Petin, VG et al. Thermocompensatory reactions of rabbits in response to microwave irradiation at different environmental temperatures. *Radiat. Biol. Radioecol.*, 2001; 41(6), 712-717.

[34] Kolganova, OI; Zhavoronkov, LP; Petin, VG et al. Rabbit heating by microwave exposure at various ambient temperatures. *Journal of Radiation Protection*, 2010; 35(3), 100-105.

[35] Konoplyannikov, AG. *Your Health and Radiation*. Moscow: IZDAT, 1993.

[36] Kumar, A; Kiefer, J; Schneider, E et al. Inhibition of recovery from potentially lethal damage by chemicals in Chinese hamster V79A cells. *Radiat. Environ. Biophys.*, 1985a; 24(2), 89-98.

[37] Kumar, A; Kiefer, J; Schneider, E et al. Enhanced cell killing, inhibition of recovery from potentially lethal damage and increased mutation frequency by 3-aminobenzamide in Chinese hamster V79 cells exposed to X-rays. *Int. J. Radiat. Biol.*, 1985b; 47, 103-112.

[38] Kushneva, VS. On the problem of the long-term effects of the combined injury to animals of silicon dioxide and radon. In: Zatukinski DI editor. *Long-term Effects of Injuries Caused by the Action of Ionizing Radiation*. State Publishers of Medical Literature, 1959. 21-28.

[39] Kusiak, RA; Ritchie, AC; Muller, J et al. Mortality from lung cancer in Ontario uranium miners. *Br. J. Ind. Med.*, 1993; 50(10), 920-928.

[40] Kuwabara, Y; Matsubara, S; Yoshimatsu, S et al. Combined effect of ultrasound and ionizing radiation on lymphocyte chromosomes. *Acta Radiol. Oncol.*, 1986 ; 25(4-6), 291-294.

[41] Little, JB. Low dose radiation effects: interaction and synergism. *Health Phys.,* 1990; 59(1), 49-55.

[42] Little, JB; Ueno, AN; Dahlberg, WK. Differential response of human and rodent cell lines to chemical inhibition of the repair of potentially lethal damage. *Radiat. Environ. Biophys.*, 1989; 28, 193-202.

[43] Lubin, JH; Boice, JD; Edling, C et al. *Radon and Lung Cancer Risk: a Joint Analysis of 11 Underground Miner Studies*. National Institutes of Health Publication: Washington, 1994. No. 94-3644.

[44] Madas, BG; Balashazy, I. Mutation induction by inhaled radon progeny modeled at the tissue level. *Radiat. Environm. Biophys.*, 2011, 50, 553-570.

[45] Magos, L. Epidemiological and experimental aspects of metal carcinogenesis: physiochemical properties, kinetics and the active species. *Environ. Health Perspect.*, 1991; 95, 157-189.

[46] Mattern, MR; Painter, RB. Dependence of mammalian DNA replication of DNA supercoiling. II. Effects of novobiocin. *Biochim. Biophys. Acta.*, 1979; 553, 306-312.

[47] McRee, DI. Determination of the absorption of microwave radiation by biological specimens in a 2450 MHz microwave field. *Health Phys.*, 1974; 26(5), 385-390.

[48] Michaelson, SM. Thermal effects of single and repeated exposures to microwaves – a review. In: Czerski P, Ostrowski K, Shore ML, Silverman Ch, Suess VJ, Waldeskog B editors. *Biological Effects and Health Hazards of Microwave Radiation*. Warsaw: Polish Medical Publishers, 1974; 1-14.

[49] Michaelson, SM. Thermoregulation in intense microwave fields. In: Adair ER, editor. *Microwaves and Thermoregulation*. New York: Academic Press, 1983; 283-295.

[50] Miyakoshi, J; Heki, S; Kano E. Cellular responses to hyperthermia and radiation in Chinese hamster cells. In: Sugahara T, editor. *Modification of Radiosensitivity in Cancer Treatment*. Tokyo: Academic Press, 1984; 335-350.

[51] Monfared, AS; Jalali, F; Sedaghat, S et al. High natural background radiation areas in Ramsar, Iran: can inhabitants feel safe? *Int. J. Low Radiat.*, 2006; 3(2/3), 171-177.

[52] Nakashima, K; Kawamata, A; Fujiki, I et al. The individual and combined effects of X-irradiation and hyperthermia on early somite mouse embryos in culture. *Teratology*, 1991; 44(6), 635-639.

[53] Ohshima, M; Ward, JM. Dietary iodine deficiency as a tumor promoter and carcinogen in male F344/NCr rats. *Cancer Res.*, 1986; 46(2), 877-883.

[54] Panov, D; Visnjic, V; Novak, L et al. Radiobiological effects of joint action of alpha radiation (Po-210) and quartz dust m the rats respiratory and renal system. *Proc. Fourth Congress Int. Radiat. Protection Association,* 1977; 4, 1165-1168.

[55] Pershagen, G; Akerblom, G; Axelson, O et al. Residential radon exposure and lung cancer in Sweden. *N. Engl. J. Med.*, 1994; 330(3), 159-164.

[56] Petin, VG; Kim, JK; Kolganova, OI et al. Synergistic interaction of environmental temperature and microwaves: prediction and optimisation. *J. Radiat. Protect.,* 2011; 36(1), 1-7.

[57] Petin, VG; Zhurakovskaya, GP; Lisowski, MA. Some radiobiological aspects of combined exposures. *Medical Radiology*, 1993; 38(3), 18-22

[58] Ponomareva, VL; Lyarskii, PP; Burykina, LN et al. The results of experimental studies of the late effects of dust combined with radiation and the problems of hygienic standardization. In: *Proceedings of the Radiobiological Conference of Socialist Countries*. Varna, 1978.

[59] Preston, RJ. The effect of cytosine arabinozide on the frequency of X-ray-induced chromosome aberrations in normal human leukocytes. *Mutat. Res.*, 1980; 69, 71-79.

[60] Purnell, MR; Whish, WJ. Novel inhibitors of poly(ADP-ribose) synthetase. *Biochem. J.*, 1980; 185, 775-777.

[61] Seyama, T; Kajitani, T; Inoh, A et al. Synergistic effect of radiation and N-nitrosoethylurea in the induction of lymphoma in mice: cellular kinetics and carcinogenesis. *Jpn. J. Cancer Res.,* 1985; 76(1), 20-27.

[62] Shakhtarin, VV; Tsyb, AF; Stepanenko, AF et al. Iodine deficiency and thyroid cancer morbidity following the accident at the Chernobyl power plant. In: Thomas G,

Karaoglou A, Williams ED editors. *Radiation and Thyroid Cancer*. Singapure, New Jersey, London, Hong Kong: World Scientific, 1999; 277-282.

[63] Sharp, JG; Grouse, DA. Apparent synergism between radiation and the carcinogen 1,2-dimethyl-hydra2ine in the induction of colonic tumors in rats. *Radiat. Res.,* 1989; 117(2), 304-317.

[64] Shellabarger, CJ; Stone, JP; Holtzman, S. Effect of interval between neutron radiation and diethylstilbestrol on mammary carcinogenesis in female ACI rats. *Environ. Health Perspect.,* 1983; 50, 227-232.

[65] Shore, RE. Overview of radiation-induced skin cancer in humans. *Int. J. Radiat. Biol.,* 1990 ; 57(4), 809-827.

[66] Streffer, C; Müller, WU. Radiation risk from combined exposures to ionizing radiation and chemicals. *Adv. Radiat. Biol.*, 1984; 11, 173-210.

[67] Taylor, PR; Quiao, YL; Schatzkin, A et al. Relation of arsenic exposure to lung cancer among tin miners in Yunnan Province, China. *Br. J. Ind. Med.*, 1989 ; 46, 881-886.

[68] Takahashi, S; Takeda, E; Kubota, Y et al. Inhibition of radiation-induced DNA double-strand breaks by nickel and arsenite. *Radiat. Res.*, 2000; 154(6), 686-691.

[69] Urano, M; Kahn, J; Majima, H et al. The cytotoxic effect of cis-diamminedichloroplatinum (II) on culture Chinese hamster ovary cells at elevated temperatures: Arrhenius plot analysis. *Int. J. Hyperthermia*, 1990; 6, 581-590.

[70] Utsumi, H; Shibuya, ML; Elkind, MM. Novobiocin inhibits the repair of potentially lethal damage but not the repair of sublethal damage. *Radiat. Res.*, 1990; 123, 55-60.

[71] Wang, Ya; Iliakis, G. Effects of 5'-iododeoxyuridine on the repair of radiation induced potentially lethal damage interphase chromatin breaks and DNA double strand breaks in Chinese hamster ovary cells. *Int. J. Radiat. Oncol. Biol. Phys.*, 1992; 23, 353-360.

[72] Weichselbaum, RR. Radioresistant and repair proficient cells may determine radiocurability in human tumors. *Int. J. Radiat. Oncol. Biol. Phys.*, 1986; 12, 637-639.

[73] Yang, L; Douple, EB; O'Hara, JA et al. Carboplatin enhances the production and persistence of radiation-induced DNA single-strand breaks. *Radiat. Res.*, 1995; 143, 302-308.

[74] Zhavoronkov, LP; Petin, VG. The dose-power approach in hygienic standardization of microwaves: the concept, biophysical modeling, experimental estimation. In: Repacholi MH, Rubtsova NB, Muc AM editors. *Electromagnetic Fields: Biological Effects and Hygienic Standardization*. Geneva: World Health Organization, 1999; 457-465.

Author Contact Information

Dr. Vladislav G. Petin,
Head of Biophysical Laboratory,
Medical Radiological Research Center,
Obninsk, Kaluga Region, Russia
Tel: +7(48439)9-70-08
vgpetin@yahoo.com

Dr. J.K. Kim,
Korea Atomic Energy Research Institute
Daedeok-Daero 989-111, Yuseong-Gu,
Daejeon, Korea
Tel: +82.42.868-2000
Fax: +82.42.682.8465

INDEX

A

B

C

D

L

M

O

P

R

S

T

U

V

X

Y